arithmetic: revision and practice

M. Bowman
Mainholm Academy, Ayr

R. Elvin
Formerly Head of Mathematics, Thomas Sumpter School, Scunthorpe

A. Ledsham
Head of Mathematics, The Abbey International College, Malvern Wells

C. Oliver
Thomas Sumpter School, Scunthorpe

Oxford University Press

Oxford University Press, Walton Street, Oxford OX2 6DP

London Glasgow New York Toronto
Delhi Bombay Calcutta Madras Karachi
Kuala Lumpur Singapore Hong Kong Tokyo
Nairobi Dar es Salaam Cape Town
Melbourne Auckland

and associates in
Beirut Berlin Ibadan Mexico City Nicosia

ISBN 0 19 914087 1
First published 1981
© M. Bowman, R. Elvin, A. Ledsham, C. Oliver 1981
Reprinted 1981, 1982

Set by Illustration Services, Oxford
Printed in Great Britain at The University Press, Oxford
by Eric Buckley, Printer to The University.

This book is written for pupils sitting the S.C.E.
O grade examination in Arithmetic and will be equally
useful for candidates preparing for similar examinations
set by CSE examination boards and by the Royal
Society of Arts.

Each section includes brief reminder notes and worked
examples followed by a large number of carefully
graded questions to suit the wide range of ability of
pupils who enter these examinations. The clear
arrangement of the contents allows the book to be
used with any scheme of teaching for appropriate
selection of material as course work or for final
revision. Particular emphasis is placed on commercial
aspects, estimation of errors, and the use of rough
estimates and the pocket calculator. The book also
includes original questions organized into test papers
similar to those of the examination.

All numerical answers are provided at the end of the
book.

The aim of the authors is to give the student confidence
and facility in all aspects of the subject. They believe
this is best gained by constant work on the fundamental
techniques which this book emphasises.

M.D.B.
R.E.
A.H.C.L.
C.O.

CONTENTS

Example 1

Find the 'odd answer out' of:
(a) $26 + 36 + 72$
(b) $34 + 62 + 38$
(c) $32 + 56 + 44$

(a)   26   (b)   34   (c)   32
      36   62   56
  + 72   + 38   + 44
  ————— ————— —————
  134   134   132

So (c) is the 'odd answer out' because its answer is 132.

Exercise 1.1a

Find the 'odd answer out' for the following:

1. (a) 6   (b) 7   (c) 4
      8   5   15
  + 10   + 12   + 6

2. (a) 9   (b) 8   (c) 7
     11   9   12
  + 12   + 14   + 13

3. (a) 5   (b) 18   (c) 9
     16   14   10
  + 15   + 3   + 16

4. (a) 13   (b) 2   (c) 6
      8   15   18
  + 12   + 17   + 10

5. (a) 19   (b) 17   (c) 15
     11   9   5
  + 8   + 12   + 17

6. (a) $25 + 7 + 11$ 7. (a) $24 + 13 + 15$
  (b) $22 + 11 + 9$ (b) $31 + 5 + 17$
  (c) $18 + 17 + 8$ (c) $29 + 12 + 11$

8. (a) $23 + 27 + 12$ 9. (a) $28 + 19 + 25$
  (b) $19 + 22 + 21$ (b) $17 + 32 + 24$
  (c) $9 + 32 + 20$ (c) $42 + 9 + 22$

10. (a) $31 + 27 + 36$ 11. (a) $53 + 17 + 63$
  (b) $42 + 13 + 40$ (b) $48 + 29 + 56$
  (c) $11 + 56 + 27$ (c) $58 + 9 + 65$

12. (a) $11 + 59 + 49$ 13. (a) $73 + 45 + 27$
  (b) $7 + 71 + 40$ (b) $69 + 55 + 22$
  (c) $19 + 39 + 60$ (c) $79 + 32 + 34$

14. (a) $87 + 36 + 29$ 15. (a) $36 + 94 + 50$
  (b) $59 + 44 + 49$ (b) $29 + 97 + 53$
  (c) $63 + 29 + 61$ (c) $45 + 82 + 52$

Example 2

Find the 'odd answer out' of:
(a) $317 + 232 + 387$
(b) $255 + 318 + 362$
(c) $297 + 421 + 218$

(a)   317   (b)   255   (c)   297
     232   318   421
  + 387   + 362   + 218
  ————— ————— —————
  936   935   936

So (b) is the 'odd answer out' because its answer is 935.

Exercise 1.1b

Find the 'odd answer out' for the following:

1. (a) 287 (b) 309 (c) 187
     326   237   412
  + 233 + 299 + 246

2. (a) 336 (b) 256 (c) 307
     422   309   236
  + 148 + 340 + 362

3. (a) 311 (b) 294 (c) 325
     307   355   402
  + 257 + 227 + 148

4. (a) 435 (b) 396 (c) 287
     269   319   376
  + 277 + 266 + 319

5. (a) 401 (b) 339 (c) 375
     355   286   319
  + 138 + 270 + 201

6. (a) $324 + 331 + 131$ 7. (a) $439 + 203 + 173$
  (b) $281 + 329 + 175$ (b) $356 + 195 + 263$
  (c) $199 + 321 + 265$ (c) $503 + 138 + 174$

8. (a) $396 + 237 + 272$ 9. (a) $517 + 218 + 215$
  (b) $435 + 187 + 284$ (b) $184 + 311 + 456$
  (c) $511 + 209 + 185$ (c) $396 + 305 + 249$

10. (a) $455 + 302 + 187$ 11. (a) $376 + 319 + 179$
  (b) $512 + 331 + 101$ (b) $295 + 318 + 262$
  (c) $432 + 217 + 296$ (c) $337 + 291 + 247$

12. (a) 1329 + 132 + 19
    (b) 1401 + 35 + 44
    (c) 1016 + 229 + 236

13. (a) 5276 + 81 + 97
    (b) 5085 + 88 + 281
    (c) 3540 + 1850 + 65

14. (a) 4327 + 424 + 155
    (b) 3215 + 1344 + 346
    (c) 3752 + 1060 + 94

15. (a) 3127 + 1346 + 362
    (b) 2955 + 1121 + 760
    (c) 2578 + 1592 + 666

10. (a) 93 − 84   11. (a) 70 − 56
    (b) 87 − 79   (b) 51 − 38
    (c) 46 − 38   (c) 63 − 49

12. (a) 45 − 9   13. (a) 92 − 59
    (b) 83 − 48   (b) 69 − 35
    (c) 62 − 27   (c) 81 − 48

14. (a) 90 − 23   15. (a) 71 − 48
    (b) 84 − 16   (b) 98 − 75
    (c) 75 − 8   (c) 80 − 56

Example 3

Find the 'odd answer out' of:
(a) 36 − 23   (b) 45 − 32   (c) 72 − 58

$$
\begin{array}{lll}
\text{(a)}\quad 36 & \text{(b)}\quad 45 & \text{(c)}\quad 72 \\
\underline{-\,23} & \underline{-\,32} & \underline{-\,58} \\
\quad 13 & \quad 13 & \quad 14
\end{array}
$$

So (c) is the 'odd answer out' because its answer is 14.

Exercise 1.1c

Find the 'odd answer out' for the following:

1. (a)   14   (b)   27   (c)   30
    $\underline{-\ 9}$ $\underline{-21}$ $\underline{-25}$

2. (a)   27   (b)   38   (c)   42
    $\underline{-13}$ $\underline{-25}$ $\underline{-28}$

3. (a)   57   (b)   61   (c)   49
    $\underline{-34}$ $\underline{-38}$ $\underline{-25}$

4. (a)   82   (b)   68   (c)   53
    $\underline{-37}$ $\underline{-23}$ $\underline{-\ 7}$

5. (a)   50   (b)   78   (c)   48
    $\underline{-19}$ $\underline{-46}$ $\underline{-17}$

6. (a) 93 − 27   7. (a) 95 − 78
   (b) 85 − 19   (b) 77 − 59
   (c) 74 − 9   (c) 53 − 36

8. (a) 80 − 42   9. (a) 90 − 58
   (b) 53 − 15   (b) 72 − 39
   (c) 61 − 24   (c) 51 − 18

Example 4

Find the 'odd answer out' of:
(a) 716 − 353
(b) 551 − 187
(c) 1175 − 811

$$
\begin{array}{lll}
\text{(a)}\quad 716 & \text{(b)}\quad 551 & \text{(c)}\quad 1175 \\
\underline{-\,353} & \underline{-\,187} & \underline{-\ 811} \\
\quad 363 & \quad 364 & \quad 364
\end{array}
$$

So (a) is the 'odd answer out' because its answer is 363.

Exercise 1.1d

Find the 'odd answer out' for the following:

1. (a)   848   (b)   763   (c)   927
    $\underline{-613}$ $\underline{-527}$ $\underline{-691}$

2. (a)   464   (b)   675   (c)   733
    $\underline{-319}$ $\underline{-529}$ $\underline{-588}$

3. (a)   754   (b)   636   (c)   984
    $\underline{-538}$ $\underline{-419}$ $\underline{-768}$

4. (a)   823   (b)   653   (c)   742
    $\underline{-468}$ $\underline{-297}$ $\underline{-386}$

5. (a)   712   (b)   835   (c)   764
    $\underline{-234}$ $\underline{-357}$ $\underline{-285}$

6. (a) 687 − 336   7. (a) 582 − 254
   (b) 538 − 186   (b) 843 − 516
   (c) 771 − 419   (c) 519 − 192

8. (a) 728 − 183   9. (a) 790 − 253
   (b) 813 − 269   (b) 802 − 264
   (c) 841 − 296   (c) 830 − 292

10. (a) $980 - 345$
    (b) $908 - 273$
    (c) $829 - 193$

11. (a) $628 - 91$
    (b) $619 - 83$
    (c) $634 - 98$

12. (a) $757 - 73$
    (b) $752 - 67$
    (c) $778 - 93$

13. (a) $3875 - 2532$
    (b) $3216 - 1872$
    (c) $2921 - 1577$

14. (a) $2346 - 1523$
    (b) $1756 - 932$
    (c) $1328 - 505$

15. (a) $1342 - 1267$
    (b) $1838 - 1762$
    (c) $1594 - 1518$

Exercise 1.1e

1. Jack has 26 marbles, Jim has 42 and Fred has 37. How many have they got altogether? If Henry has a large collection of 150, how many more has he got than the others put together?

2. Robert is 138 centimetres tall. If he is 39 centimetres shorter than his father, how tall is his father? If the father is 68 centimetres taller than Robert's younger brother Paul, how tall is Paul?

3. Halesowen is 126 metres above sea level and Adam's Hill is 304 metres above sea level. By how many metres does Adam's Hill stand above Halesowen? If Walton Hill stands 189 metres above Halesowen, how high is Walton Hill above sea level?

4. The population of Storrington is 2750. If this is 850 more than that of West Chiltington, what is the population of West Chiltington? If the population of West Chiltington is 3550 less than that of Steyning, what is the population of Steyning?

5. Ludlow is 87 metres above sea level and Titterstone Clee Hill stands 446 metres above Ludlow. How high is Titterstone Clee Hill above sea level? If Brown Clee Hill is 7 metres higher than Titterstone Clee Hill, by how many metres does Brown Clee Hill stand above Ludlow?

6. June is 15 years old, and Jane is 6 years older than June. How old is Jane? If Jane is 13 years older than Jill, how old is Jill?

7. There are 218 customers in Alice's Restaurant, and 79 fewer in Joe's Diner. How many customers are in Joe's Diner? If there are 194 more customers in The Big Bite snack bar than in Joe's Diner, how many customers are in the snack bar?

8. Hay's Building is 392 metres high, and is 139 metres taller than the Victoria Hotel. How high is the Victoria Hotel? If the Allied Insurance House is 228 metres taller than the Victoria Hotel, how high is the Allied Insurance House?

9. A ship is sailing 472 metres above the sea bed, and a whale is swimming 296 metres below the ship. How far above the sea bed is the whale? A scuba diver is swimming 275 metres above the whale. How far above the sea bed is the diver?

10. Bill takes 5850 steps during a walking race. If Jack takes 1130 fewer steps than Bill, how many steps does Jack take? If Fred takes 164 steps more than Jack, how many steps does Fred take?

11. A Jumbo Jet is flying at a height of 5280 metres, and a Starfighter is flying 1990 metres below. At what height is the Starfighter? A Cessna is flying 2655 metres below the Starfighter. At what height is the Cessna?

12. On a seaside golf course, the first tee is 32 metres above sea level, and the fifth tee is 57 metres above sea level. By how many metres does the fifth tee stand above the first tee? If the twelfth tee is 46 metres above the first tee, how many metres above sea level is the twelfth tee? By how many metres does the twelfth tee stand above the fifth tee?

13. A newsagent sells 56 magazines one Friday. If he sells 17 more newspapers than magazines, how many newspapers does he sell? If he sells 13 more magazines than paperback books, how many paperbacks does he sell? How many more newspapers than paperback books does he sell?

14. Charlie sells 197 ice cream cones from his van. Compared to his ice cream sales, he sells 47 more ice lollies and 28 fewer choc ices. How many choc ices does he sell? How many ice lollies does he sell? How many more ice lollies than choc ices does he sell?

15. There are 97 more patrons in the Royal Cinema than in the Ritz. If there are 285 patrons in the Ritz, how many are in the Royal Cinema? If there are 106 more patrons in the Palace Cinema than in the Ritz, how many are in the Palace Cinema? How many more patrons are in the Palace Cinema than in the Royal Cinema?

To multiply by 10, move the figures one place to the left.

Example 1

$$36 \times 10 = 360$$

Hundreds	Tens	Units
	3	6

Hundreds	Tens	Units
3	6	0

To multiply by 100, move the figures two places to the left.

Example 2

$$36 \times 100 = 3600$$

Th	H	T	U		Th	H	T	U
		3	6		3	6	0	0

Exercise 1.2a

Multiply each of the following
(a) by 10;   (b) by 100:

1. 7   2. 14   3. 25   4. 37
5. 10   6. 70   7. 220   8. 400
9. 100   10. 786

Copy the following and fill in the empty spaces.
11. 8 × 10 =   16. 30 ×   = 3000
12.   × 10 = 160   17.   × 100 = 12 000
13. 32 × 100 =   18. 500 × 10 =
14. 46 ×   = 4600   19.   × 10 = 9000
15.   × 100 = 4900   20.   × 100 = 30 000

To divide by 10, move the figures one place to the right.

Example 3

$$250 \div 10 = 25$$

H	T	U		H	T	U
2	5	0			2	5

To divide by 100, move the figures two places to the right.

Example 4

$$4600 \div 100 = 46$$

Th	H	T	U		Th	H	T	U
4	6	0	0				4	6

Exercise 1.2b

Divide each of the following
(a) by 10;   (b) by 100:

1. 6600   2. 9800   3. 10 500
4. 12 000   5. 20 000   6. 125 000
7. 301 000   8. 1000   9. 200 000
10. 10 000

Copy the following and fill in the empty spaces.
11. 720 ÷ 10 =   16. 7000 ÷   = 70
12.   ÷ 10 = 36   17.   ÷ 100 = 180
13. 2800 ÷ 100 =   18. 4000 ÷ 10 =
14. 5300 ÷   = 53   19.   ÷ 10 = 300
15.   ÷ 100 = 65   20.   ÷ 100 = 900

To multiply by a whole number ending in 0, multiply by 10, then multiply the answer by the number in front of the 0.

Example 5

(a) $12 \times 20 = 120 \times 2 = 240$
(b) $64 \times 40 = 640 \times 4 = 2560$
(c) $15 \times 120 = 150 \times 12 = 1800$

To multiply by a whole number ending in 00, multiply by 100, then multiply the answer by the number in front of the 00.

Example 6

(a) $12 \times 200 = 1200 \times 2 = 2400$
(b) $64 \times 400 = 6400 \times 4 = 25600$
(c) $15 \times 1200 = 1500 \times 12 = 18000$

Exercise 1.2c

Multiply each of the following by 20:
1. 6   2. 13   3. 21   4. 27
5. 112   6. 163   7. 225   8. 384
9. 97   10. 500

Multiply each of the following by 70:
11. 4   12. 15   13. 19   14. 26
15. 50   16. 112   17. 208   18. 139
19. 350   20. 600

Multiply each of the following by 300:
21. 8   22. 11   23. 17   24. 29
25. 47   26. 60   27. 121   28. 150
29. 700   30. 3000

Copy the following and fill in the empty spaces.
31. $18 \times 30 =$ 36. $42 \times \quad = 8400$
32. $\quad \times 30 = 270$ 37. $\quad \times 40 = 1680$
33. $15 \times 500 =$ 38. $63 \times 90 =$
34. $36 \times \quad = 720$ 39. $\quad \times 60 = 1260$
35. $\quad \times 200 = 1800$ 40. $\quad \times 400 = 28\,000$

To divide by a whole number ending in 0,
divide by 10, then divide the answer by the
number in front of the 0.

Example 7

(a) $1200 \div 20 = 120 \div 2 = 60$
(b) $640 \div 40 = 64 \div 4 = 16$
(c) $9600 \div 120 = 960 \div 12 = 80$

To divide by a whole number ending in 00,
divide by 100, then divide the answer by the
number in front of the 00.

Example 8

(a) $1200 \div 200 = 12 \div 2 = 6$
(b) $3900 \div 1300 = 39 \div 13 = 3$
(c) $96\,000 \div 1200 = 960 \div 12 = 80$

Exercise 1.2d

Divide the following by 20:
1. 120   2. 300   3. 240   4. 4000
5. 5000   6. 6800   7. 80 000   8. 12 000
9. 7200   10. 108 000

Divide each of the following by 50:
11. 750   12. 950   13. 4000   14. 7000
15. 11 000 16. 60 000   17. 85 000   18. 5700
19. 24 000 20. 125 000

Divide each of the following by 700:
21. 1400   22. 3500   23. 7700   24. 21 000
25. 49 000 26. 43 400   27. 62 300   28. 5600
29. 88 200 30. 175 000

Copy the following and fill in the empty spaces.
31. $280 \div 20$ 36. $14\,000 \div \quad = 70$
32. $\quad \div 20 = 15$ 37. $\quad \div 50 = 25$
33. $63\,000 \div 700 =$ 38. $7500 \div 150 =$
34. $3300 \div \quad = 11$ 39. $\quad \div 200 = 350$
35. $840 \div \quad = 21$ 40. $\quad \div 60 = 35$

Example 9

Find the 'odd answer out' of:
(a) 232×4 (b) 34×27 (c) 51×18

(a)   232   (b)   34
      $\times$ 4 $\times$ 27
      ─────   ─────
         928   238   (34×7)
                               680   (34×20)
                               ─────
                                918
                               ═════

(c)   51
      $\times$ 18
      ─────
         408   (51×8)
         510   (51×10)
      ─────
         918

So (a) is the 'odd answer out' because its
answer is 928.

Exercise 1.2e

Find the 'odd answer out' for the following:
1. (a)   6   (b)   12   (c)   7
       $\times$ 8 $\times$ 4 $\times$ 7

2. (a)   36   (b)   29   (c)   16
       $\times$ 4 $\times$ 5 $\times$ 9

3. (a)   75   (b)   124   (c)   93
       $\times$ 5 $\times$ 3 $\times$ 4

4. (a)   117   (b)   146   (c)   195
       $\times$ 5 $\times$ 4 $\times$ 3

5. (a)   394   (b)   709   (c)   591
       $\times$ 9 $\times$ 5 $\times$ 6

6. (a)   459   (b)   328   (c)   287
              × 5   × 7   × 8

7. (a) 81 × 11   **8.** (a) 63 × 12
(b) 223 × 4   (b) 69 × 11
(c) 99 × 9   (c) 84 × 9

9. (a) 21 × 18   **10.** (a) 32 × 15
(b) 22 × 17   (b) 30 × 16
(c) 27 × 14   (c) 27 × 18

11. (a) 56 × 24   **12.** (a) 49 × 36
(b) 64 × 21   (b) 104 × 17
(c) 52 × 26   (c) 126 × 14

Example 10

Find the 'odd answer out' of:
(a) 161 ÷ 7   (b) 207 ÷ 9   (c) 192 ÷ 8

$$\text{(a) } 7\overline{)161}^{\;23} \qquad \text{(b) } 9\overline{)207}^{\;23} \qquad \text{(c) } 8\overline{)192}^{\;24}$$

So (c) is the 'odd answer out' because its answer is 24.

Exercise 1.2f

Find the 'odd answer out' for the following:

1. (a) 330 ÷ 6   **2.** (a) 220 ÷ 5
(b) 270 ÷ 5   (b) 352 ÷ 8
(c) 495 ÷ 9   (c) 270 ÷ 6

3. (a) 2630 ÷ 5   **4.** (a) 2275 ÷ 7
(b) 3162 ÷ 6   (b) 2916 ÷ 9
(c) 2108 ÷ 4   (c) 2600 ÷ 8

5. (a) 944 ÷ 4   **6.** (a) 1309 ÷ 7
(b) 1180 ÷ 5   (b) 1488 ÷ 8
(c) 711 ÷ 3   (c) 930 ÷ 5

7. (a) 3708 ÷ 3   **8.** (a) 7242 ÷ 6
(b) 4940 ÷ 4   (b) 6030 ÷ 5
(c) 6180 ÷ 5   (c) 8442 ÷ 7

9. (a) 6510 ÷ 5   **10.** (a) 2928 ÷ 4
(b) 5212 ÷ 4   (b) 2196 ÷ 3
(c) 9121 ÷ 7   (c) 4386 ÷ 6

11. (a) 2808 ÷ 12   **12.** (a) 1023 ÷ 11
(b) 2585 ÷ 11   (b) 1104 ÷ 12
(c) 2106 ÷ 9   (c) 736 ÷ 8

Example 11

Find the 'odd answer out' of:
(a) 1357 ÷ 23
(b) 1020 ÷ 17
(c) 4602 ÷ 78

```
           59
(a)  23)1357
         115 ¦   (5 × 23)
         207
         207     (9 × 23)
```

```
           60
(b)  17)1020
         102 ¦   (17 × 6)
           0
           0     (17 × 0)
```

```
           59
(c)  78)4602
         390 ¦   (78 × 5)
         702
         702     (78 × 9)
```

So (b) is the 'odd answer out' because its answer is 60.

Exercise 1.2g

Find the 'odd answer out' for the following:

1. (a) 540 ÷ 15   **2.** (a) 602 ÷ 14
(b) 630 ÷ 18   (b) 688 ÷ 16
(c) 468 ÷ 13   (c) 756 ÷ 18

3. (a) 715 ÷ 13   **4.** (a) 736 ÷ 23
(b) 1026 ÷ 19   (b) 672 ÷ 21
(c) 935 ÷ 17   (c) 858 ÷ 26

5. (a) 1056 ÷ 24   **6.** (a) 925 ÷ 25
(b) 990 ÷ 22   (b) 1008 ÷ 28
(c) 1100 ÷ 25   (c) 936 ÷ 26

7. (a) 1921 ÷ 17   **8.** (a) 2546 ÷ 19
(b) 2147 ÷ 19   (b) 2144 ÷ 16
(c) 1792 ÷ 16   (c) 2430 ÷ 18

9. (a) 3749 ÷ 23   **10.** (a) 1664 ÷ 32
(b) 4727 ÷ 29   (b) 1908 ÷ 36
(c) 3444 ÷ 21   (c) 1855 ÷ 35

11. (a) 1107 ÷ 41   **12.** (a) 2013 ÷ 61
(b) 980 ÷ 35   (b) 1760 ÷ 55
(c) 1512 ÷ 54   (c) 1485 ÷ 45

Exercise 1.2h

1. John weighs 21 kilograms. If his father is four times heavier, how much does his father weigh? If the father is three times heavier than John's older brother David, how heavy is David?

2. Mary is four times heavier than her baby sister Julie. If Mary weighs 60 kilograms, how much does Julie weigh? If their mother is five times heavier than Julie, how much does their mother weigh?

3. The distance from Birmingham to Evesham is 48 kilometres. If Bristol is three times further away from Birmingham than Evesham, how far is it from Birmingham to Bristol? If Bristol is eight times further away from Birmingham than Alverchurch, how far is it from Birmingham to Alverchurch?

4. Ann, Janet and Christine have picked some flowers. Ann has picked 54, three times as many as Janet has. How many has Janet picked? How many has Christine picked if she has picked five times as many as Janet? How many less than 200 have they picked together?

5. A wooden crate weighs 56 kilograms. If a steel container is three times heavier, how much does the container weigh? If the steel container is seven times heavier than a plastic box, how heavy is the plastic box?

6. A mobile library contains 162 fiction books. If a public library contains twenty times as many fiction books, how many does it contain? If the public library contains six times as many fiction books as a school library, how many are in the school library?

7. One Thursday, there are 96 visitors to an ancient castle. If there are fifteen times as many visitors to a leisure centre, how many people visited the leisure centre? If the leisure centre had 12 times as many visitors as an art exhibition, how many visitors went to the exhibition?

8. A jeweller has 65 watches. If he has eighteen times as many rings as watches, how many rings has he? If he has thirty times fewer clocks than rings, how many clocks has he?

9. A confectioner sells 87 packets of sweets. If she sells 25 times as many ice creams, how many ice creams does she sell? If she sells 15 times as many ice creams as she does boxes of chocolates, how many boxes does she sell?

10. A farmer owns 18 cows. If he owns 40 times as many sheep, how many sheep does he own? If he owns 24 times as many sheep as hens, how many hens has he?

11. A magazine has 78 pages, and a book has 9 times as many pages as the magazine. How many pages has the book? If a pamphlet has 54 times fewer pages than the book, how many pages has the pamphlet?

12. A laboratory has 27 Bunsen burners and 28 times as many test tubes. How many test tubes are there? If there are 21 times as many test tubes as there are measuring cylinders, how many measuring cylinders are there?

13. A man travelled 8000 kilometres in his car last year, and a businessman travelled 6 times as far. How far did the businessman travel? If the businessman travelled 32 times as far as a young woman, how far did the young woman travel?

14. A wholesaler sells 1800 eggs. If he sells 400 times as many bags of crisps, how many bags does he sell? If he sells 60 times as many eggs as he does boxes of chocolates, how many boxes does he sell?

15. There are 306 cars in a parking area. If there are 18 times fewer cars parked in a lane, how many cars are there in the lane? A multi-storey car park holds 280 times as many cars as are in the lane. How many cars does the multi-storey hold?

To add and subtract a series of numbers, add the positive numbers first; then do the subtractions.

Example 1

(a) $12 - 8 + 6 = 12 + 6 - 8$
$= 18 - 8 = 10$
(b) $12 - 20 + 19 = 12 + 19 - 20$
$= 31 - 20 = 11$
(c) $14 - 16 + 8 - 4 = 14 + 8 - 16 - 4$
$= 22 - 16 - 4 = 6 - 4$
$= 2$

Exercise 1.3a

1. $4 + 3 - 2$ 2. $4 - 3 + 2$
3. $6 - 8 + 12$ 4. $12 + 8 - 10$
5. $12 - 14 + 5$ 6. $22 + 14 - 28$
7. $13 - 25 + 12$ 8. $33 - 14 + 17$
9. $15 - 40 + 53$ 10. $16 - 35 + 22$
11. $1 - 9 + 14$ 12. $26 - 26 + 14$
13. $11 - 35 + 42$ 14. $9 - 18 + 13$
15. $23 + 15 - 32$ 16. $37 - 42 + 16$
17. $29 - 18 + 7$ 18. $85 - 94 + 23$
19. $63 - 44 + 17$ 20. $108 + 54 - 130$
21. $5 + 3 - 2 - 3$ 22. $8 - 9 + 4 - 2$
23. $16 - 4 + 5 - 13$ 24. $12 - 20 - 4 + 32$
25. $5 - 19 - 21 + 43$ 26. $17 - 13 + 12 + 14$
27. $18 - 23 + 19 - 14$ 28. $31 - 42 + 16 - 3$
29. $22 + 8 - 16 - 12$ 30. $13 - 39 - 27 + 63$
31. $36 - 27 + 9 - 14$ 32. $86 - 57 - 39 + 22$
33. $74 - 92 - 27 + 58$ 34. $12 + 28 - 13 - 19$
35. $16 - 4 - 9 - 2$ 36. $84 - 108 + 16 + 23$
37. $118 - 240 + 165 - 36$ 38. $360 + 482 - 561 - 17$
39. $42 - 328 + 218 + 96$ 40. $456 - 172 - 19 - 104$

A calculation in brackets must be done before any other; and division and multiplication must be done before addition or subtraction.

Example 2

(a) $3 + 14 \times 2 = 3 + 28 = 31$
(b) $(3 + 14) \times 2 = 17 \times 2 = 34$
(c) $5 \times 3 + 4 \div 2 = 15 + 2 = 17$

Exercise 1.3b

1. $6 + 12 \times 3$ 2. $14 + 6 \times 8$
3. $9 + 7 \times 11$ 4. $10 \times 9 + 2$
5. $15 \times 3 + 16$ 6. $9 - 4 \times 2$
7. $7 \times 2 - 2$ 8. $33 - 9 \times 3$
9. $11 \times 6 - 5$ 10. $28 - 14 \times 2$
11. $12 + 4 \div 2$ 12. $9 + 6 - 3$
13. $72 - 8 + 1$ 14. $30 \div 3 + 7$
15. $24 + 48 \div 12$ 16. $18 - 6 \div 3$
17. $24 - 8 \div 4$ 18. $40 \div 8 - 4$
19. $50 \div 10 - 5$ 20. $276 - 108 \div 12$
21. $(4 + 2) \times 3$ 22. $5 \times (3 + 2)$
23. $(12 + 5) \times 2$ 24. $(4 + 2) \times 13$
25. $12 \times (14 + 27)$ 26. $(12 - 5) \times 2$
27. $(34 - 16) \times 4$ 28. $(26 - 13) \times 5$
29. $9 \times (58 - 41)$ 30. $(90 - 45) \times 3$
31. $(32 + 24) \div 7$ 32. $(19 + 89) \div 12$
33. $39 \div (9 + 4)$ 34. $(25 - 5) \div 5$
35. $(92 - 29) \div 9$ 36. $132 \div (27 - 15)$
37. $6 \times 4 + 3 \times 9$ 38. $9 \times 7 - 4 \times 8$
39. $14 \times 6 + 72 \div 8$ 40. $484 \div 22 - 6 \times 3$

1.4 NUMBERS AND FACTORS

The *multiples* of 6 are:
6, (1×6); 12, (2×6); 18; 24; etc.

Example 1

List the first five multiples of 12.
The multiples are:
12, (1×12); 24, (2×12); 36; 48; 60.

Exercise 1.4a

List the first five multiples of:
1. 2   2. 5   3. 8   4. 7
5. 11   6. 20   7. 30   8. 60
9. 15   10. 25   11. 16   12. 18
13. 14   14. 13   15. 24   16. 21
17. 45   18. 36   19. 51   20. 72

6 can be divided by 1, 2, 3, and 6.
These numbers are the *factors* of 6.

Example 2

Find the factors of 72.
72 = 1 × 72 or 2 × 36 or 3 × 24
  or 4 × 18 or 6 × 12 or 8 × 9.
So the factors of 72 are
1, 2, 3, 4, 6, 8, 9, 12, 18, 24, 36, and 72.

Exercise 1.4b

Find the factors of:

1. 3	**2.** 8	**3.** 10	**4.** 12
5. 15	**6.** 18	**7.** 30	**8.** 27
9. 24	**10.** 32	**11.** 45	**12.** 40
13. 54	**14.** 42	**15.** 60	**16.** 48
17. 63	**18.** 84	**19.** 66	**20.** 72

A *square* number is the product of two identical factors.

Example 3

(a) 16 = 4 × 4, so 16 is a square number.
(b) 144 = 12 × 12, so 12 is a square number.
Note that 1 is also a square number.
Square numbers are sometimes called *perfect squares*.

Exercise 1.4c

List the factors of the following numbers and state which are square numbers.

1. 2	**2.** 5	**3.** 6	**4.** 7
5. 11	**6.** 20	**7.** 35	**8.** 4
9. 50	**10.** 25	**11.** 16	**12.** 22
13. 14	**14.** 13	**15.** 28	**16.** 21
17. 49	**18.** 36	**19.** 51	**20.** 70

A *prime* number has only two different factors.

Example 4

(a) 2 = 1 × 2, so 2 is a prime number.
(b) 13 = 1 × 13, so 13 is a prime number.
(c) 51 = 1 × 51 or 3 × 17.
  So 51 is not a prime number.

(d) 1 has only one factor, so 1 is *not* a prime number.

Exercise 1.4d

List the factors of the following numbers and state which are prime numbers.

1. 17	**2.** 33	**3.** 26	**4.** 29
5. 39	**6.** 57	**7.** 37	**8.** 91
9. 87	**10.** 53	**11.** 111	**12.** 97
13. 61	**14.** 67	**15.** 119	**16.** 103
17. 73	**18.** 123	**19.** 101	**20.** 117

The *prime factors* of 6 are 2 and 3; these are the prime numbers which are also factors of 6.

Example 5

Find the prime factors of 72.
72 = 2 × 36
  = 2 × 2 × 18
  = 2 × 2 × 2 × 9
  = 2 × 2 × 2 × 3 × 3
So the prime factors of 72 are
2 × 2 × 2 × 3 × 3.

Exercise 1.4e

Find the prime factors of:

1. 10	**2.** 15	**3.** 28	**4.** 24
5. 36	**6.** 40	**7.** 45	**8.** 54
9. 70	**10.** 63	**11.** 78	**12.** 60
13. 66	**14.** 112	**15.** 96	**16.** 162
17. 154	**18.** 180	**19.** 168	**20.** 216

The largest factor of both 16 and 28 is 4; 4 is called the *highest common factor* (H.C.F.) of 16 and 28.

Example 6

Find the H.C.F. of 48 and 60.
48 = 2 × 24
  = 2 × 2 × 12
  = 2 × 2 × 2 × 6
  = 2 × 2 × 2 × 2 × 3

$60 = 2 \times 30$
$\quad = 2 \times 2 \times 15$
$\quad = 2 \times 2 \times 3 \times 5$

The common factors (heavy type) are
2, 2, and 3.
So the H.C.F. of 48 and 60 is
$2 \times 2 \times 3 = 12$.

Exercise 1.4f

Find the H.C.F. of:

1. 6 and 8	2. 12 and 18
3. 30 and 24	4. 36 and 27
5. 56 and 42	6. 72 and 48
7. 54 and 36	8. 63 and 84
9. 75 and 105	10. 64 and 80
11. 54, 42, and 78	12. 36, 44, and 68
13. 70, 84, and 98	14. 64, 96, and 112
15. 56, 72, and 104	16. 66, 84, and 90
17. 108, 132, and 156	18. 52, 65, and 91
19. 96, 128, and 144	20. 24, 36, and 52

The smallest multiple of both 16 and 28
is 112; 112 is called the *lowest common
multiple* (L.C.M.) of 16 and 28.

Example 7

Find the L.C.M. of 48 and 60.

$1 \times 48 = 48$	$1 \times 60 = 60$
$2 \times 48 = 96$	$2 \times 60 = 120$
$3 \times 48 = 144$	$3 \times 60 = 180$
$4 \times 48 = 192$	$4 \times 60 = 240$
$5 \times 48 = 240$	

So 240 is the L.C.M. of 48 and 60.

Exercise 1.4g

Find the L.C.M. of:

1. 6 and 8	2. 10 and 15
3. 8 and 12	4. 6 and 9
5. 9 and 12	6. 10 and 12
7. 6 and 15	8. 8 and 20
9. 9 and 15	10. 12 and 16
11. 3, 4, and 8	12. 3, 9, and 12
13. 5, 8, and 10	14. 4, 5, and 6
15. 6, 9, and 12	16. 10, 12, and 15
17. 9, 12, and 16	18. 15, 20, and 24
19. 15, 18, and 30	20. 24, 36, and 48

Exercise 1.4h

Find the answer to each part of the following and
name the 'odd answer out'.

1. (a) Sum of the first six prime numbers.
   (b) L.C.M. of 5 and 8.
   (c) Sum of both square numbers between 10
       and 30.
2. (a) Sum of all prime numbers less than 10.
   (b) H.C.F. of 56 and 70.
   (c) Sum of the first three square numbers.
3. (a) Sum of both prime numbers between 15 and
       20.
   (b) L.C.M. of 12 and 9.
   (c) Sum of the 3rd and 5th square numbers.
4. (a) Difference between both prime numbers
       between 30 and 40.
   (b) H.C.F. of 36, 42, and 54.
   (c) Difference between the 4th and the 3rd
       square numbers.
5. (a) Sum of the nearest two prime numbers to 30.
   (b) L.C.M. of 6, 9, and 27.
   (c) Sum of all square numbers which are greater
       than 1 but less than 30.
6. (a) Sum of all prime numbers between 20 and 40.
   (b) L.C.M. of 8, 15, and 20.
   (c) Sum of all square numbers between 15 and 50.
7. (a) Product of the 1st and the 4th prime numbers.
   (b) H.C.F. of 60, 90, and 105.
   (c) Difference between the 8th and the 7th
       square numbers.
8. (a) Sum of the first three prime numbers that end
       with a 7.
   (b) L.C.M. of 15, 20, and 30.
   (c) Difference between the first two square num-
       bers that end with a 4.
9. (a) Product of the highest prime number which
       is less than 30 and the only even prime
       number.
   (b) L.C.M. of 7, 8, and 14.
   (c) Sum of the first two square numbers that
       end with a 9.
10. (a) Difference between the highest prime num-
        ber which is less than 40 and the lowest one
        which is greater than 20.
    (b) H.C.F. of 80, 96, and 144.
    (c) Difference between the 5th and the 3rd
        square numbers.

The decimal point separates whole numbers from fractions.

Example 1

3·5 is 3 units and 5 tenths.
3·55 is 3 units, 5 tenths, and 5 hundredths,
(or 3 units and 55 hundredths).
3·555 is 3 units, 5 tenths, 5 hundredths
and 5 thousandths,
(or 3 units and 555 thousandths).

Exercise 1.5a

Give the value of the underlined figures:
1. 2·7$\underline{5}$
2. 10·6$\underline{4}$
3. 5$\underline{2}$3·92
4. 5·38$\underline{5}$
5. 9·23$\underline{6}$
6. 31·5$\underline{6}$7
7. 25·3$\underline{2}$8
8. 28$\underline{6}$·4
9. 32$\underline{5}$·272
10. 200·00$\underline{3}$

When adding or subtracting decimal fractions, the decimal points are put underneath each other to make sure that each figure is in its correct column.

Example 2

Find the 'odd answer out' of:
(a) 13·46 + 1·2 + 75·36
(b) 7·34 + 24 + 58·78
(c) 32·14 + 25·08 + 32·9

$$
\begin{array}{lll}
\text{(a)} \quad 13{\cdot}46 & \text{(b)} \quad 7{\cdot}34 & \text{(c)} \quad 32{\cdot}14 \\
\quad\;\; 1{\cdot}2 & \quad\;\; 24 & \quad\;\; 25{\cdot}08 \\
+ 75{\cdot}36 & + 58{\cdot}78 & + 32{\cdot}9 \\
\hline
\;\; 90{\cdot}02 & \;\; 90{\cdot}12 & \;\; 90{\cdot}12
\end{array}
$$

So (a) is the 'odd answer out' because its answer is 90·02.

Example 3

Find the 'odd answer out' of:
(a) 24·68 − 13·24
(b) 39 − 27·46   (c) 59·8 − 48·36

$$
\begin{array}{lll}
\text{(a)} \quad 24{\cdot}68 & \text{(b)} \quad 39{\cdot}00 & \text{(c)} \quad 59{\cdot}80 \\
- 13{\cdot}24 & - 27{\cdot}46 & - 48{\cdot}36 \\
\hline
\;\; 11{\cdot}44 & \;\; 11{\cdot}54 & \;\; 11{\cdot}44
\end{array}
$$

So (b) is the 'odd answer out' because its answer is 11·54.

Exercise 1.5b

$$
\begin{array}{ll}
\text{1.} \quad 36{\cdot}2 & \text{2.} \quad 27{\cdot}8 \\
\quad + 1{\cdot}4 & \quad + 3{\cdot}4 \\
\end{array}
$$

$$
\begin{array}{ll}
\text{3.} \quad 32{\cdot}1 & \text{4.} \quad 53{\cdot}25 \\
\quad + 3{\cdot}75 & \quad + 18{\cdot}86 \\
\end{array}
$$

5. 81·25 + 6·584
6. 39·4 + 3·785
7. 21·6 + 9·3 + 7·26
8. 23·4 + 5·87 + 32·115
9. 12·06 + 4·5 + 0·375
10. 1·26 + 4·735 + 19

$$
\begin{array}{ll}
\text{11.} \quad 3{\cdot}8 & \text{12.} \quad 4{\cdot}3 \\
\quad\; - 2{\cdot}4 & \quad\; - 2{\cdot}5 \\
\end{array}
$$

$$
\begin{array}{ll}
\text{13.} \quad 6{\cdot}75 & \text{14.} \quad 8{\cdot}37 \\
\quad\; - 2{\cdot}9 & \quad\; - 4{\cdot}65 \\
\end{array}
$$

15. 9·2 − 5·45
16. 20·32 − 13·17
17. 13·76 − 9·38
18. 38·2 − 19·35
19. 135·35 − 28·7
20. 101·26 − 59·37

Find the 'odd answer out' for the following:
21. (a) 38·7 + 1·9 + 3·25
    (b) 25 + 1·65 + 17·1
    (c) 32·35 + 1·95 + 9·55
22. (a) 21·25 + 12 + 3·85
    (b) 24·325 + 9·2 + 3·575
    (c) 19·3 + 4·35 + 13·55
23. (a) 4·375 + 0·585 + 12·32
    (b) 8·36 + 5·325 + 3·495
    (c) 9·8 + 5 + 2·38
24. (a) 3·27 + 5·19 + 11·38
    (b) 8 + 1·345 + 10·395
    (c) 12·3 + 0·975 + 6·465
25. (a) 10·2 + 1·065 + 15·005
    (b) 11·32 + 5 + 9·85
    (c) 6·005 + 0·845 + 19·42
26. (a) 25·29 − 11·56
    (b) 37·13 − 23·5
    (c) 45·2 − 31·47

27. (a) $8·325 - 3·65$
    (b) $9·34 - 4·665$
    (c) $8 - 3·425$
28. (a) $13·285 - 9·29$
    (b) $7·6 - 3·615$
    (c) $10·36 - 6·375$
29. (a) $14·27 - 9·145$ 30. (a) $153·23 - 28·78$
    (b) $13·065 - 7·84$ (b) $159·4 - 35·05$
    (c) $5·6 - 0·475$ (c) $132 - 7·65$

To multiply by 10, move the figures one place to the left.

Example 4

$$3·6 \times 10 = 36$$

Units	tenths		T	U	t
3	6		3	6	

$$0·36 \times 10 = 3·6$$

U	t	h		U	t
0	3	6		3	6

To multiply by 100, move the figures two places to the left.

Example 5

$$3·6 \times 100 = 360$$
$$0·36 \times 100 = 36$$

Exercise 1.5c

Multiply each of the following:
(a) by 10; (b) by 100:
1. $2·4$ 2. $12·2$ 3. $3·75$
4. $15·36$ 5. $2·135$ 6. $18·576$
7. $0·85$ 8. $0·7$ 9. $0·2368$
10. $0·0139$

Copy the following and fill in the empty spaces.
11. $3·6 \times 10 =$
12. $\quad \times 10 = 45$
13. $2·9 \times 100 =$
14. $\quad \times 100 = 320$
15. $0·53 \times \quad = 53$
16. $0·9 \times 100 =$
17. $\quad \times 100 = 80$
18. $0·02 \times \quad = 2$

19. $0·004 \times 10 =$
20. $\quad \times 10 = 0·015$

To divide by 10, move the figures one place to the right.

Example 6

$$36 \div 10 = 3·6$$
$$3·6 \div 10 = 0·36$$

To divide by 100, move the figures two places to the right.

Example·7

$$360 \div 100 = 3·6$$
$$3·6 \div 100 = 0·036$$

Exercise 1.5d

Divide each of the following
(a) by 10; (b) by 100:
1. $25·3$ 2. $38·16$ 3. $6·25$
4. $7·35$ 5. 36 6. 60
7. $20·4$ 8. $100·3$ 9. $0·85$
10. $0·032$

Copy the following and fill in the empty spaces.
11. $4·2 \div 10 =$
12. $\quad \div 10 = 0·51$
13. $36 \div 100 =$
14. $\quad \div 100 = 4·8$
15. $65 \div \quad = 0·65$
16. $60 \div 100 =$
17. $\quad \div 100 = 0·3$
18. $5 \div \quad = 0·05$
19. $0·09 \div 10 =$
20. $\quad \div 10 = 0·0012$

Multiply a decimal by a whole number as follows.

Example 8

$$3·6 \times 3 = 10·8$$

(a) Multiply 36 by 3 to give 108.
(b) Count the number of decimal places, i.e. one.
(c) Put this number of places (one) in the answer to give 10·8.

Exercise 1.5e

1. $1 \cdot 2 \times 3$	**2.** $0 \cdot 8 \times 2$
3. $3 \cdot 8 \times 4$	**4.** $4 \cdot 3 \times 8$
5. $7 \cdot 5 \times 5$	**6.** $2 \cdot 16 \times 7$
7. $3 \cdot 12 \times 9$	**8.** $5 \cdot 15 \times 12$
9. $0 \cdot 24 \times 8$	**10.** $1 \cdot 36 \times 16$
11. $2 \cdot 34 \times 15$	**12.** $0 \cdot 92 \times 16$
13. $0 \cdot 015 \times 9$	**14.** $0 \cdot 018 \times 7$
15. $0 \cdot 028 \times 5$	**16.** $0 \cdot 0036 \times 8$
17. $0 \cdot 0045 \times 6$	**18.** $0 \cdot 24 \times 30$
19. $0 \cdot 018 \times 50$	**20.** $0 \cdot 026 \times 15$

Multiply a decimal by a decimal as follows.

Example 9

$$0 \cdot 136 \times 0 \cdot 23 = 0 \cdot 031\ 28$$

(a) Multiply 136 by 23 to give 3128.
(b) Count the number of decimal places, i.e. five.
(c) Put this number of places (five) in the answer to give $0 \cdot 031\ 28$.

Exercise 1.5f

1. $3 \cdot 7 \times 0 \cdot 6$	**2.** $9 \cdot 2 \times 0 \cdot 3$
3. $8 \cdot 6 \times 0 \cdot 7$	**4.** $3 \cdot 12 \times 0 \cdot 8$
5. $8 \cdot 25 \times 0 \cdot 4$	**6.** $6 \cdot 31 \times 0 \cdot 12$
7. $51 \cdot 5 \times 0 \cdot 9$	**8.** $32 \cdot 4 \times 0 \cdot 6$
9. $86 \cdot 5 \times 0 \cdot 12$	**10.** $4 \cdot 25 \times 0 \cdot 13$
11. $5 \cdot 12 \times 0 \cdot 16$	**12.** $4 \cdot 32 \times 0 \cdot 15$
13. $23 \cdot 2 \times 0 \cdot 14$	**14.** $52 \cdot 1 \times 0 \cdot 15$
15. $0 \cdot 246 \times 0 \cdot 4$	**16.** $0 \cdot 113 \times 0 \cdot 8$
17. $1 \cdot 24 \times 0 \cdot 05$	**18.** $5 \cdot 25 \times 0 \cdot 012$
19. $6 \cdot 34 \times 0 \cdot 015$	**20.** $4 \cdot 82 \times 0 \cdot 016$
21. $1 \cdot 25 \times 2 \cdot 3$	**22.** $7 \cdot 48 \times 1 \cdot 9$
23. $4 \cdot 2 \times 3 \cdot 05$	**24.** $6 \cdot 9 \times 4 \cdot 72$
25. $1 \cdot 05 \times 9 \cdot 4$	**26.** $3 \cdot 35 \times 2 \cdot 72$
27. $64 \cdot 2 \times 1 \cdot 3$	**28.** $17 \cdot 9 \times 5 \cdot 4$
29. $96 \cdot 7 \times 12 \cdot 4$	**30.** $10 \cdot 4 \times 87 \cdot 2$

Divide a decimal by a whole number as follows.

Example 10

$$6 \cdot 25 \div 25 = 0 \cdot 25$$

```
        0·25
    25)6·25
        5 0
        1 25
        1 25
```

Exercise 1.5g

1. $0 \cdot 6 \div 2$	**2.** $0 \cdot 8 \div 4$
3. $0 \cdot 72 \div 3$	**4.** $0 \cdot 98 \div 7$
5. $0 \cdot 858 \div 6$	**6.** $1 \cdot 274 \div 7$
7. $2 \cdot 448 \div 9$	**8.** $3 \cdot 675 \div 15$
9. $6 \cdot 048 \div 24$	**10.** $7 \cdot 072 \div 32$
11. $1 \cdot 032 \div 12$	**12.** $1 \cdot 245 \div 15$
13. $0 \cdot 095 \div 5$	**14.** $0 \cdot 0868 \div 7$
15. $0 \cdot 336 \div 8$	**16.** $0 \cdot 837 \div 9$
17. $0 \cdot 585 \div 13$	**18.** $0 \cdot 756 \div 18$
19. $0 \cdot 325 \div 25$	**20.** $0 \cdot 451 \div 41$

Divide a decimal by a decimal as follows.

Example 11

$$3 \cdot 128 \div 2 \cdot 3$$

(a) Turn the divisor into a whole number by multiplying it by 10, 100, etc:
$$2 \cdot 3 \times 10 = 23$$
(b) Multiply the dividend by the same number: $\quad 3 \cdot 128 \times 10 = 31 \cdot 28$
(c) Then divide as shown in Example 10.
$3 \cdot 128 \div 2 \cdot 3 = 31 \cdot 28 \div 23 = 1 \cdot 36$.

Exercise 1.5h

1. $8 \cdot 48 \div 0 \cdot 4$	**2.** $8 \cdot 37 \div 0 \cdot 9$
3. $7 \cdot 02 \div 0 \cdot 6$	**4.** $1 \cdot 05 \div 0 \cdot 7$
5. $5 \cdot 45 \div 0 \cdot 5$	**6.** $14 \cdot 4 \div 0 \cdot 8$
7. $11 \cdot 7 \div 0 \cdot 9$	**8.** $0 \cdot 42 \div 0 \cdot 03$
9. $1 \cdot 26 \div 0 \cdot 09$	**10.** $3 \cdot 25 \div 0 \cdot 05$
11. $252 \div 0 \cdot 4$	**12.** $365 \div 0 \cdot 5$
13. $828 \div 0 \cdot 9$	**14.** $325 \div 0 \cdot 05$
15. $0 \cdot 285 \div 0 \cdot 15$	**16.** $0 \cdot 196 \div 0 \cdot 14$
17. $2 \cdot 16 \div 0 \cdot 18$	**18.** $6 \cdot 75 \div 0 \cdot 25$
19. $22 \cdot 4 \div 0 \cdot 16$	**20.** $19 \cdot 5 \div 0 \cdot 15$

21. $6\cdot3 \div 2\cdot1$
22. $27\cdot5 \div 2\cdot5$
23. $13\cdot2 \div 1\cdot2$
24. $6\cdot25 \div 1\cdot25$
25. $264 \div 2\cdot4$
26. $4\cdot68 \div 1\cdot8$
27. $19\cdot84 \div 3\cdot2$
28. $0\cdot516 \div 0\cdot24$
29. $0\cdot0391 \div 1\cdot7$
30. $122\cdot55 \div 4\cdot3$

To multiply by a whole number ending in 0, multiply by 10, then multiply the answer by the number in front of the 0.

Example 12

(a) $1\cdot2 \times 30 = 12 \times 3 = 36$
(b) $0\cdot23 \times 70 = 2\cdot3 \times 7 = 16\cdot1$
(c) $2\cdot5 \times 120 = 25 \times 12 = 300$

To multiply by a whole number ending in 00, multiply by 100, then multiply the answer by the number in front of the 00.

Example 13

(a) $1\cdot2 \times 300 = 120 \times 3 = 360$
(b) $0\cdot23 \times 700 = 23 \times 7 = 161$
(c) $2\cdot5 \times 1200 = 250 \times 12 = 3000$

Exercise 1.5i

Multiply each of the following by 20:
1. $1\cdot6$
2. $2\cdot4$
3. $6\cdot8$
4. $17\cdot2$
5. $25\cdot2$
6. $30\cdot5$
7. $6\cdot02$
8. $0\cdot52$
9. $0\cdot08$
10. $0\cdot00075$

Multiply each of the following by 90:
11. $3\cdot4$
12. $5\cdot7$
13. $6\cdot3$
14. $13\cdot5$
15. $21\cdot6$
16. $40\cdot9$
17. $4\cdot02$
18. $0\cdot38$
19. $0\cdot06$
20. $0\cdot0025$

Multiply each of the following by 400:
21. $2\cdot9$
22. $7\cdot8$
23. $9\cdot5$
24. $17\cdot7$
25. $35\cdot6$
26. $64\cdot2$
27. $3\cdot09$
28. $0\cdot71$
29. $0\cdot003$
30. $0\cdot00034$

Copy the following and fill in the empty spaces:
31. $2\cdot2 \times 20 =$
32. $\quad \times 20 = 64$
33. $7\cdot2 \times 30 =$
34. $\quad \times 30 = 393$
35. $0\cdot21 \times \quad = 8\cdot4$
36. $0\cdot77 \times 500 =$
37. $0\cdot62 \times \quad = 124$
38. $0\cdot004 \times 600 =$
39. $0\cdot023 \times \quad = 6\cdot9$
40. $\quad \times 90 = 189$

To divide by a whole number ending in 0, divide by 10, then divide the answer by the number in front of the 0.

Example 14

(a) $66\cdot3 \div 30 = 6\cdot63 \div 3 = 2\cdot21$
(b) $0\cdot154 \div 70 = 0\cdot0154 \div 7 = 0\cdot0022$
(c) $3\cdot84 \div 120 = 0\cdot384 \div 12 = 0\cdot032$

To divide by a whole number ending in 00, divide by 100, then divide the answer by the number in front of the 00.

Example 15

(a) $68\cdot4 \div 200 = 0\cdot684 \div 2 = 0\cdot342$
(b) $15\cdot5 \div 500 = 0\cdot155 \div 5 = 0\cdot031$
(c) $2760 \div 1200 = 27\cdot6 \div 12 = 2\cdot3$

Exercise 1.5j

Divide each of the following by 30:
1. $12\cdot0$
2. $3\cdot6$
3. $52\cdot8$
4. $6\cdot72$
5. 2412
6. $25\cdot2$
7. $0\cdot72$
8. $0\cdot0852$
9. $0\cdot024$
10. $0\cdot108$

Divide each of the following by 40:
11. $24\cdot0$
12. $8\cdot4$
13. $68\cdot4$
14. $6\cdot24$
15. 3224
16. $27\cdot6$
17. $0\cdot56$
18. $0\cdot0936$
19. $0\cdot032$
20. $0\cdot308$

Divide each of the following by 600:
21. $120\cdot0$
22. $78\cdot6$
23. $95\cdot4$
24. $258\cdot0$
25. $48\cdot6$
26. $237\cdot6$
27. $0\cdot78$
28. $0\cdot0852$
29. $0\cdot972$
30. $0\cdot0588$

Copy the following and fill in the empty spaces:
31. $1\cdot6 \div 20 =$
32. $6\cdot8 \div \quad = 0\cdot34$
33. $12\cdot6 \div 30 =$
34. $\quad \div 30 = 3\cdot2$
35. $2\cdot64 \div 40 =$
36. $8\cdot7 \div \quad = 0\cdot29$
37. $1\cdot28 \div 200 =$
38. $10\cdot5 \div \quad = 0\cdot021$
39. $0\cdot644 \div 70 =$
40. $\quad \div 90 = 0\cdot022$

To add and subtract a series of numbers, add the positive numbers first, then do the subtractions.

Example 1

(a) $8{\cdot}35 - 3{\cdot}1 + 2{\cdot}07 = 8{\cdot}35 + 2{\cdot}07 - 3{\cdot}1$
$= 10{\cdot}42 - 3{\cdot}1 = 7{\cdot}32$

(b) $4{\cdot}6 - 8{\cdot}91 + 6{\cdot}52 = 4{\cdot}6 + 6{\cdot}52 - 8{\cdot}91$
$= 11{\cdot}12 - 8{\cdot}91 = 2{\cdot}21$

(c) $2{\cdot}33 - 4{\cdot}7 + 5{\cdot}84 - 2 = 2{\cdot}33 + 5{\cdot}84 - 4{\cdot}7 - 2$
$= 8{\cdot}17 - 4{\cdot}7 - 2$
$= 3{\cdot}47 - 2 = 1{\cdot}47$

Exercise 1.6a

1. $7{\cdot}4 + 9{\cdot}8 - 2{\cdot}21$
2. $6{\cdot}8 - 5{\cdot}95 + 2{\cdot}07$
3. $3{\cdot}72 - 6{\cdot}1 + 4{\cdot}8$
4. $12{\cdot}4 + 7{\cdot}9 - 16$
5. $0{\cdot}85 - 3{\cdot}2 + 5{\cdot}15$
6. $3{\cdot}9 + 0{\cdot}8 - 1{\cdot}7$
7. $9{\cdot}8 - 12 + 8{\cdot}6$
8. $11{\cdot}4 - 3{\cdot}65 + 7{\cdot}2$
9. $1{\cdot}4 - 6{\cdot}3 + 9{\cdot}02$
10. $9{\cdot}8 - 9{\cdot}8 + 2{\cdot}3$
11. $0{\cdot}68 - 7{\cdot}2 + 9$
12. $7{\cdot}4 - 12{\cdot}6 + 7{\cdot}4$
13. $9 - 6{\cdot}48 + 2{\cdot}7$
14. $8{\cdot}4 - 13 + 5{\cdot}09$
15. $6{\cdot}3 - 8{\cdot}5 + 9{\cdot}02$
16. $8{\cdot}2 + 7{\cdot}9 - 6{\cdot}3$
17. $12{\cdot}6 - 15{\cdot}9 + 13{\cdot}5$
18. $0{\cdot}7 - 0{\cdot}2 + 0{\cdot}5$
19. $1{\cdot}9 - 2{\cdot}2 + 0{\cdot}8$
20. $0{\cdot}3 - 1{\cdot}7 + 1{\cdot}5$
21. $0{\cdot}8 - 0{\cdot}2 + 3{\cdot}1 - 1{\cdot}4$
22. $6{\cdot}1 - 8{\cdot}2 + 3{\cdot}1 - 0{\cdot}5$
23. $12{\cdot}4 - 0{\cdot}8 + 6 - 3{\cdot}05$
24. $2{\cdot}7 - 4{\cdot}6 - 5 + 8{\cdot}9$
25. $1{\cdot}04 - 6{\cdot}3 - 12{\cdot}7 + 21{\cdot}5$
26. $13{\cdot}2 + 4{\cdot}8 - 6{\cdot}9 - 7{\cdot}1$
27. $18{\cdot}1 - 9{\cdot}9 + 10{\cdot}3 - 8{\cdot}6$
28. $5{\cdot}4 - 7{\cdot}6 + 4{\cdot}2 - 0{\cdot}7$
29. $32{\cdot}2 + 4{\cdot}2 - 26{\cdot}6 - 3{\cdot}9$
30. $48{\cdot}4 - 50{\cdot}1 + 26{\cdot}3 - 19{\cdot}8$
31. $5{\cdot}02 - 4{\cdot}87 + 1{\cdot}35 - 0{\cdot}96$
32. $1{\cdot}93 - 4{\cdot}7 - 1{\cdot}5 + 5{\cdot}6$
33. $27{\cdot}3 - 30 + 19{\cdot}4 - 12{\cdot}6$
34. $5{\cdot}2 - 7 - 12{\cdot}6 + 19{\cdot}7$
35. $4{\cdot}88 + 9{\cdot}35 - 1{\cdot}64 - 7{\cdot}25$
36. $3{\cdot}1 - 7{\cdot}2 - 8{\cdot}3 + 13{\cdot}2$
37. $19{\cdot}4 - 21{\cdot}6 + 4{\cdot}7 + 5{\cdot}1$
38. $22{\cdot}4 + 14{\cdot}3 - 19{\cdot}9 - 15{\cdot}6$
39. $7 - 9{\cdot}2 - 4{\cdot}8 + 14{\cdot}9$
40. $14{\cdot}2 - 7{\cdot}4 - 6{\cdot}3 - 0{\cdot}5$

A calculation in brackets must be done before any other; and division and multiplication must be done before addition and subtraction

Example 2

(a) $2{\cdot}14 + 3{\cdot}5 \times 2{\cdot}1 = 2{\cdot}14 + 7{\cdot}35 = 9{\cdot}49$

(b) $(2{\cdot}14 + 3{\cdot}5) \times 2{\cdot}1 = 5{\cdot}64 \times 2{\cdot}1 = 11{\cdot}844$

(c) $2{\cdot}6 \times 1{\cdot}5 + 6{\cdot}4 \div 1{\cdot}6 = 3{\cdot}9 + 4{\cdot}0 = 7{\cdot}9$

Exercise 1.6b

1. $2{\cdot}8 + 0{\cdot}8 \times 3{\cdot}5$
2. $4{\cdot}2 + 2{\cdot}4 \times 2{\cdot}5$
3. $5{\cdot}0 \times 0{\cdot}7 + 3{\cdot}9$
4. $6{\cdot}5 \times 2{\cdot}2 + 9{\cdot}1$
5. $12{\cdot}6 + 2{\cdot}1 \times 1{\cdot}3$
6. $15{\cdot}8 - 6{\cdot}5 \times 2$
7. $3{\cdot}4 \times 3{\cdot}5 - 3{\cdot}5$
8. $3{\cdot}8 \times 2{\cdot}3 - 1{\cdot}98$
9. $12{\cdot}4 - 4{\cdot}3 \times 2$
10. $0{\cdot}9 \times 4 - 0{\cdot}28$
11. $6{\cdot}04 + 0{\cdot}8 \div 0{\cdot}4$
12. $3{\cdot}7 + 2{\cdot}4 \div 0{\cdot}5$
13. $4{\cdot}2 \div 0{\cdot}6 + 3{\cdot}3$
14. $2{\cdot}76 \div 0{\cdot}3 + 5{\cdot}2$
15. $6{\cdot}2 + 5{\cdot}4 \div 1{\cdot}8$
16. $4{\cdot}9 + 9 \div 1{\cdot}5$
17. $6{\cdot}3 \div 2{\cdot}1 - 1{\cdot}1$
18. $8{\cdot}1 - 4{\cdot}32 \div 1{\cdot}2$
19. $19{\cdot}8 \div 1{\cdot}8 - 0{\cdot}9$
20. $2{\cdot}1 - 0{\cdot}7 \div 0{\cdot}35$
21. $2{\cdot}5 \times (1{\cdot}6 + 3{\cdot}4)$
22. $(0{\cdot}83 + 2{\cdot}17) \times 0{\cdot}4$
23. $3{\cdot}1 \times (4{\cdot}2 + 0{\cdot}8)$
24. $(2{\cdot}1 + 1{\cdot}4) \times 2{\cdot}4$
25. $(6{\cdot}2 + 0{\cdot}7) \times 3{\cdot}3$
26. $(3{\cdot}5 - 1{\cdot}5) \times 4{\cdot}5$
27. $(4{\cdot}8 - 3{\cdot}4) \times 2{\cdot}5$
28. $0{\cdot}7 \times (6{\cdot}2 - 4{\cdot}3)$
29. $1{\cdot}6 \times (2{\cdot}48 - 0{\cdot}98)$
30. $(6{\cdot}2 - 3{\cdot}9) \times 1{\cdot}5$
31. $(3{\cdot}1 + 2{\cdot}1) \div 0{\cdot}8$
32. $(3{\cdot}11 + 1{\cdot}21) \div 1{\cdot}2$
33. $80 \div (0{\cdot}9 + 1{\cdot}6)$
34. $3{\cdot}78 \div (0{\cdot}7 + 0{\cdot}35)$
35. $(7{\cdot}72 - 0{\cdot}88) \div 0{\cdot}3$
36. $(0{\cdot}63 - 0{\cdot}18) \div 0{\cdot}09$
37. $0{\cdot}924 \div (5{\cdot}64 - 4{\cdot}54)$
38. $2{\cdot}52 \div (6{\cdot}34 - 5{\cdot}99)$
39. $7{\cdot}2 \times 9{\cdot}5 + 5{\cdot}2 \div 0{\cdot}13$
40. $4{\cdot}44 \div 0{\cdot}24 - 7{\cdot}5 \times 0{\cdot}32$

Decimal places

To write a number to a given number of decimal places, look at the digit following the figure in the required decimal place;
(a) if this digit is less than 5, remove it;
(b) if it is more than 5, add 1 to the figure in front of it;
(c) if it is exactly 5, and is followed by no other digits (except zeroes), make the figure in front an even number (if this figure is already even, leave it alone; if it is odd, increase it by 1).

Example 1

$3.724 = 3.72$ (to two decimal places)
$4.6867 = 4.687$ (3 D.P.)
$0.75 = 0.8$ (1 D.P.)
$3.650 = 3.6$ (1 D.P.)
$1.6851 = 1.69$ (2 D.P.)

Exercise 1.7a

Correct each of the following to the number of decimal places indicated.

1. 2·643 (2 D.P.)	2. 1·338 (2 D.P.)
3. 17·64 (1 D.P.)	4. 42·79 (1 D.P.)
5. 1·7342 (3 D.P.)	6. 1·5628 (3 D.P.)
7. 13·65 (1 D.P.)	8. 25·375 (2 D.P.)
9. 8·054 (2 D.P.)	10. 24·03 (1 D.P.)
11. 5·104 (2 D.P.)	12. 4·507 (2 D.P.)
13. 27·08 (1 D.P.)	14. 9·509 (2 D.P.)
15. 6·305 (2 D.P.)	16. 3·899 (2 D.P.)
17. 5·099 (2 D.P.)	18. 14·799 (2 D.P.)
19. 13·99 (1 D.P.)	20. 1·999 (2 D.P.)
21. 0·685 (2 D.P.)	22. 91·801 (1 D.P.)
23. 0·7450 (2 D.P.)	24. 3·899 (1 D.P.)
25. 5·55 (1 D.P.)	26. 13·615 (2 D.P.)
27. 1·4725 (3 D.P.)	28. 0·0451 (2 D.P.)
29. 9·99 (1 D.P.)	30. 1·8501 (1 D.P.)

Nearest unit

A number can be written correct to a given unit.

Example 2

$14.7 = 15$ correct to the nearest whole number
$6.023 = 6.0$ correct to the nearest tenth
$6450 = 6400$ correct to the nearest hundred

Exercise 1.7b

Correct each of the following to the nearest whole number.

1. 26·4	2. 6·03
3. 5·96	4. 0·804
5. 138·5	6. 12·35
7. 0·64	8. 35·5
9. 4·27	10. 104·7

Correct each of the following to the nearest tenth.

11. 4·76	12. 12·84
13. 0·85	14. 7·914
15. 16·75	16. 0·07
17. 6·351	18. 1·251
19. 3·99	20. 4·2731

Correct each of the following to the nearest ten.

21. 4763	22. 947
23. 6325	24. 1004
25. 127	26. 368
27. 74	28. 4096
29. 142·3	30. 1788·421

Significant figures

In any number, the first non-zero figure is the first significant figure.

Example 13

1·006 has 4 significant figures
0·006 has only 1 significant figure.
1·006 written correct to 3 significant figures is 1·01.
5246 written correct to 2 significant figures (2 S.F.) is 5200.

Exercise 1.7c

State the number of significant figures in each of the following.

1. 1·325	2. 3·26	3. 0·853
4. 0·057	5. 5·01	6. 4·250
7. 320	8. 200	9. 5·0
10. 8·00	11. 5·24	12. 8·357

Correct each of the following to the number of significant figures indicated.

13. 3·223 (2 S.F.)	14. 7·574 (2 S.F.)
15. 13·36 (3 S.F.)	16. 17·21 (3 S.F.)
17. 31·4 (2 S.F.)	18. 36·9 (2 S.F.)
19. 15·19 (2 S.F.)	20. 19·87 (2 S.F.)
21. 3574 (3 S.F.)	22. 4285 (3 S.F.)

23. 5486 (2 S.F.)   24. 20·04 (3 S.F.)
25. 3·99 (2 S.F.)   26. 9·999 (3 S.F.)
27. 0·0637 (2 S.F.)   28. 0·00724 (2 S.F.)
29. 0·0088 (1 S.F.)   30. 0·00769 (1 S.F.)

27. (a) $13·1 \div 2·4$ 28. (a) $77·8 \div 4·9$
    (b) $36·9 \div 9·8$ (b) $31·6 \div 2·7$
    (c) $19·7 \div 4·9$ (c) $87·1 \div 9·2$

29. (a) $9·63 \div 0·48$ 30. (a) $0·43 \div 1·95$
    (b) $34·4 \div 0·63$ (b) $0·38 \div 0·77$
    (c) $5·91 \div 0·26$ (c) $0·99 \div 1·62$

Rough estimates

A rough estimate is often useful as a check on your work. Write each number correct to 1 significant figure, and then do the calculation.

$\approx$ means 'approximately equal to', but the sign $\simeq$ is also used.

Example 4

Find rough estimates for:
(a) $8·9 \times 3·2$ (b) $23·6 \div 0·38$

(a) $8·9 \times 3·2 \approx 9 \times 3 \approx 27$
(b) $23·6 \div 0·38 \approx 20 \div 0·4$
$\approx 200 \div 4 \approx 50$

Exercise 1.7d

Find rough estimates for:
1. $9·1 \times 1·9$ 2. $8·2 \times 2·9$
3. $7·1 \times 3·8$ 4. $5·8 \times 2·3$
5. $8·9 \times 4·2$ 6. $12·9 \times 5·1$
7. $14·7 \times 3·3$ 8. $16·2 \times 2·8$
9. $18·1 \times 0·49$ 10. $15·4 \times 0·36$
11. $6·2 \div 2·1$ 12. $9·3 \div 3·2$
13. $8·4 \div 4·1$ 14. $8·9 \div 2·8$
15. $5·7 \div 2·9$ 16. $9·6 \div 4·6$
17. $14·8 \div 4·9$ 18. $17·6 \div 3·7$
19. $16·2 \div 0·41$ 20. $20·3 \div 0·52$

Find rough estimates for each of the following, and for each question find which estimate is the 'odd one out'.

21. (a) $12·1 \times 4·8$ 22. (a) $15·2 \times 4·9$
    (b) $14·3 \times 3·8$ (b) $12·3 \times 5·7$
    (c) $7·71 \times 5·4$ (c) $13·4 \times 9·6$

23. (a) $13·8 \times 7·1$ 24. (a) $15·9 \times 3·4$
    (b) $15·7 \times 6·3$ (b) $24·2 \times 1·9$
    (c) $28·6 \times 3·7$ (c) $13·3 \times 3·6$

25. (a) $12·2 \times 0·67$ 26. (a) $18·2 \div 3·9$
    (b) $11·3 \times 0·78$ (b) $24·1 \div 9·6$
    (c) $36·4 \times 0·19$ (c) $28·3 \div 5·7$

Example 5

Find rough estimates for:

(a) $\dfrac{123}{0·039 \times 22}$ (b) $\dfrac{4·1 \times 0·38}{20·4}$

(a) $\dfrac{123}{0·039 \times 22} \approx \dfrac{\cancel{100}^{\,5}}{0·04 \times \cancel{20}_1}$

$\approx \dfrac{5}{0·04} \approx \dfrac{500}{4} \approx 125$

(b) $\dfrac{4·1 \times 0·38}{20·4} \approx \dfrac{{}^1\cancel{4} \times 0·4}{\cancel{20}_5}$

$\approx \dfrac{0·4}{5} \approx 0·08$

Exercise 1.7e

Find rough estimates for:

1. $\dfrac{84}{2·1 \times 17·8}$ 2. $\dfrac{109}{49 \times 0·97}$

3. $\dfrac{39·3}{3·6 \times 2·2}$ 4. $\dfrac{26·7}{2·7 \times 11·4}$

5. $\dfrac{58·6}{9·8 \times 3·4}$ 6. $\dfrac{0·77}{4·2 \times 1·9}$

7. $\dfrac{36}{12·3 \times 0·43}$ 8. $\dfrac{6·3}{2·43 \times 33}$

9. $\dfrac{462}{0·23 \times 54·6}$ 10. $\dfrac{0·37}{5·42 \times 7·5}$

11. $\dfrac{4·2 \times 12·3}{21·9}$ 12. $\dfrac{10·8 \times 9·6}{4·76}$

13. $\dfrac{0·76 \times 54}{84·2}$ 14. $\dfrac{94·5 \times 0·56}{2·51}$

15. $\dfrac{0·13 \times 99}{0·49}$ 16. $\dfrac{5·4 \times 6·31}{58·2}$

17. $\dfrac{8·6 \times 14·3}{0·34}$ 18. $\dfrac{7·56 \times 23}{0·39}$

19. $\dfrac{16 \times 0·24}{0·46}$ 20. $\dfrac{77·6 \times 0·053}{0·38}$

The unit of length is the metre.
1000 millimetres (mm) = 1 metre (m)
100 centimetres (cm)   = 1 metre (m)
1000 metres (m)   = 1 kilometre (km)

Example 1

Change (a) 426 cm to m
       (b) 2·64 km to m
(a) 426 cm = 426 ÷ 100 = 4·26 m
(b) 2·64 km = 2·64 × 1000 = 2640 m

Exercise 1.8a

Change the following as indicated.
1. 357 cm to m   2. 5329 cm to m
3. 3760 cm to m   4. 49 cm to m
5. 60 cm to m   6. 9 cm to m
7. 5276 mm to m   8. 752 mm to m
9. 80 mm to m   10. 7 mm to m
11. 9137 m to km   12. 830 m to km
13. 3·372 km to m   14. 2·49 km to m
15. 19·6 km to m   16. 3·45 m to cm
17. 9·2 m to cm   18. 5·936 m to mm
19. 8·21 m to mm   20. 7·9 m to mm

The unit of mass is the kilogram
1000 milligrams (mg) = 1 gram (g)
1000 grams (g)   = 1 kilogram (kg)
1000 kilograms (kg)   = 1 tonne (t)

Example 2

Change (a) 276 mg to g
       (b) 52·63 g to mg
(a) 276 mg = 276 ÷ 1000 = 0·276 g
(b) 52·63 g = 52·63 × 1000 = 52 630 mg

Exercise 1.8b

Change the following as indicated.
1. 1328 mg to g   2. 536 mg to g
3. 780 mg to g   4. 90 mg to g
5. 8 mg to g   6. 1500 g to kg
7. 590 g to kg   8. 30 g to kg
9. 2 g to kg   10. 1320 kg to t
11. 800 kg to t   12. 4·536 g to mg
13. 8·98 g to mg   14. 3·4 g to mg
15. 5·26 kg to g   16. 8·5 kg to g
17. 0·7 kg to g   18. 3·71 t to kg
19. 5·6 t to kg   20. 0·3 t to kg

The unit for the measurement of capacity
is the litre.
1000 millilitres (ml) = 1 litre (l)

Example 3

Change 2646 ml to litres.
2646 ml = 2646 ÷ 1000 = 2·646 litres

Exercise 1.8c

Change the following as indicated
1. 3278 ml to litres   2. 8250 ml to litres
3. 9300 ml to litres   4. 6035 ml to litres
5. 5020 ml to litres   6. 1·332 litres to ml
7. 7·6 litres to ml   8. 0·755 litres to ml
9. 0·32 litres to ml   10. 0·1 litres to ml
11. 470 ml to litres   12. 300 ml to litres
13. 25 ml to litres   14. 1470 ml to litres
15. 9 litres to ml   16. 84 litres to ml
17. 0·5 litres to ml   18. 0·008 litres to ml
19. 186 000 ml to litres 20. 370 000 ml to litres

Example 4

Find the cost of 650 g of bacon at £2
per kg.

$650 \text{ g} = \frac{650}{1000}$ kg at £2

$= \frac{13}{20}$ kg at £2   = £1·30

Exercise 1.8d

Find the cost of the following.
1. 400 g of soap powder at 50p per kg.
2. 600 g of flour at 25p per kg.
3. 450 g of sugar at 40p per kg.
4. 350 g of jam at 60p per kg.
5. 800 g of cheese at £1·50 per kg.
6. 600 g of butter at £1·25 per kg.
7. 45 cm of dress cloth at 80p per m.
8. 50 m of thread at 60p per km.
9. 600 ml of milk at 25p per l.
10. A 400 ml can of beer at 65p per l.
11. 0·5 litres of petrol at 32p per l.
12. 0·25 kg of peas at 60p per kg.
13. 1500 ml of methylated spirit at £1·96 per l.
14. 3 m of wire at £1·25 for 5 m.
15. 4·2 litres of paint at £2·80 per l.

A fraction keeps the same value when its numerator (top line) and denominator (bottom line) are both either:

(i) multiplied by the same number,

or (ii) divided by the same number.

Example 1

(a) $\frac{1}{2} = \frac{1 \times 3}{2 \times 3} = \frac{3}{6}$

(b) $\frac{10}{12} = \frac{10 \div 2}{12 \div 2} = \frac{5}{6}$

Exercise 2.1a

Copy and complete the following.

1. $\frac{1}{6} = \frac{}{12}$ 2. $\frac{5}{8} = \frac{10}{}$

3. $\frac{}{3} = \frac{4}{6}$ 4. $\frac{9}{} = \frac{18}{26}$

5. $\frac{17}{51} = \frac{1}{}$ 6. $\frac{24}{36} = \frac{}{3}$

Write each of the following in its simplest form.

7. $\frac{3}{6}$ 8. $\frac{4}{12}$ 9. $\frac{5}{20}$

10. $\frac{4}{6}$ 11. $\frac{9}{12}$ 12. $\frac{18}{24}$

13. $\frac{27}{45}$ 14. $\frac{54}{63}$ 15. $\frac{26}{39}$

An *improper* fraction has its numerator bigger than its denominator.

Example 2

(a) $\frac{7}{5}$ (b) $\frac{14}{3}$ (c) $\frac{15}{6}$

Improper fractions can be written as *mixed numbers*.

Example 3

(a) $\frac{7}{5} = 1\frac{2}{5}$ (b) $\frac{14}{3} = 4\frac{2}{3}$

(c) $\frac{15}{6} = 2\frac{3}{6} = 2\frac{1}{2}$

Mixed numbers can be changed into improper fractions.

Example 4

(a) $1\frac{1}{4} = \frac{5}{4}$ (b) $3\frac{2}{7} = \frac{23}{7}$

(c) $8\frac{1}{10} = \frac{81}{10}$

Exercise 2.1b

Write as mixed numbers:

1. $\frac{3}{2}$ 2. $\frac{4}{3}$ 3. $\frac{7}{4}$ 4. $\frac{5}{2}$

5. $\frac{7}{3}$ 6. $\frac{18}{7}$ 7. $\frac{9}{6}$ 8. $\frac{32}{24}$

Write as improper fractions:

9. $1\frac{1}{4}$ 10. $1\frac{1}{5}$ 11. $1\frac{3}{7}$ 12. $1\frac{2}{5}$

13. $2\frac{1}{2}$ 14. $3\frac{1}{4}$ 15. $6\frac{1}{4}$ 16. $3\frac{3}{6}$

17. $3\frac{2}{3}$ 18. $7\frac{3}{5}$ 19. $3\frac{5}{7}$ 20. $7\frac{6}{9}$

Fractions can only be added or subtracted when their denominators are the same: only their numerators are then added or subtracted.

Example 5

(a) $\frac{1}{3} + \frac{1}{3} = \frac{2}{3}$.   (b) $\frac{3}{5} + \frac{4}{5} = \frac{7}{5} = 1\frac{2}{5}$

(c) $\frac{3}{7} - \frac{2}{7} = \frac{1}{7}$ (d) $\frac{5}{6} - \frac{1}{6} = \frac{4}{6} = \frac{2}{3}$

Exercise 2.1c

1. $\frac{1}{5} + \frac{1}{5}$ 2. $\frac{2}{5} + \frac{1}{5}$ 3. $\frac{3}{10} + \frac{4}{10}$

4. $\frac{3}{8} + \frac{1}{8}$ 5. $\frac{2}{3} + \frac{2}{3}$ 6. $\frac{3}{10} + \frac{7}{10}$

7. $\frac{5}{7} + \frac{6}{7}$ 8. $\frac{3}{8} + \frac{7}{8}$ 9. $\frac{2}{3} - \frac{1}{3}$

10. $\frac{4}{5} - \frac{2}{5}$ 11. $\frac{5}{7} - \frac{2}{7}$ 12. $\frac{7}{8} - \frac{1}{8}$

13. $\frac{9}{10} - \frac{3}{10}$ 14. $\frac{7}{15} - \frac{2}{15}$ 15. $\frac{15}{16} - \frac{9}{16}$

To add or subtract fractions with different denominators, a common denominator is found.

A common denominator of $\frac{1}{2}$, $\frac{1}{3}$ and $\frac{1}{4}$ is 12, this is the smallest number which can be divided by 2, 3 and 4.

Example 6

(a) $\frac{1}{2} + \frac{2}{5} = \frac{5}{10} + \frac{4}{10} = \frac{9}{10}$

(b) $\frac{5}{6} - \frac{1}{2} = \frac{5}{6} - \frac{3}{6} = \frac{2}{6} = \frac{1}{3}$

Exercise 2.1d

1. $\frac{1}{3} + \frac{1}{5}$ 2. $\frac{1}{3} + \frac{1}{4}$ 3. $\frac{1}{5} + \frac{1}{10}$

4. $\frac{2}{3} + \frac{1}{4}$ 5. $\frac{3}{4} + \frac{1}{8}$ 6. $\frac{1}{3} + \frac{1}{6}$

7. $\frac{1}{4} + \frac{1}{12}$ 8. $\frac{2}{3} + \frac{1}{12}$ 9. $\frac{2}{9} + \frac{5}{18}$

10. $\frac{1}{8} + \frac{1}{4} + \frac{3}{16}$ 11. $\frac{1}{2} - \frac{1}{4}$ 12. $\frac{1}{5} - \frac{1}{10}$

13. $\frac{1}{2} - \frac{1}{3}$ 14. $\frac{1}{4} - \frac{1}{5}$ 15. $\frac{2}{3} - \frac{1}{4}$

16. $\frac{7}{8} - \frac{3}{4}$ 17. $\frac{5}{6} - \frac{1}{2}$ 18. $\frac{11}{12} - \frac{3}{4}$

19. $\frac{5}{6} - \frac{1}{4}$ 20. $\frac{9}{10} - \frac{11}{15}$

To add or subtract mixed numbers, first change them to improper fractions.

Example 7

(a) $1\frac{3}{4} + 1\frac{1}{2} = \frac{7}{4} + \frac{3}{2}$

$\qquad = \frac{7}{4} + \frac{6}{4} = \frac{13}{4} = 3\frac{1}{4}$

(b) $2\frac{2}{3} + 2\frac{5}{6} = \frac{8}{3} + \frac{17}{6}$

$\qquad = \frac{16}{6} + \frac{17}{6} = \frac{33}{6} = 5\frac{3}{6} = 5\frac{1}{2}$

(c) $4\frac{2}{3} - 2\frac{1}{2} = \frac{14}{3} - \frac{5}{2}$

$\qquad = \frac{28}{6} - \frac{15}{6} = \frac{13}{6} = 2\frac{1}{6}$

Exercise 2.1e

1. $2\frac{1}{4} + 1\frac{1}{4}$ 2. $3\frac{1}{3} + 1\frac{1}{5}$

3. $2\frac{1}{3} + 3\frac{1}{4}$ 4. $4\frac{1}{5} + 1\frac{1}{10}$

5. $6\frac{3}{4} + 1\frac{1}{8}$ 6. $1\frac{7}{10} + 1\frac{3}{5}$

7. $3\frac{2}{3} + 1\frac{3}{4}$ 8. $2\frac{2}{5} + 3\frac{7}{10}$

9. $1\frac{2}{3} + 1\frac{5}{6}$ 10. $1\frac{5}{12} + 1\frac{1}{3}$

11. $1\frac{3}{4} + 1\frac{7}{12}$ 12. $2\frac{9}{10} + 1\frac{1}{2}$

13. $4\frac{3}{4} - 3\frac{1}{2}$ 14. $2\frac{5}{6} - 1\frac{2}{3}$

15. $3\frac{7}{8} - 2\frac{1}{2}$ 16. $8\frac{4}{5} - 3\frac{7}{10}$

17. $8 - 2\frac{3}{4}$ 18. $2\frac{1}{2} - 1\frac{2}{3}$

19. $4\frac{1}{5} - 2\frac{3}{5}$ 20. $5\frac{1}{6} - 2\frac{2}{3}$

21. $4\frac{1}{2} - 3\frac{5}{6}$ 22. $4\frac{3}{4} - 1\frac{11}{12}$

23. $1\frac{9}{10} - \frac{14}{15}$ 24. $3\frac{1}{4} - 1\frac{5}{6}$

25. $7\frac{2}{3} - 4\frac{7}{9}$

To multiply fractions, multiply the numerators together and the denominators together.

Example 8

$$\frac{2}{3} \times \frac{5}{9} = \frac{2 \times 5}{3 \times 9} = \frac{10}{27}$$

Cancelling can make this easier.

Example 9

$$\frac{7}{9} \times \frac{12}{35} = \frac{\cancel{7}^1 \times \cancel{12}^4}{\cancel{9}_3 \times \cancel{35}_5} = \frac{1 \times 4}{3 \times 5} = \frac{4}{15}$$

Exercise 2.1f

1. $\frac{1}{2} \times \frac{1}{3}$ 2. $\frac{1}{4} \times \frac{2}{5}$ 3. $\frac{3}{4} \times \frac{1}{2}$

4. $\frac{3}{7} \times \frac{1}{2}$ 5. $\frac{2}{3} \times \frac{4}{5}$ 6. $\frac{1}{3} \times \frac{3}{5}$

7. $\frac{1}{3} \times \frac{6}{7}$ 8. $\frac{3}{5} \times \frac{10}{21}$ 9. $\frac{4}{7} \times \frac{21}{32}$

10. $\frac{5}{6} \times \frac{9}{11}$ 11. $\frac{1}{9} \times \frac{12}{13}$ 12. $\frac{3}{8} \times \frac{4}{21}$

13. $\frac{5}{16} \times \frac{6}{25}$ 14. $\frac{5}{7} \times \frac{14}{15}$ 15. $\frac{12}{13} \times \frac{39}{48}$

To multiply mixed numbers, first change them to improper fractions.

Example 10

(a) $1\frac{2}{5} \times 1\frac{1}{2} = \frac{7}{5} \times \frac{3}{2}$

$\qquad = \frac{7 \times 3}{5 \times 2} = \frac{21}{10} = 2\frac{1}{10}$

(b) $3\frac{1}{3} \times 1\frac{1}{5} = \frac{10}{3} \times \frac{6}{5}$

$\qquad = \frac{\cancel{10}^2 \times \cancel{6}^2}{\cancel{3}_1 \times \cancel{5}_1} = \frac{2 \times 2}{1 \times 1} = 4$

Exercise 2.1g

1. $1\frac{1}{4} \times 2\frac{1}{3}$ 2. $1\frac{2}{3} \times 1\frac{1}{4}$

3. $2\frac{1}{2} \times 2\frac{1}{2}$ 4. $1\frac{3}{4} \times 1\frac{2}{3}$

5. $3\frac{1}{4} \times 1\frac{1}{5}$ 6. $1\frac{1}{4} \times 2\frac{2}{3}$

7. $1\frac{1}{15} \times 2\frac{1}{2}$ 8. $3\frac{3}{4} \times 1\frac{1}{5}$

9. $2\frac{1}{2} \times 5$ 10. $7\frac{1}{2} \times 4$

11. $2\frac{1}{7} \times 1\frac{1}{3}$ 12. $2\frac{5}{8} \times 3\frac{2}{7}$

13. $4\frac{4}{7} \times 2\frac{5}{8}$ 14. $3\frac{3}{5} \times 3\frac{1}{3}$

15. $1\frac{1}{4} \times 1\frac{1}{2} \times 1\frac{1}{3}$

To divide by a fraction, multiply by its inverse; e.g. the inverse of $\frac{1}{2}$ is $\frac{2}{1}$ and the inverse of $4\frac{1}{2}$ (i.e. $\frac{9}{2}$) is $\frac{2}{9}$.

Example 11

(a) $\frac{2}{3} \div \frac{3}{4} = \frac{2}{3} \times \frac{4}{3}$

$\qquad = \frac{2 \times 4}{3 \times 3} = \frac{8}{9}$

(b) $2\frac{1}{2} \div 1\frac{1}{4} = \frac{5}{2} \div \frac{5}{4}$

$\qquad = \frac{5}{2} \times \frac{4}{5}$

$\qquad = \frac{{}^{1}\cancel{5} \times \cancel{4}^{2}}{\cancel{2} \times \cancel{5}_{1}} = \frac{1 \times 2}{1 \times 1} = 2$

Exercise 2.1h

1. $\frac{1}{4} \div \frac{1}{3}$ 2. $\frac{2}{5} \div \frac{2}{7}$

3. $\frac{4}{5} \div \frac{3}{4}$ 4. $\frac{3}{7} \div \frac{2}{5}$

5. $\frac{5}{12} \div \frac{3}{5}$ 6. $\frac{1}{3} \div \frac{5}{9}$

7. $\frac{2}{5} \div \frac{9}{10}$ 8. $\frac{3}{7} \div \frac{11}{14}$

9. $\frac{4}{9} \div \frac{2}{3}$ 10. $\frac{2}{5} \div \frac{4}{5}$

11. $5 \div 1\frac{1}{4}$ 12. $6 \div 1\frac{1}{2}$

13. $7\frac{1}{2} \div 2\frac{1}{2}$ 14. $3\frac{1}{2} \div 1\frac{3}{4}$

15. $1\frac{1}{10} \div 1\frac{1}{5}$ 16. $1\frac{3}{8} \div 2\frac{1}{4}$

17. $2\frac{6}{7} \div 1\frac{1}{14}$ 18. $2\frac{2}{3} \div 1\frac{7}{9}$

19. $1\frac{5}{12} \div 3\frac{3}{16}$ 20. $3\frac{3}{5} \div 2\frac{1}{4}$

Exercise 2.1i

In questions **1** to **10** find the 'odd answer out'.

1. (a) $\frac{4}{5} + \frac{1}{30}$ 2. (a) $\frac{9}{10} - \frac{11}{40}$

(b) $\frac{9}{20} + \frac{7}{15}$ (b) $\frac{1}{3} + \frac{7}{24}$

(c) $\frac{4}{9} + \frac{7}{18}$ (c) $\frac{3}{5} + \frac{3}{20}$

3. (a) $\frac{5}{12} - \frac{4}{15}$ 4. (a) $\frac{4}{15} - \frac{1}{6}$

(b) $\frac{1}{15} + \frac{1}{10}$ (b) $\frac{1}{5} - \frac{3}{40}$

(c) $\frac{2}{5} - \frac{1}{4}$ (c) $\frac{5}{6} - \frac{17}{24}$

5. (a) $1\frac{1}{2} + \frac{2}{3}$ 6. (a) $1\frac{9}{20} + 1\frac{3}{10}$

(b) $2\frac{5}{8} - \frac{11}{24}$ (b) $2\frac{1}{12} + \frac{2}{3}$

(c) $3\frac{3}{4} - 1\frac{5}{12}$ (c) $3\frac{1}{6} - \frac{2}{3}$

7. (a) $2\frac{7}{10} - \frac{3}{40}$ 8. (a) $2\frac{3}{4} - 1\frac{11}{20}$

(b) $1\frac{3}{5} + \frac{9}{10}$ (b) $\frac{1}{2} + \frac{1}{4} + \frac{5}{12}$

(c) $4\frac{1}{4} - 1\frac{5}{8}$ (c) $\frac{1}{5} + \frac{2}{3} + \frac{3}{10}$

9. (a) $3\frac{1}{3} - \frac{8}{15}$ 10. (a) $1\frac{1}{6} + \frac{1}{5} + \frac{1}{3}$

(b) $1\frac{3}{20} + \frac{7}{10} + \frac{3}{4}$ (b) $1\frac{4}{5} + \frac{3}{4} - \frac{17}{20}$

(c) $1\frac{1}{2} + 1\frac{1}{10} + \frac{1}{5}$ (c) $1\frac{1}{4} + \frac{11}{15} - \frac{1}{12}$

11. The distance from Halesowen to Quinton is $2\frac{1}{2}$ miles. I am given a lift for part of the way and the car's mileometer records $1\frac{3}{10}$ miles. If I walk the rest of the way, how far do I walk?

12. The distance from Halesowen to Stourbridge is $4\frac{1}{2}$ miles. I am given a lift for part of the way and the car's mileometer records $2\frac{9}{10}$ miles. If I walk the rest of the way, how far do I walk?

13. A one-litre flask is filled with milk, and it is used to fill two glasses, one of capacity $\frac{1}{2}$ of a litre and the other of capacity $\frac{1}{6}$ of a litre. What fraction of a litre will remain in the flask?

14. In a certain class $\frac{3}{4}$ of the children have dark hair, $\frac{1}{12}$ have ginger hair and the remainder are blonde. What fraction of the class have blonde hair?

15. At a polling-station $\frac{1}{2}$ of the people voted Labour, $\frac{1}{3}$ voted Conservative, $\frac{1}{15}$ voted Liberal and the remainder voted for an independent candidate. What fraction voted for the independent candidate?

16. A group of people travelled from Stourbridge to Kidderminster, $\frac{1}{20}$ of them decided to walk, $\frac{1}{12}$ went by car, $\frac{2}{5}$ went by train and all the rest travelled by bus. What fraction went by bus?

17. Two tonnes of coal are shared between three men. The first receives $\frac{3}{4}$ of a tonne, the second $\frac{4}{5}$ of a tonne. What fraction of a tonne does the third receive?

18. Three tonnes of sand are shared between three builders. The first receives $1\frac{1}{2}$ tonnes, the second $\frac{5}{6}$ of a tonne. What fraction of a tonne does the third builder receive?

In questions **19** to **30** find the 'odd answer out'.

19. (a) $\frac{20}{33} \times \frac{44}{45}$ 20. (a) $\frac{6}{7} \times \frac{21}{32}$

(b) $\frac{27}{40} \times \frac{32}{45}$ (b) $\frac{7}{9} \times \frac{24}{35}$

(c) $\frac{14}{15} \times \frac{18}{35}$ (c) $\frac{6}{25} \div \frac{9}{20}$

21. (a) $\frac{24}{35} \div \frac{20}{21}$ 22. (a) $\frac{8}{21} \div \frac{9}{14}$

(b) $\frac{5}{6} \times \frac{44}{45}$ (b) $\frac{10}{21} \div \frac{8}{9}$

(c) $\frac{33}{40} \times \frac{48}{55}$ (c) $\frac{14}{15} \times \frac{40}{63}$

23. (a) $1\frac{7}{20} \times \frac{35}{36}$ **24.** (a) $1\frac{1}{20} \times 2\frac{1}{12}$

(b) $1\frac{11}{24} \div \frac{14}{15}$ (b) $\frac{11}{18} \times 3\frac{6}{7}$

(c) $3\frac{3}{4} \times \frac{5}{12}$ (c) $1\frac{9}{35} \div \frac{8}{15}$

25. (a) $3\frac{11}{15} \div 1\frac{7}{25}$ **26.** (a) $1\frac{7}{20} \div 4\frac{4}{5}$

(b) $2\frac{2}{9} \times 1\frac{5}{16}$ (b) $\frac{25}{27} \times \frac{9}{22} \times \frac{4}{5}$

(c) $2\frac{7}{24} \times 1\frac{7}{33}$ (c) $\frac{18}{25} \times \frac{15}{16} \times \frac{5}{12}$

27. (a) $4\frac{4}{9} \div 2\frac{1}{12}$ **28.** (a) $1\frac{13}{27} \times 2\frac{11}{12} \div 3\frac{1}{9}$

(b) $2\frac{2}{9} \times 2\frac{1}{10} \times \frac{16}{35}$ (b) $3\frac{3}{14} \times \frac{15}{36} \div 1\frac{1}{8}$

(c) $1\frac{1}{24} \times 2\frac{1}{4} \times \frac{14}{15}$ (c) $1\frac{1}{4} \times 1\frac{8}{27} \div 1\frac{1}{6}$

29. (a) $\frac{7}{8} \times \frac{4}{5} \div \frac{7}{10}$ **30.** (a) $3\frac{1}{8} \times 1\frac{1}{15} \div \frac{5}{9}$

(b) $1\frac{1}{2} \times \frac{6}{7} \times \frac{7}{8}$ (b) $5\frac{1}{4} \times 1\frac{1}{3} \times 1\frac{5}{7}$

(c) $\frac{3}{4} \times 1\frac{3}{5} \div 1\frac{1}{5}$ (c) $\frac{4}{17} \times 4\frac{1}{2} \times 11\frac{1}{3}$

2.2 FRACTIONS AND DECIMALS

Change a decimal fraction into a common fraction or a mixed number as follows.

(i) Write each separate figure after the decimal point as a common fraction.
(ii) Add these fractions together.
(iii) Make sure this sum is given in its simplest form.

Example 1

(a) $0 \cdot 27 = \frac{2}{10} + \frac{7}{100} = \frac{20}{100} + \frac{7}{100} = \frac{27}{100}$

(b) $0 \cdot 35 = \frac{3}{10} + \frac{5}{100} = \frac{30}{100} + \frac{5}{100}$

$= \frac{35}{100} = \frac{7}{20}$

(c) $1 \cdot 025 = 1 + \frac{0}{10} + \frac{2}{100} + \frac{5}{1000}$

$= 1 + \frac{20}{1000} + \frac{5}{1000}$

$= 1\frac{25}{1000} = 1\frac{1}{40}$

Exercise 2.2a

Change the following into common fractions or mixed numbers.

1. $0 \cdot 7$	2. $0 \cdot 5$	3. $0 \cdot 4$
4. $0 \cdot 9$	5. $0 \cdot 45$	6. $0 \cdot 26$
7. $0 \cdot 85$	8. $0 \cdot 32$	9. $0 \cdot 58$
10. $0 \cdot 72$	11. $0 \cdot 875$	12. $0 \cdot 325$
13. $0 \cdot 625$	14. $0 \cdot 475$	15. $0 \cdot 03$
16. $0 \cdot 08$	17. $0 \cdot 05$	18. $0 \cdot 07$
19. $0 \cdot 055$	20. $0 \cdot 024$	21. $0 \cdot 062$
22. $0 \cdot 0275$	23. $0 \cdot 0625$	24. $0 \cdot 003$
25. $0 \cdot 006$	26. $0 \cdot 002$	27. $0 \cdot 0035$
28. $0 \cdot 0048$	29. $3 \cdot 2$	30. $8 \cdot 25$
31. $12 \cdot 16$	32. $41 \cdot 02$	33. $5 \cdot 018$
34. $13 \cdot 012$	35. $2 \cdot 0125$	36. $8 \cdot 0175$
37. $14 \cdot 004$	38. $11 \cdot 005$	39. $16 \cdot 0025$
40. $21 \cdot 0015$		

Change a common fraction into a decimal fraction by dividing the numerator (top line) by the denominator (bottom line). The answer will be either

(i) an exact decimal (see example 2),
or (ii) a recurring decimal (see example 3),
or (iii) a non-terminating decimal when the answer is usually written to a given number of decimal places (see example 4).

Example 2

Change (a) $\frac{3}{4}$ into a decimal fraction

(b) $\frac{7}{40}$ into a decimal fraction

(a) $\frac{3}{4} = 3 \div 4$

$= 0 \cdot 75$

$$\begin{array}{r} 0{\cdot}75 \\ 4\overline{)\ 3{\cdot}00} \\ \underline{2\ 8} \\ 20 \\ \underline{20} \\ \cdot\ \cdot \end{array}$$

(b) $\frac{7}{40} = 7 \div 40$

$= 0 \cdot 175$

$$\begin{array}{r} 0{\cdot}175 \\ 40\overline{)\ 7{\cdot}000} \\ \underline{40} \\ 300 \\ \underline{280} \\ 200 \\ \underline{200} \\ \cdot\ \cdot\ \cdot \end{array}$$

Exercise 2.2b

Change the following to decimal fractions.

1. $\frac{3}{8}$ 2. $\frac{1}{8}$ 3. $\frac{1}{4}$ 4. $\frac{3}{20}$

5. $\frac{11}{20}$ 6. $\frac{7}{20}$ 7. $\frac{13}{20}$ 8. $\frac{19}{20}$

9. $\frac{3}{5}$ 10. $\frac{1}{5}$ 11. $\frac{4}{5}$ 12. $\frac{11}{40}$

13. $\frac{9}{40}$ 14. $\frac{17}{40}$ 15. $\frac{21}{40}$ 16. $\frac{3}{16}$

17. $\frac{11}{16}$ 18. $\frac{7}{50}$ 19. $\frac{9}{50}$ 20. $\frac{31}{50}$

Example 3

Change (a) $\frac{1}{3}$ into a decimal fraction

(b) $\frac{2}{11}$ into a decimal fraction

(a) $\frac{1}{3} = 1 \div 3$

$= 0.333$

$= 0.\dot{3}$

$$\begin{array}{r} 0.333 \\ 3\overline{)1.000} \end{array}$$

(b) $\frac{2}{11} = 2 \div 11$

$= 0.1818$

$= 0.\dot{1}\dot{8}$

$$\begin{array}{r} 0.1818 \\ 11\overline{)2.0000} \end{array}$$

Exercise 2.2c

Change the following to decimal fractions.

1. $\frac{2}{3}$ 2. $\frac{5}{6}$ 3. $\frac{1}{6}$ 4. $\frac{5}{12}$

5. $\frac{7}{12}$ 6. $\frac{11}{12}$ 7. $\frac{5}{11}$ 8. $\frac{9}{11}$

9. $\frac{3}{11}$ 10. $\frac{7}{11}$ 11. $\frac{4}{15}$ 12. $\frac{11}{15}$

13. $\frac{7}{15}$ 14. $\frac{13}{15}$ 15. $\frac{8}{15}$ 16. $\frac{7}{30}$

17. $\frac{11}{30}$ 18. $\frac{13}{30}$ 19. $\frac{5}{18}$ 20. $\frac{7}{18}$

Example 4

Change $\frac{5}{17}$ into a decimal fraction correct to 3 decimal places

$\frac{5}{17} = 5 \div 17$

$= 0.2941 \ldots$

$= 0.294$ (to 3 D.P.)

$$\begin{array}{r} 0.2941 \\ 17\overline{)5.0000} \\ 3\,4 \\ \hline 1\,60 \\ 1\,53 \\ \hline 70 \\ 68 \\ \hline 20 \\ 13 \\ \hline 7 \end{array}$$

Exercise 2.2d

Change the following to decimal fractions correct to 3 decimal places.

1. $\frac{5}{7}$ 2. $\frac{6}{7}$ 3. $\frac{4}{7}$ 4. $\frac{2}{7}$

5. $\frac{1}{7}$ 6. $\frac{3}{7}$ 7. $\frac{7}{13}$ 8. $\frac{10}{13}$

9. $\frac{9}{13}$ 10. $\frac{11}{13}$ 11. $\frac{10}{21}$ 12. $\frac{2}{21}$

13. $\frac{16}{21}$ 14. $\frac{9}{17}$ 15. $\frac{4}{17}$ 16. $\frac{3}{17}$

17. $\frac{6}{17}$ 18. $\frac{3}{19}$ 19. $\frac{5}{19}$ 20. $\frac{7}{19}$

Example 5

Find the difference between the fractions $\frac{1}{3}$ and $\frac{19}{300}$ giving the answer (a) as a fraction (b) as a decimal.

(a) $\frac{1}{3} - \frac{19}{300} = \frac{100}{300} - \frac{19}{300} = \frac{81}{300} = \frac{27}{100}$

(b) $\frac{27}{100} = 0.27$

Exercise 2.2e

1. A club has 56 members, of whom 35 are men. What fraction of the membership are men? Express this fraction also as a decimal.
2. One day 120 trains arrive at a station and 42 of them are late. What fraction of the trains arrive late? Express this fraction also as a decimal.

3. A village school has 150 pupils and 135 are present on a certain day. What fraction of the pupils are present? Express this fraction also as a decimal.

4. Find the sum of the fractions $\frac{1}{3}$ and $\frac{17}{300}$, giving the answer (a) as a fraction, (b) as a decimal.

5. Find the difference between the fractions $\frac{1}{7}$ and $\frac{3}{70}$, giving the answer (a) as a fraction, (b) as a decimal.

6. Find the sum of the fractions $\frac{2}{3}$ and $\frac{7}{30}$, giving the answer (a) as a fraction, (b) as a decimal.

7. Find the difference between the fractions $\frac{2}{3}$ and $\frac{53}{300}$, giving the answer (a) as a fraction, (b) as a decimal.

8. A piece of string is $\frac{1}{3}$ of a metre in length, if a piece equal to $\frac{7}{30}$ of a metre in length is cut off, what length remains? Express the answer (a) as a fraction, (b) as a decimal.

9. I have $\frac{2}{3}$ of a kilogram of sand. If $\frac{1}{30}$ of a kilogram is then added, how much will I then have? Express the answer (a) as a fraction, (b) as a decimal.

10. A flask holds $\frac{2}{3}$ of a litre of coffee and some is poured out so as to fill a cup of capacity $\frac{11}{30}$ of a litre. How much remains in the flask? Express the answer (a) as a fraction, (b) as a decimal.

In questions 11 to 15, find which is the greater and by how much. Express the answer (a) as a fraction, (b) as a decimal.

11. (a) the sum of $\frac{3}{7}$ and $\frac{19}{70}$
    (b) the sum of $\frac{2}{3}$ and $\frac{7}{300}$

12. (a) the sum of $\frac{2}{7}$ and $\frac{3}{700}$
    (b) the sum of $\frac{2}{9}$ and $\frac{7}{90}$

13. (a) the difference of $\frac{8}{9}$ and $\frac{17}{90}$
    (b) the difference of $\frac{5}{7}$ and $\frac{3}{700}$

14. (a) the sum of $\frac{5}{9}$ and $\frac{13}{900}$
    (b) the difference of $\frac{2}{3}$ and $\frac{11}{300}$

15. (a) the sum of $\frac{2}{3}$ and $\frac{61}{300}$
    (b) the difference of $\frac{6}{7}$ and $\frac{19}{700}$

To arrange numbers in order of size, first change each to a decimal fraction.

Example 6

Arrange the following numbers in order of size beginning with the smallest.

$\frac{1}{2}, 0.25, 1\frac{3}{4}, \frac{4}{10}, 0.64$

$\frac{1}{2} = 1 \div 2 = 0.5$

$1\frac{3}{4} = \frac{7}{4} = 7 \div 4 = 1.75$

$\frac{4}{10} = 4 \div 10 = 0.4$

0·25 and 0·64 are already decimal fractions. The order of the decimal fractions is
0·25, 0·4, 0·5, 0·64, 1·75
So, the order of the given numbers is
$0.25, \frac{4}{10}, \frac{1}{2}, 0.64, 1\frac{3}{4}$

Exercise 2.2f

Write the numbers in each list in order of size, beginning with the smallest.

1. $2, 0.48, \frac{3}{4}, 0.7, 1\frac{1}{5}$
2. $\frac{3}{8}, 0.6, \frac{3}{4}, \frac{7}{10}, 0.85$
3. $0.65, \frac{2}{5}, 0.8, \frac{5}{8}, 0.76$
4. $\frac{3}{4}, \frac{3}{5}, 0.71, \frac{7}{8}, 0.26$
5. $1.2, 1\frac{3}{5}, 1.39, 1\frac{5}{8}, 1\frac{7}{20}$
6. $\frac{6}{5}, \frac{3}{4}, 1\frac{1}{8}, \frac{11}{20}, \frac{1}{2}$
7. $\frac{3}{10}, 0.28, \frac{3}{5}, \frac{1}{4}, \frac{3}{8}$
8. $0.6, \frac{1}{2}, \frac{7}{10}, \frac{2}{5}, \frac{9}{20}$
9. $2\frac{1}{4}, 2.3, 2\frac{2}{5}, 2, 2.41$
10. $0.1, 0.11, \frac{1}{8}, \frac{3}{20}, 0.12$

Write the numbers in each list in order of size, beginning with the largest.

11. $1, 0.8, \frac{3}{5}, \frac{13}{20}, \frac{1}{2}$
12. $\frac{3}{4}, \frac{4}{5}, 0.7, \frac{3}{8}, 0.81$
13. $\frac{5}{8}, 0.71, \frac{7}{10}, 0.69, \frac{17}{25}$
14. $\frac{11}{25}, 0.41, \frac{9}{20}, 0.39, \frac{4}{10}$
15. $0.51, \frac{9}{16}, \frac{1}{2}, \frac{11}{20}, 0.52$
16. $1.02, 1\frac{1}{10}, 1.2, 1\frac{2}{5}, 1.3$
17. $\frac{7}{25}, \frac{1}{4}, 0.26, \frac{27}{100}, 0.29$
18. $2.21, 2\frac{1}{5}, 2\frac{3}{10}, 2\frac{9}{50}, 2.31$
19. $0.01, \frac{11}{100}, 0.10, \frac{7}{50}, \frac{4}{25}$
20. $\frac{5}{16}, 0.31, \frac{3}{10}, \frac{9}{25}, 0.313$

To add and subtract a series of numbers, add the positive numbers first; then do the subtractions.

A calculation in brackets must be done before any other; and division and multiplication must be done before addition or subtraction.

Example 1

(a) $3\frac{1}{2} - 1\frac{1}{4} + 2\frac{1}{3} = 3\frac{1}{2} + 2\frac{1}{3} - 1\frac{1}{4}$

$$= \frac{7}{2} + \frac{7}{3} - \frac{5}{4}$$

$$= \frac{42}{12} + \frac{28}{12} - \frac{15}{12}$$

$$= \frac{70}{12} - \frac{15}{12} = \frac{55}{12} = 4\frac{7}{12}$$

(b) $\frac{4}{5} - 2\frac{1}{4} + 3\frac{1}{2} = \frac{4}{5} + 3\frac{1}{2} - 2\frac{1}{4}$

$$= \frac{4}{5} + \frac{7}{2} - \frac{9}{4}$$

$$= \frac{16}{20} + \frac{70}{20} - \frac{45}{20}$$

$$= \frac{86}{20} - \frac{45}{20} = \frac{41}{20} = 2\frac{1}{20}$$

(c) $1\frac{1}{2} - 2\frac{1}{10} + 3 - \frac{3}{5} = 1\frac{1}{2} + 3 - 2\frac{1}{10} - \frac{3}{5}$

$$= \frac{3}{2} + 3 - \frac{21}{10} - \frac{3}{5}$$

$$= \frac{30}{20} + \frac{60}{20} - \frac{42}{20} - \frac{12}{20}$$

$$= \frac{90}{20} - \frac{42}{20} - \frac{12}{20}$$

$$= \frac{48}{20} - \frac{12}{20} = \frac{36}{20} = 1\frac{16}{20}$$

$$= 1\frac{4}{5}$$

Example 2

(a) $1\frac{1}{2} + 2\frac{1}{4} \times \frac{2}{3} = \frac{3}{2} + \frac{\overset{3}{\cancel{9}}}{\underset{2}{\cancel{4}}} \times \frac{\overset{1}{\cancel{2}}}{\underset{1}{\cancel{3}}}$

$$= \frac{3}{2} + \frac{3 \times 1}{2 \times 1}$$

$$= \frac{3}{2} + \frac{3}{2} = \frac{6}{2} = 3$$

(b) $(1\frac{1}{2} + 2\frac{1}{4}) \times \frac{2}{3} = (\frac{3}{2} + \frac{9}{4}) \times \frac{2}{3}$

$$= (\frac{6}{4} + \frac{9}{4}) \times \frac{2}{3}$$

$$= \frac{\overset{5}{\cancel{15}}}{\underset{2}{\cancel{4}}} \times \frac{\overset{1}{\cancel{2}}}{\underset{1}{\cancel{3}}}$$

$$= \frac{5}{2} = 2\frac{1}{2}$$

(c) $2\frac{3}{5} \times 1\frac{2}{3} + \frac{9}{10} \div 3\frac{3}{5} = \frac{13}{5} \times \frac{5}{3} + \frac{9}{10} \div \frac{18}{5}$

$$= \frac{13}{\underset{1}{\cancel{5}}} \times \frac{\overset{1}{\cancel{5}}}{3} + \frac{\overset{1}{\cancel{9}}}{\underset{2}{\cancel{10}}} \times \frac{\overset{1}{\cancel{5}}}{\underset{2}{\cancel{18}}}$$

$$= \frac{13}{3} + \frac{1}{4} = \frac{52}{12} + \frac{3}{12}$$

$$= \frac{55}{12} = 4\frac{7}{12}$$

Exercise 2.3a

1. $2\frac{1}{4} + 1\frac{1}{2} - 1\frac{3}{4}$
2. $3\frac{1}{3} - 4\frac{1}{2} + 5$
3. $1\frac{5}{8} - \frac{1}{16} + 2\frac{3}{4}$
4. $3\frac{2}{5} - 4\frac{1}{2} + 2$
5. $\frac{1}{8} - 1\frac{5}{6} + 1\frac{3}{4}$
6. $\frac{3}{7} + 4 - 3\frac{2}{3}$
7. $\frac{5}{6} - \frac{7}{8} + \frac{3}{4}$
8. $3 - 4\frac{5}{6} + 1\frac{9}{10}$
9. $8\frac{1}{5} - 1\frac{1}{2} - 2\frac{3}{4}$
10. $7\frac{7}{8} - 3\frac{5}{6} - 1\frac{11}{12}$
11. $\frac{5}{9} + \frac{2}{3} - \frac{1}{6} - \frac{5}{18}$
12. $1\frac{3}{4} + 2\frac{1}{2} - 1\frac{3}{8} + 1\frac{1}{2}$
13. $3\frac{1}{7} + 1\frac{1}{4} - \frac{13}{14} + 3$
14. $2\frac{5}{8} - 3\frac{1}{2} + 1\frac{1}{4} + \frac{1}{2}$
15. $1\frac{5}{6} - \frac{1}{4} - \frac{2}{3} + \frac{5}{12}$
16. $\frac{7}{8} - 2\frac{1}{4} + 2\frac{3}{16} - \frac{1}{2}$
17. $3\frac{1}{5} - 4\frac{3}{4} + 3\frac{1}{2} - 1\frac{2}{3}$
18. $\frac{9}{10} - 1\frac{3}{5} - 2\frac{1}{3} + 3\frac{1}{2}$
19. $3\frac{7}{8} - 1\frac{1}{4} + \frac{5}{8} - 2\frac{1}{3}$
20. $2\frac{3}{4} - 1\frac{1}{3} - \frac{3}{20} - \frac{4}{5}$

Exercise 2.3b

1. $1\frac{3}{4} + 2\frac{1}{2} \times 1\frac{3}{5}$
2. $1\frac{2}{3} \times \frac{3}{10} + 1\frac{3}{4}$
3. $(1\frac{7}{10} + 2\frac{1}{2}) \div \frac{3}{10}$
4. $8\frac{3}{4} - 4\frac{2}{5} \times 1\frac{4}{11}$
5. $14\frac{1}{4} \div (\frac{5}{8} + 1\frac{1}{4})$
6. $(2\frac{7}{8} - 1\frac{9}{16}) \div \frac{7}{8}$
7. $1\frac{4}{5} \times 2\frac{1}{2} \div \frac{3}{4}$
8. $2\frac{3}{5} \times 2\frac{1}{7} - 1\frac{2}{3}$
9. $1\frac{5}{6} - 2\frac{1}{2} + 3\frac{3}{4}$
10. $(1\frac{1}{6} + 2\frac{2}{3}) \times 1\frac{1}{8}$
11. $12\frac{3}{5} - \frac{14}{15} \div \frac{2}{9}$
12. $(\frac{9}{10} - \frac{3}{4}) \times 4\frac{4}{9}$
13. $3\frac{7}{8} \times 3\frac{1}{5} - 9\frac{9}{10}$
14. $\frac{9}{16} \times \frac{4}{7} \div 1\frac{15}{21}$
15. $2\frac{2}{3} + 1\frac{3}{7} \times 6\frac{1}{8}$
16. $(5\frac{2}{3} + 1\frac{1}{7}) \div \frac{5}{12}$
17. $3\frac{1}{4} \div 1\frac{1}{8} + \frac{5}{9} \div \frac{2}{15}$
18. $3\frac{1}{2} + 2(\frac{1}{3} + \frac{7}{8})$
19. $(2\frac{5}{7} - 1\frac{2}{3}) \div (5\frac{1}{3} + 2\frac{11}{15})$
20. $\frac{3}{4}(3\frac{17}{18} - 1\frac{1}{6}) - 1\frac{5}{12}$

When calculating a fraction of any quantity, 'of' means multiply.

Example 1

(a) Find $\frac{3}{5}$ of 20 litres.

(b) Annette and Lorna share £2·48. Annette receives five-eighths of the money, and Lorna takes the remainder. How much does each girl receive?

(a) $\quad \frac{3}{{}_1\cancel{5}} \times \frac{\cancel{20}^4}{1} = 12$ litres

(b) Annette's share $= \dfrac{5}{{}_1\cancel{8}} \times \dfrac{\cancel{284}^{31}}{1}$ p

$\qquad\qquad\qquad = 155\text{p} = £1·55$

Lorna's share $\quad = £2·48 - £1·55$

$\qquad\qquad\qquad = £0·93$

Exercise 2.4

1. A car's petrol tank can hold 40 litres. If the tank is $\frac{1}{4}$ full, how many litres of petrol are in it?
2. Three-quarters of a class of 24 pupils are girls. How many are girls? How many are boys?
3. A strip of brass is 64 cm long, and $\frac{7}{8}$ of this strip is used in making a model. What length is used? What length is left?
4. Hugo earns £2·55. He spends $\frac{4}{15}$ of this money, and saves the remainder. How much does he spend? How much does he save?
5. Two machinists must sew 56 dresses before going home. If Violet sews $\frac{3}{7}$ of the dresses, how many does she sew? How many must Rose sew?
6. A bingo machine contains 45 numbered balls. If $\frac{4}{5}$ of these balls carry even numbers, how many have even numbers? How many have odd numbers?
7. There are 96 keys on a computer keyboard, and $\frac{3}{8}$ of them are yellow. How many are yellow? If all the other keys are blue, how many are blue?

8. Anne is given £4·50, and gives $\frac{3}{10}$ of this money to her friend Karen. How much does Karen receive? How much has Anne left?
9. In a mixture weighing 738 kg, $\frac{7}{9}$ is cement, and the remainder is sand. What weight of cement is in the mixture? What weight of sand is in the mixture?
10. $\frac{11}{12}$ of the tadpoles in a tank grow into frogs. If there are 192 tadpoles, how many grow into frogs? How many do not?
11. The distance from my house to the airport is $5\frac{1}{4}$ kilometres. If I walk $\frac{2}{3}$ of the way, how far do I walk?
12. The distance from Troon Marina to a house in Prestwick is $11\frac{1}{4}$ kilometres. If I walk $\frac{2}{5}$ of the way, how far do I walk?
13. $8\frac{1}{4}$ tonnes of coal are to be shared between a number of men. If one man receives $\frac{6}{11}$ of the load, how many tonnes does he get?
14. $10\frac{1}{2}$ tonnes of sand are to be shared between a number of builders. If one of them receives $\frac{4}{7}$ of the load, how many tonnes does he get?
15. An urn contains $12\frac{1}{2}$ litres of lemonade. How many glasses, each of capacity $\frac{5}{16}$ litre, will it be able to fill?
16. A grain merchant has only $13\frac{1}{2}$ tonnes in his stock. If he has several customers who are all ordering $\frac{3}{4}$ of a tonne, how many can he supply?
17. A plastic tank contains $15\frac{3}{4}$ litres of fruit juice. If a man drains $\frac{2}{9}$ of the contents from the tank, what quantity of juice does he take?
18. $\frac{3}{4}$ of the paint in a tin is used to paint a garage. If the tin contained $5\frac{1}{2}$ litres before painting began, how many litres were used?
19. The distance from a holiday camp to a town is $9\frac{1}{3}$ kilometres. A boy cycles $\frac{9}{14}$ of this distance, and then rests. How far does he cycle? How far has he still to go?
20. Four-fifths of a consignment of liquid cleaner is sold to a farmer. If the whole consignment weighs $11\frac{2}{3}$ kilograms, what weight is sold to the farmer? What weight remains?

Ratio compares quantities of the *same* kind and is found by writing one as a fraction of the other in its simplest form.

Example 1

Give each of the following ratios in their simplest form.

(a) 15 to 75

$$\frac{15}{75} = \frac{1}{5}$$

∴ ratio is 1 : 5

(b) 50 cm to 2 m

$$\frac{50 \text{ cm}}{2 \text{ m}} = \frac{50 \text{ cm}}{200 \text{ cm}}$$

$$= \frac{50}{200} = \frac{1}{4}$$

∴ ratio is 1 : 4

(c) £2·75 to £1·25

$$\frac{£2·75}{£1·25} = \frac{275p}{125p}$$

$$= \frac{275}{125} = \frac{11}{5}$$

∴ ratio is 11 : 5

Exercise 2.5a

Give each of the following ratios in their simplest form.

1. 24 to 96	2. 18 to 108
3. 25 to 75	4. 16 to 40
5. 36 to 54	6. 45 to 60
7. 60 to 72	8. 32 to 56
9. 60 to 24	10. 32 to 12
11. 28 to 16	12. 45 to 25
13. 36 to 21	14. 60 cm to 4 m
15. 80 cm to 6 m	16. 120 m to 2 km
17. 450 kg to 1 t	18. 250 g to 2 kg
19. 450 ml to 6 litres	20. 24 mm to 30 cm
21. 75p to £2	22. 3 m to 45 cm
23. 3 km to 800 m	24. 4 kg to 720 g
25. 2 l to 150 ml	26. £5 to 40p
27. £1·35 to £1·80	28. £1·26 to £1·44
29. £1·20 to £2·80	30. 19p to £1·14
31. 48p to £1·32	32. £4·32 to £1·80
33. £2·52 to £1·96	34. £2·70 to 36p
35. £2·25 to £1·00	36. 80 cm to 1·28 m
37. 2·4 mm to 1·6 cm	38. 900 g to 1·26 kg
39. 1·08 m to 24 cm	40. 1·32 litres to 550 ml

Example 2

In a class there are 27 boys and 15 girls. Find the ratio of the number of boys to the number of girls.

$$\frac{\text{number of boys}}{\text{number of girls}} = \frac{27}{15} = \frac{9}{5}$$

∴ the ratio of the number of boys to the number of girls is 9 : 5.

Exercise 2.5b

1. In a certain month there were 12 wet days and 18 fine days. Find the ratio of wet days to fine days.
2. Amongst a group of boys there were 20 Ayr United supporters and 24 Rangers supporters. Find the ratio of Ayr United supporters to Rangers supporters.
3. On an overcrowded bus there were 9 standing passengers and 30 seated passengers. Find the ratio of standing passengers to seated passengers.
4. In a certain street there were 20 bungalows and 35 houses. Find the ratio of bungalows to houses.
5. In a cricket match a batsman scored 54 runs in his first innings and 96 runs in his second innings. Find the ratio of his first score relative to the second.
6. There are 20 people present in a class, but 12 are absent because they have influenza. Find the ratio of those present to those absent.
7. Amongst a group of girls, 24 chose to play hockey and 21 chose to play netball. Find the ratio of those who played hockey to those who played netball.
8. A group of people travelled from Glasgow to Edinburgh; 54 went by train and 12 went by bus. Find the ratio of those who used the train to those who used the bus.
9. In a group of men there were 56 smokers and 40 non-smokers. Find the ratio of smokers to non-smokers.
10. A local bus from Birmingham to Kidderminster takes 64 minutes whereas an express bus takes only 48 minutes. Find the ratio of the slow time to the fast time.

Example 3

Divide £50 in the ratio 2:3:5
Number of shares = 2 + 3 + 5 = 10

So 1 share = $\frac{£50}{10}$ = £5

The amounts are
  2 X £5 = £10
  3 X £5 = £15
  5 X £5 = £25
Check: £10 + £15 + £25 = £50.

Exercise 2.5c

1. Divide £90 in the ratio 3:7.
2. Divide £84 in the ratio 2:5.
3. Divide £45 in the ratio 4:5.
4. Share out 60 sweets between Michael and Margaret in the ratio 5:7.
5. Share out 48 kg of sand between Mr. Johnson and Mr. Walker in the ratio 3:5.
6. Mary and Julie receive £66 from a rich uncle. They share the money between them in the ratio 5:6. How much does each receive?
7. Divide £120 in the ratio 3:4:5.
8. Divide £72 in the ratio 2:3:4.
9. Divide £132 in the ratio 2:3:6.
10. Divide £56 in the ratio 1:2:5.
11. Share out 60 marbles between Tom, Jack and Bill in the ratio 2:3:7.
12. Share out 45 beads between Ann, Jane and Susan in the ratio 1:3:5.
13. Share out 96 kg coal between Mr. Smith, Mr. Jones and Mr. Thompson in the ratio 1:5:6.
14. David, Peter and John organize a fête which raises £110. The money is shared between charities for the old, for the blind, and for animals in the ratio 5:4:2. How much does each charity receive?
15. At a local election 120 people voted. They voted Liberal, Conservative and Labour in the respective ratio 1:3:6. How many people voted for each party?

Example 4

£185 is to be shared between John, Mary and Jane (whose ages are 10, 12, and 15 years respectively) in the same ratio as their ages. How much does each receive?

Ratio of ages = 10:12:15
Number of shares = 10 + 12 + 15 = 37

∴ 1 share = £185 ÷ 37 = £5
  John gets £5 X 10 = £50
  Mary gets £5 X 12 = £60
  Jane gets £5 X 15 = £75
Check: £50 + £60 + £75 = £185.

Exercise 2.5d

1. £160 is to be shared between Bill, Jane and Wendy (whose ages are 9, 10, and 13 years respectively) in the same ratio as their ages. How much does each receive?
2. Mr. Andrews, Mr. Bailey and Mr. Carter own 4, 5, and 6 parts respectively of the same business. If the business makes a profit of £120 in a certain week, how much does each receive?
3. In a television quiz contest the prize money is £108. It is shared out between the top three contestants in the same ratio as that of the marks they obtain. If their marks are 15, 13, and 8, how much does each receive?
4. A football club pays out £140 in bonus money to its three top-scoring players in the same ratio as that of the goals they have scored in a certain tournament. If they score 16, 11, and 8 goals, how much does each receive?
5. A cricket club pays £120 in bonus money to its three most successful batsmen in the same ratio as the number of centuries they have scored during the season. If they have scored 11, 7, and 6 centuries, how much does each receive?
6. 72 kg of compost is to be spread over three small garden plots in a ratio equal to that of their areas. If their areas are 7 m², 8 m², and 9 m², how much is used on each plot?
7. 540 kg of dry concrete mixture is to be spread over three surfaces in a ratio equal to that of their areas. If their areas are 9 m², 11 m², and 16 m², how much is used on each surface?
8. £10 is to be shared between three fruit pickers in a ratio equal to that of the number of bags they are able to fill. If they fill 5, 6, and 9 bags, how much does each receive?
9. Three boys sell some magazines for a total profit of £2. The profit is shared between them in a ratio equal to that of the number of magazines they sell. If they sell 7, 8, and 10 magazines, how much does each receive?
10. 15 t of surface soil is to be spread over three small lawns in a ratio equal to that of their areas. If their areas are 7 m², 11 m² and 12 m², how much is used on each surface?

Two quantities are in *direct proportion* when an increase in one is in the same ratio as an increase in the other. e.g. if 5 oranges cost 20p, 10 oranges cost 40p: the cost is in direct proportion to the quantity. The cost varies directly as the quantity.

Example 5

If 4 m of cloth cost £10, what will 10 m cost?
The number of metres of cloth is increased in the ratio $\frac{10}{4}$.
So the cost is increased in the ratio $\frac{10}{4}$.
∴ 10 m costs $\frac{10}{4}$ × £10 = £25.

Example 6

If 12 eggs cost 54p, how many eggs could be bought for 18p?
The cost of the eggs decreases in the ratio $\frac{18}{54}$.
So the number of eggs bought decreases in the ratio $\frac{18}{54}$.
∴ number of eggs bought = $\frac{18}{54}$ × 12
= 4 eggs.

Exercise 2.5e

1. Curtain hooks cost 30p for 12. Find the cost of
   (a) 40 curtain hooks,
   (b) 8 curtain hooks.
2. If 12 oranges cost 88p, find the cost of
   (a) 15 oranges,
   (b) 9 oranges.
3. If 20 m² of carpet costs £36, find the cost of
   (a) 35 m²,
   (b) 15 m².
4. Ceiling tiles cost £1·56 for 12. Find the cost of
   (a) 7, (b) 30, (c) 9, (d) 20.
5. It takes me 44 minutes to cycle from Pitlochry to Pitagowan, a distance of 16 km. At the same rate, how long will it take to cycle
   (a) from Pitlochry to Perth (40 km)?
   (b) from Pitlochry to Blair Atholl (12 km)?
6. If 12 brass screws cost 16p, how many can be bought for (a) 40p? (b) 12p?
7. 12 grapefruits cost 90p. How many can be bought for (a) £2·10? (b) 75p?
8. If 20 m² of flooring cost £15, what area of the same flooring can be bought for (a) £33? (b) £9?

9. Ceramic tiles cost £3·24 for 12. How many can be bought for
   (a) £8·64?   (b) £5·94?
   (c) £2·70?   (d) £2·43?
10. It takes me 40 minutes to cycle from Ayr to Patna, a distance of 15 km. Which of the distances below can I cycle in
   (a) 48 minutes?
   (b) 32 minutes?
   Ayr to Annbank Station (7·5 km)
   Ayr to Loans (10·5 km)
   Ayr to Maybole (12 km)
   Ayr to Ochiltree (18 km)
11. 8 boxes of chocolate soldiers weigh 240 g. What is the weight of
   (a) 6 boxes? (b) 10 boxes?
12. 11 bags of frozen peas cost £6·93. Find the cost of (a) 7 bags, (b) 20 bags.
13. A furnace burns 42 tonnes of coal every 5 hours. How much coal does it burn in
   (a) 3 hours? (b) 12 hours?
14. 420 small nails cost 84p.
   (a) What is the cost of 600 nails?
   (b) How many nails can be bought for £2·20?
15. 14 ferrets can be bought for £57·68.
   (a) What is the cost of 3 ferrets?
   (b) How many ferrets can be bought for £32·96?
16. 28 crates of luggage weigh 4·20 tonnes.
   (a) What is the weight of 13 crates?
   (b) How many crates weigh 2·55 tonnes?
17. At a steady speed, a motorcyclist travels 15 km in 20 minutes. At the same speed, how long will he take to travel
   (a) 18 km? (b) 6 km? (c) 24 km?
18. 5 kg of coffee costs £32
   Find the cost of
   (a) $2\frac{1}{2}$ kg, (b) $3\frac{1}{4}$ kg, (c) $1\frac{3}{8}$ kg.
19. $1\frac{3}{4}$ tonnes of plastic costs £770.
   Find the cost of
   (a) $3\frac{1}{2}$ tonnes, (b) $2\frac{1}{5}$ tonnes, (c) $1\frac{1}{8}$ tonnes.
20. $4\frac{1}{2}$ metres of steel rod costs £11·25.
   Find the cost of
   (a) 3 m, (b) $5\frac{1}{2}$ m, (c) $2\frac{1}{2}$ m.

Two quantities are in *inverse proportion* when an increase in one is in the same ratio as a *decrease* in the other.
e.g. if 5 combine harvesters reap a field in 2 days, then 10 combine harvesters would reap the same field in 1 day.
The time for harvesting varies inversely as the number of harvesters used.

Example 7

A car travelling at an average speed of
60 km/h completes a journey in 38
minutes. How long would the same journey
take at an average speed of 40 km/h?
The speed decreases in the ratio $\frac{40}{60}$,

so the time increases in the ratio $\frac{60}{40}$.

$\therefore$ time taken $= \frac{60}{40} \times 38 = 57$ minutes.

Example 8

If 5 men can unload a lorry in 18 minutes,
how long would it take 9 men to unload
the same lorry?
The number of men increases in the ratio $\frac{9}{5}$,

so the time taken decreases in the ratio $\frac{5}{9}$.

$\therefore$ time taken $= \frac{5}{9} \times 18 = 10$ minutes.

Exercise 2.5f

1. If 12 dockers can load a barge in 75 minutes,
   how long will it take (a) 10 dockers, (b) 25
   dockers to load the same barge?
2. If 15 men can unload a ship's cargo in 100
   minutes, how long will it take (a) 6 men, (b) 20
   men to unload the same cargo?
3. If 15 bricklayers can build a wall in 30 hours of
   working time, how long will it take (a) 9 bricklayers, (b) 25 bricklayers to build the same wall?
4. An electric heater raises the temperature of 8
   kilograms of water by 18 degrees in a certain
   amount of time. By how many degrees would it
   raise the temperature of
   (a) 6 kilograms, (b) 9 kilograms of water in the
   same time?
5. An electric heater raises the temperature of 900
   grams of water by 20 degrees in a certain amount
   of time. By how many degrees would it raise the
   temperature of (a) 750 grams, (b) 1·2 kilograms
   of water in the same time?
6. A 500 watt electric heater boils a given amount of
   water in 12 minutes. How long would it take
   (a) a less powerful heater of 300 watts, (b) a
   more powerful heater of 750 watts to boil the
   same amount of water?
7. A 1·2 kilowatt electric heater boils a given amount
   of water in 10 minutes. How long would it take
   (a) a less powerful heater of 800 watts, (b) a
   more powerful heater of 2 kilowatts to boil the
   same amount of water?

8. Water flows constantly into a cylindrical beaker
   of base area 24 cm^2. The water level rises by
   10 cm every second. How quickly would the
   water level rise in a beaker of base area
   (a) 20 cm^2, (b) 30 cm^2 if the flow rate remained
   the same?
9. A car travelling at 90 km/h completes a certain
   journey in 40 minutes. How long would it take a
   car travelling at (a) 100 km/h, (b) 75 km/h to
   complete the same journey?
10. A car travelling at 96 km/h takes 35 minutes to
    travel between two service areas on a motorway.
    How long would it take a car travelling at
    (a) 112 km/h, (b) 80 km/h to travel between
    the same two service areas?
11. Charles has been given money for his birthday,
    and can buy exactly 50 Mammoth Ices at 24p
    each. With the same amount of money, how
    many of each of the following can he buy?
    (a) Giant Lollies, at 20p each.
    (b) Atomic Coaster choc ices, at 15p each.
12. Lottie has enough hay to feed 12 horses for
    10 days. How many days will the same quantity
    of hay last if she feeds
    (a) 6 horses? (b) 15 horses? (c) 40 horses?
13. A man types at the rate of 68 words per minute,
    for 5 hours. How long will he take to type the
    same material at
    (a) 34 words per minute?
    (b) 102 words per minute?
14. A spaceship has sufficient food to feed 5
    crewmen for 6 days. How many days will the
    same quantity of food last if there are
    (a) 3 crewmen? (b) 10 crewmen?
    (c) 30 crewmen?
15. A taxi driver charges 25 pence per kilometre,
    and a girl can afford to travel 12 kilometres.
    For the same total cost, how far could she
    travel if the cost per kilometre was
    (a) 30 pence? (b) 15 pence? (c) $37\frac{1}{2}$ pence?
16. The cost of a coach, shared equally between
    30 people, is £1·20 per person. If 10 extra
    people agree to help share the cost, what is the
    new cost per person?
17. When set at 75 characters per second, a
    facsimile machine prints a page in 15 seconds.
    How long will it take to print the same page if
    the number of characters per second is
    (a) increased to 125?
    (b) decreased to 45?
18. 4 joiners build a garage in $5\frac{1}{2}$ days. Working at
    the same rate, how long would the same job take
    (a) 2 joiners? (b) 11 joiners?

The *scale* of a map is the ratio of the distance between two points on the map to the actual distance between those points on the ground. Use the *same units* for both distances. e.g. mm and mm; cm and cm; etc.

Example 1

Two towns, 5 km apart, are 10 cm apart on a map.
Find the scale of the map.
Actual distance = 5 km = 500 000 cm
Distance on map = 10 cm
Scale = 10:500 000
$\quad$ = 1:50 000

Exercise 2.6a

1. Two windows are 5 cm apart on a plan. If the windows are actually 3 m apart, what is the scale of the plan?
2. Two towns, 20 km apart, are drawn 10 cm apart on a map. What is the scale of the map?
3. A cottage and a farmhouse are really 2 km apart, but are 5 cm apart on a map. What is the scale of the map?
4. A road is 8 cm long, on a map. If the road is actually 12 km long, what is the scale of the map?
5. The distance from a hotel to a town centre on a map is found to be 15 cm. If the true distance is 3 km, find the scale of the map.
6. The distance from a fire door to a staircase is 4 cm on a plan. If the real distance is 20 m, what is the scale of the plan?
7. A chimney, 150 m high, measures 5 cm on a plan. Find the scale of the plan.
8. The distance between two ferry ports is 48 km. If this is represented by a distance of 8 cm on a map, what is the scale of the map?
9. A farmer's field is 1·5 cm long, on a map. If the field is actually 300 m long, find the scale of the map.

10. A café and a disco are 750 m apart, but the distance between them on a map is 0·75 cm. What is the scale of the map?

Example 2

A map has a scale of 1:20 000.
Find the distance on the ground represented by a distance of 6·5 cm on the map.
Distance on ground = 6·5 × 20 000 cm
$\qquad\qquad\qquad$ = 130 000 cm = 1·3 km

Exercise 2.6b

1. A plan has a scale of 1:200. If the distance between a door and a wall is 4 cm on this plan, find the true distance from door to wall.
2. On a plan of scale 1:800, the front of a house measures 7 cm. Find the true length of the front of the house.
3. The distance from Ayburgh to Groaton is 10 cm on a map. If the scale of the map is 1:200 000, find the real distance between these towns.
4. A plan shows a distance of 9 mm between two lights. If the scale of the plan is 1:10 000, find the actual distance between these two lights.
5. Two chimneys are 4 mm apart on a plan. If the scale of the plan is 1:50 000, find the actual distance between the chimneys.
6. The distance from an airport to a harbour is 1·5 cm on a map. Find the real distance between the airport and harbour if the scale of the map is 1:200 000.
7. On a plan the height of a tower is 8 cm. If the scale is 1:250, find the true height of the tower.
8. A test tank is 2·5 cm high on a scale drawing. If the scale is 1:400, find the actual height of the tank.
9. On a plan of scale 1:450 the thickness of a steel beam is 2·6 mm.
Find the real thickness of the beam.
10. The scale of a map is 1:250 000. If the width of a peat bog is 5·6 cm on this map, find the actual width of the bog.

Example 3

On a map of scale 1:25 000, find the distance which represents 2 km.

$$\text{Distance on map} = \frac{1}{25\,000} \times \frac{2}{1}$$

$$= \frac{1}{\underset{1}{25\,000}} \times \frac{\overset{8}{200\,000}}{1} \text{ cm}$$

$$= 8 \text{ cm}$$

Exercise 2.6c

1. The distance between two houses is 100 m. How far apart should the houses be drawn on a plan of scale 1:500?
2. St. Evert and Pludlow are 4 km apart. How far apart will they be on a map of scale 1:200 000?
3. The distance between two chimneys is 750 m. How far apart should they be drawn on a plan of scale 1:2500?

4. The height of a block of flats is 680 m. On a plan of scale 1:4000 how high will the flats appear?
5. The distance from a harbour to an island is 5 km. On a map of scale 1:250 000 how far apart are they?
6. The distance from a generator to a set of floodlights is 350 m. How far apart should they be on a plan of scale 1:70 000?
7. In a sports stadium the distance from the entrance to the changing rooms is 72 m. What is the corresponding distance on a plan of scale 1:1200?
8. The distance from one railway station to another is 16·8 km. What distance will separate the stations on a map of scale 1 : 300 000?
9. A road is 4·5 m wide. How wide will it be on a plan of scale 1:3000?
10. On a map of scale 1:60 000 how long must I draw a road bridge if its real length is 1·2 km?

2.7 PERCENTAGE

A percentage is a fraction with a denominator of 100.

Example 1

Write each percentage as a fraction in its simplest form.

(a) $7\% \quad = \frac{7}{100}$

(b) $25\% \; = \frac{25}{100} = \frac{25 \div 25}{100 \div 25} = \frac{1}{4}$

(c) $84\% \quad = \frac{84}{100} = \frac{84 \div 4}{100 \div 4} = \frac{21}{25}$

(d) $160\% = \frac{160}{100} = \frac{8}{5} = 1\frac{3}{5}$

Example 2

Write each percentage as a decimal fraction.

(a) $17\% \quad = \frac{17}{100} = 0\cdot17$

(b) $75\% \quad = \frac{75}{100} = 0\cdot75$

(c) $3\% \quad = \frac{3}{100} = 0\cdot03$

(d) $283\% = \frac{283}{100} = 2\cdot83$

Exercise 2.7a

Write each percentage as a fraction in its simplest form.

1. 9%	2. 15%	3. 45%	4. 20%
5. 70%	6. 32%	7. 68%	8. 56%
9. 42%	10. 6%	11. 140%	12. 180%
13. 130%	14. 210%	15. 275%	16. 335%
17. 236%	18. 308%	19. 405%	20. 465%

Write each percentage as a decimal fraction.

21. 15%	22. 35%	23. 90%	24. 80%
25. 60%	26. 5%	27. 16%	28. 72%
29. 44%	30. 58%	31. 125%	32. 156%
33. 170%	34. 230%	35. 225%	36. 340%
37. 212%	38. 304%	39. 408%	40. 401%

Example 3

Write each percentage as a fraction in its simplest form.

(a) $12\frac{1}{2}\% \quad = \frac{12\frac{1}{2}}{100} = \frac{12\frac{1}{2} \times 2}{100 \times 2} = \frac{25}{200} = \frac{1}{8}$

(b) $32\frac{1}{4}\% \quad = \frac{32\frac{1}{4}}{100} = \frac{32\frac{1}{4} \times 4}{100 \times 4} = \frac{129}{400}$

(c) $6\frac{2}{3}\% = \dfrac{6\frac{2}{3}}{100} = \dfrac{20}{300} = \dfrac{1}{15}$

(d) $128\frac{4}{7}\% = \dfrac{128\frac{4}{7}}{100} = \dfrac{900}{700} = \dfrac{9}{7} = 1\frac{2}{7}$

Example 4

(a) $37\frac{1}{2}\% = \dfrac{37\cdot5}{100}$ (as $\frac{1}{2} = 0\cdot5$) $= 0\cdot375$

(b) $6\frac{1}{4}\% = \dfrac{6\cdot25}{100}$ (as $\frac{1}{4} = 0\cdot25$) $= 0\cdot0625$

(c) $133\frac{1}{3}\% = \dfrac{133\cdot\dot{3}}{100}$ (as $\frac{1}{3} = 0\cdot\dot{3}$) $= 1\cdot33\dot{3}$

Exercise 2.7b

Write each percentage as a fraction in its simplest form.

1. $17\frac{1}{2}\%$ 2. $18\frac{3}{4}\%$ 3. $63\frac{1}{3}\%$ 4. $53\frac{1}{3}\%$

5. $20\frac{5}{6}\%$ 6. $41\frac{2}{3}\%$ 7. $42\frac{1}{2}\%$ 8. $5\frac{5}{9}\%$

9. $61\frac{1}{9}\%$ 10. $21\frac{7}{8}\%$ 11. $102\frac{1}{2}\%$ 12. $147\frac{1}{2}\%$

13. $123\frac{1}{3}\%$ 14. $204\frac{1}{6}\%$ 15. $427\frac{1}{2}\%$ 16. $308\frac{1}{3}\%$

17. $313\frac{1}{3}\%$ 18. $206\frac{2}{3}\%$ 19. $129\frac{1}{6}\%$ 20. $115\frac{5}{8}\%$

Write each percentage as a decimal fraction.

21. $62\frac{1}{2}\%$ 22. $87\frac{1}{2}\%$ 23. $32\frac{1}{2}\%$ 24. $47\frac{1}{2}\%$

25. $56\frac{1}{4}\%$ 26. $81\frac{1}{4}\%$ 27. $9\frac{3}{8}\%$ 28. $66\frac{2}{3}\%$

29. $11\frac{1}{9}\%$ 30. $83\frac{1}{3}\%$ 31. $112\frac{1}{2}\%$ 32. $137\frac{1}{2}\%$

33. $107\frac{1}{2}\%$ 34. $222\frac{1}{2}\%$ 35. $131\frac{1}{4}\%$ 36. $306\frac{1}{4}\%$

37. $203\frac{1}{8}\%$ 38. $333\frac{1}{3}\%$ 39. $403\frac{1}{3}\%$ 40. $216\frac{2}{3}\%$

To find the value of a percentage of a quantity, change the percentage to a common fraction; then find that fraction of the quantity.

Example 5

(a) 28% of 50 $= \dfrac{28}{100} \times \dfrac{50}{1} = \dfrac{14}{1} = 14$

(b) 66% of £1·25 $= \dfrac{66}{100} \times \dfrac{125}{1}$ pence

$= \dfrac{165}{2} = 82\frac{1}{2}$ pence

(c) $6\frac{1}{4}\%$ of 1 m 44 cm $= \dfrac{6\frac{1}{4}}{100} \times 144$ cm

$= \dfrac{25}{400} \times \dfrac{144}{1} = 9$ cm

Exercise 2.7c

Find the 'odd answer out'.

1. (a) 20% of £400 2. (a) 24% of £150
   (b) 25% of £300   (b) 25% of £140
   (c) 30% of £250   (c) 15% of £240
3. (a) 16% of £250 4. (a) 32% of £7·50
   (b) 36% of £125   (b) 15% of £16·00
   (c) 25% of £160   (c) 20% of £12·50
5. (a) 64% of £5·00 6. (a) 25% of 2 m 16 cm
   (b) 44% of £7·50   (b) 40% of 1 m 30 cm
   (c) 60% of £5·50   (c) 30% of 1 m 80 cm
7. (a) 15% of 11 m   8. (a) 15% of 4 m
   (b) 35% of 4 m 80 cm (b) 12% of 5 m 50 cm
   (c) 12% of 14 m   (c) 8% of 7 m 50 cm
9. (a) $12\frac{1}{2}\%$ of £272 10. (a) $62\frac{1}{2}\%$ of £192
   (b) $16\frac{2}{3}\%$ of £198   (b) $58\frac{1}{3}\%$ of £204
   (c) $18\frac{3}{4}\%$ of £176   (c) $66\frac{2}{3}\%$ of £180

To write one quantity as a percentage of another quantity, first put them as a fraction. Each quantity must be in the same units. Then change the fraction into a percentage by multiplying by 100.

Example 6

Find (a) 15 as a percentage of 60.
(b) £1·20 as a percentage of £6·00.
(c) 225 mm as a percentage of 20 cm.

(a) Fraction is $\frac{15}{60}$.

Percentage is $\frac{\cancel{15}^{1}}{\cancel{60}_{4}} \times \frac{\cancel{100}^{25}}{1}\% = \frac{25}{1} = 25\%$

(b) Fraction is $\frac{£1·20}{£6·00} = \frac{120p}{600p}$

Percentage is $\frac{\cancel{120}^{1}}{\cancel{600}_{5}} \times \frac{\cancel{100}^{20}}{1}\% = \frac{20}{1} = 20\%$

(c) Fraction is $\frac{225\ mm}{20\ cm} = \frac{225\ mm}{200\ mm}$

Percentage is $\frac{225}{\cancel{200}_{2}} \times \frac{\cancel{100}^{1}}{1}\% = \frac{225}{2}$

$$= 112\tfrac{1}{2}\%$$

Exercise 2.7d

Find the 'odd answer out' by changing each to a percentage.

1. (a) £42 of £168
   (b) £45 of £225
   (c) £48 of £192
2. (a) £112 of £160
   (b) £120 of £150
   (c) £144 of £180
3. (a) £4·20 of £5·60
   (b) £4·50 of £6·00
   (c) £3·60 of £5·00
4. (a) £1·89 of £4·20
   (b) £2·10 of £5·25
   (c) £1·96 of £4·90
5. (a) 2 m 16 cm of 3 m 60 cm
   (b) 2 m 24 cm of 4 m
   (c) 2 m 22 cm of 3 m 70 cm
6. (a) 3 m 40 cm of 4 m
   (b) 3 m 24 cm of 3 m 60 cm
   (c) 3 m 42 cm of 3 m 80 cm
7. (a) £35 of £315
   (b) £36 of £288
   (c) £33 of £264
8. (a) £54 of £144
   (b) £49 of £147
   (c) £48 of £128
9. (a) 32 cm of 1 m 92 cm
   (b) 30 cm of 2 m 25 cm
   (c) 33 cm of 1 m 98 cm
10. (a) 33 cm of 3 m 96 cm
    (b) 28 cm of 3 m 36 cm
    (c) 25 cm of 4 m
11. (a) 1 km 680 m of 2 km 520 m
    (b) 1 km 750 m of 2 km 800 m
    (c) 1 km 800 m of 2 km 880 m

Exercise 2.7e

1. In a box of 150 eggs, 20% were broken. How many whole eggs were there?
2. A factory has 1600 workers, but during a bus strike 15% were absent. Find the number of workers absent.
3. A school has 1850 pupils, and one day 4% of them arrived late. Find the number of those arriving late.
4. 2400 people live in the village of Bilton, and the results of a local election are given below. 25% voted Conservative, 15% voted Liberal, 40% voted Labour, 20% did not vote. Convert these percentages to actual numbers.
5. A school has 1200 pupils and the percentage attendance records for a certain week are given below.

Monday	90%	Thursday	98%
Tuesday	85%	Friday	95%
Wednesday	92%		

   Find the number of pupils present on each day.
6. A girl wins £648 000 on the pools, and donates $12\tfrac{1}{2}\%$ of her winnings to charity. How much does she donate?
7. $6\tfrac{1}{4}\%$ of a town's population are over 80 years old. If the population of the town is 38 000, how many are over 80 years old?
8. $16\tfrac{2}{3}\%$ of the cows on a farm are Jerseys. If there are 270 cows on the farm, how many are Jerseys?
9. 720 buses operate from a depot, but 144 are being repaired. What percentage of the total are in service?
10. One day 132 trains arrive at a railway station and 99 of them are on time. What percentage of them are late?
11. An examination is marked out of 144 and one pupil gets 108. Find his mark as a percentage.
12. 48 000 live in Scarborough and 7200 are council house dwellers. What percentage of the population live in council houses?
13. 3200 people live in Swanland and the results of a local election are given below. 1280 voted Conservative, 800 voted Liberal 960 voted Labour, 160 did not vote. Convert the figures to percentages.
14. A glue consists of 20 g resin, 8 g activator, and 2 g hardener. Find the percentage each chemical is of the total.
15. A shop sells 12 tape recorders, 15 radios, 30 television sets, and 3 video recorders in one month. What percentage of the total does each type of article represent?

Before beginning any new calculations on your calculator, press the $\boxed{\text{AC}}$ key. This clears the calculator.
If your calculator does not have an $\boxed{\text{AC}}$ key, press $\boxed{\text{C}}$ instead.
To add, subtract, multiply or divide numbers, enter the numbers and signs in their given order, then press $\boxed{=}$.
Always check your answer with a rough estimate (see page 17), as shown in the example.

Example 1

Calculate (a) $49 + 32 + 61$ (b) $78 - 32$
  (c) 43×27 (d) $99 \div 18$
and check your answers with a rough estimate.

(a) $\boxed{\text{AC}}\boxed{4}\boxed{9}\boxed{+}\boxed{3}\boxed{2}\boxed{+}\boxed{6}\boxed{1}\boxed{=}$ 142

(b) $\boxed{\text{AC}}\boxed{7}\boxed{8}\boxed{-}\boxed{3}\boxed{2}\boxed{=}$ 46

(c) $\boxed{\text{AC}}\boxed{4}\boxed{3}\boxed{\times}\boxed{2}\boxed{7}\boxed{=}$ 1161

(d) $\boxed{\text{AC}}\boxed{9}\boxed{9}\boxed{\div}\boxed{1}\boxed{8}\boxed{=}$ 5·5

Check: write each number correct to 1 significant figure and then repeat the calculation.

(a) $49 + 32 + 61 \approx 50 + 30 + 60 = 140$ ✓
(b) $78 - 32 \approx 80 - 30 = 50$ ✓
(c) $43 \times 27 \approx 40 \times 30 = 1200$ ✓
(d) $99 \div 18 \approx 100 \div 20 = 5$ ✓

Exercise 3.1a

Use your calculator to find each answer. Check your answers with a rough estimate.

1. $87 + 42 + 33$	**2.** $78 + 62 + 51$	**3.** $39 + 31 + 64$
4. $26 + 43 + 92$	**5.** $79 + 54 + 81$	**6.** $51 + 37 + 28$
7. $92 + 49 + 56$	**8.** $33 + 58 + 57$	**9.** $84 + 18 + 99$
10. $42 + 48 + 9$		

11. $93 - 21$	**12.** $82 - 53$	**13.** $74 - 31$	**14.** $68 - 39$
15. $87 - 58$	**16.** $79 - 16$	**17.** $62 - 24$	**18.** $91 - 12$
19. $97 - 45$	**20.** $56 - 8$		

21. 82×29	**22.** 71×37	**23.** 94×18	**24.** 53×26
25. 82×9	**26.** 58×21	**27.** 87×34	**28.** 66×43
29. 75×52	**30.** 97×22	**31.** $91 \div 28$	**32.** $81 \div 36$

33. $92 \div 32$	**34.** $63 \div 24$	**35.** $93 \div 12$	**36.** $56 \div 32$
37. $78 \div 24$	**38.** $58 \div 16$	**39.** $98 \div 16$	**40.** $99 \div 48$

Example 2

Calculate (a) $390 + 180 + 210$ (b) $570 - 390$
            (c) 340×160 (d) $810 \div 216$

and check your answers with a rough estimate.

(a) |AC| |3| |9| |0| |+| |1| |8| |0| |+| |2| |1| |0| |=| 780

(b) |AC| |5| |7| |0| |-| |3| |9| |4| |=| 176

(c) |AC| |3| |4| |0| |×| |1| |6| |0| |=| 54 400

(d) |AC| |8| |1| |0| |÷| |2| |1| |6| |=| 3·75

Check: write each number correct to 1 significant figure.

(a) $390 + 180 + 210 \approx 400 + 200 + 200 = 800$ ✓
(b) $570 - 394 \approx 600 - 400 = 200$ ✓
(c) $340 \times 160 \approx 300 \times 200 = 60\,000$ ✓
(d) $810 \div 216 \approx 800 \div 200 = 4$ ✓

Exercise 3.1b

Use your calculator to find each answer. Check your answers with a rough estimate.

1. $230 + 320 + 280$	**2.** $110 + 420 + 190$	**3.** $430 + 240 + 270$
4. $390 + 280 + 210$	**5.** $370 + 190 + 120$	**6.** $360 + 90 + 330$
7. $297 + 389 + 214$	**8.** $192 + 481 + 119$	**9.** $378 + 266 + 204$
10. $312 + 131 + 96$	**11.** $243 + 125 + 85$	**12.** $437 + 209 + 72$
13. $149 + 251 + 55$		

14. $510 - 230$	**15.** $620 - 430$	**16.** $730 - 120$	**17.** $480 - 170$
18. $660 - 380$	**19.** $870 - 80$	**20.** $698 - 276$	**21.** $567 - 389$
22. $874 - 191$	**23.** $721 - 432$	**24.** $516 - 345$	**25.** $628 - 209$

26. 280×120	**27.** 370×310	**28.** 460×230	**29.** 410×190
30. 640×270	**31.** 540×180	**32.** 321×197	**33.** 434×286
34. 548×173	**35.** 479×124	**36.** 387×231	**37.** 564×329
38. 651×95			

39. $825 \div 220$	**40.** $420 \div 224$	**41.** $742 \div 112$	**42.** $585 \div 180$
43. $564 \div 96$	**44.** $858 \div 312$	**45.** $567 \div 216$	**46.** $408 \div 192$
47. $638 \div 176$	**48.** $936 \div 288$	**49.** $833 \div 196$	**50.** $945 \div 252$

Example 3

Calculate (a) $6·8 + 2·1 + 4·3$ (b) $8·9 - 5·6$
            (c) $9·1 \times 4·8$ (d) $3·9 \div 2·4$

and check your answers with a rough estimate.

(a) |AC| |6| |.| |8| |+| |2| |.| |1| |+| |4| |.| |3| |=| 13·2

(b) |AC| |8| |.| |9| |-| |5| |.| |6| |=| 3·3

(c) |AC| |9| |.| |1| |×| |4| |.| |8| |=| 43·68

(d) |AC| |3| |.| |9| |+| |2| |.| |4| |=| 1·625

Check: write each number correct to 1 significant figure.

(a) $6 \cdot 8 + 2 \cdot 1 + 4 \cdot 3 \approx 7 + 2 + 4 = 13 \checkmark$
(b) $8 \cdot 9 - 5 \cdot 6 \approx 9 - 6 = 3 \checkmark$
(c) $9 \cdot 1 \times 4 \cdot 8 \approx 9 \times 5 = 45 \checkmark$
(d) $3 \cdot 9 \div 2 \cdot 4 \approx 4 \div 2 = 2 \checkmark$

Exercise 3.1c

Use your calculator to find each answer. Check your answers with a rough estimate.

1. $5 \cdot 9 + 3 \cdot 2 + 4 \cdot 1$
2. $8 \cdot 7 + 2 \cdot 3 + 5 \cdot 2$
3. $13 \cdot 8 + 9 \cdot 1 + 7 \cdot 4$
4. $15 \cdot 9 + 12 \cdot 2 + 8 \cdot 3$
5. $11 \cdot 6 + 20 \cdot 4 + 6 \cdot 1$
6. $18 \cdot 2 + 23 \cdot 9 + 5 \cdot 7$
7. $27 \cdot 3 + 14 \cdot 8 + 8 \cdot 6$
8. $26 \cdot 4 + 32 \cdot 6 + 9 \cdot 9$
9. $37 \cdot 1 + 21 \cdot 9 + 13 \cdot 8$
10. $40 \cdot 2 + 19 \cdot 8 + 10 \cdot 7$

11. $8 \cdot 2 - 3 \cdot 4$
12. $9 \cdot 3 - 2 \cdot 4$
13. $17 \cdot 1 - 5 \cdot 3$
14. $15 \cdot 8 - 3 \cdot 9$
15. $13 \cdot 7 - 4 \cdot 6$
16. $24 \cdot 8 - 9 \cdot 7$
17. $27 \cdot 2 - 8 \cdot 4$
18. $32 \cdot 1 - 12 \cdot 3$
19. $35 \cdot 9 - 18 \cdot 6$
20. $44 \cdot 7 - 25 \cdot 9$

21. $8 \cdot 2 \times 3 \cdot 9$
22. $9 \cdot 3 \times 5 \cdot 8$
23. $12 \cdot 1 \times 2 \cdot 7$
24. $15 \cdot 4 \times 4 \cdot 6$
25. $20 \cdot 2 \times 7 \cdot 8$
26. $14 \cdot 9 \times 6 \cdot 2$
27. $17 \cdot 8 \times 5 \cdot 3$
28. $39 \cdot 6 \times 7 \cdot 4$
29. $24 \cdot 7 \times 8 \cdot 1$
30. $49 \cdot 5 \times 5 \cdot 2$

31. $4 \cdot 2 \div 2 \cdot 4$
32. $25 \cdot 2 \div 3 \cdot 2$
33. $3 \cdot 6 \div 1 \cdot 6$
34. $11 \cdot 5 \div 4 \cdot 6$
35. $24 \cdot 7 \div 5 \cdot 2$
36. $14 \cdot 3 \div 2 \cdot 2$
37. $23 \cdot 1 \div 4 \cdot 2$
38. $6 \cdot 3 \div 2 \cdot 8$
39. $29 \cdot 4 \div 5 \cdot 6$
40. $10 \cdot 2 \div 4 \cdot 8$

Example 4

Calculate (a) $3 \cdot 89 + 4 \cdot 21 + 1 \cdot 34$ (b) $8 \cdot 22 - 2 \cdot 78$
            (c) $5 \cdot 84 \times 2 \cdot 92$ (d) $19 \cdot 57 \div 4 \cdot 12$

and check your answers with a rough estimate.

(a) $\boxed{AC}\ \boxed{3}\ \boxed{.}\ \boxed{8}\ \boxed{9}\ \boxed{+}\ \boxed{4}\ \boxed{.}\ \boxed{2}\ \boxed{1}\ \boxed{+}\ \boxed{1}\ \boxed{.}\ \boxed{3}\ \boxed{4}\ \boxed{=}$ 9·44

(b) $\boxed{AC}\ \boxed{8}\ \boxed{.}\ \boxed{2}\ \boxed{2}\ \boxed{-}\ \boxed{2}\ \boxed{.}\ \boxed{7}\ \boxed{8}\ \boxed{=}$ 5·44

(c) $\boxed{AC}\ \boxed{5}\ \boxed{.}\ \boxed{8}\ \boxed{4}\ \boxed{\times}\ \boxed{2}\ \boxed{.}\ \boxed{9}\ \boxed{2}\ \boxed{=}$ 17·0528

(d) $\boxed{AC}\ \boxed{1}\ \boxed{9}\ \boxed{.}\ \boxed{5}\ \boxed{7}\ \boxed{\div}\ \boxed{4}\ \boxed{.}\ \boxed{1}\ \boxed{2}\ \boxed{=}$ 5·25

Check: write each number correct to 1 significant figure.

(a) $3 \cdot 89 + 4 \cdot 21 + 1 \cdot 34 \approx 4 + 4 + 1 = 9 \checkmark$
(b) $8 \cdot 22 - 2 \cdot 78 \approx 8 - 3 = 5 \checkmark$
(c) $5 \cdot 84 \times 2 \cdot 92 \approx 6 \times 3 = 18 \checkmark$
(d) $19 \cdot 57 \div 4 \cdot 12 \approx 20 \div 4 = 5 \checkmark$

Exercise 3.1d

Use your calculator to find each answer. Check your answers with a rough estimate.

1. $4 \cdot 97 + 3 \cdot 86 + 2 \cdot 13$
2. $8 \cdot 78 + 6 \cdot 93 + 5 \cdot 32$
3. $15 \cdot 72 + 7 \cdot 68 + 4 \cdot 28$
4. $18 \cdot 91 + 9 \cdot 63 + 3 \cdot 31$
5. $23 \cdot 74 + 15 \cdot 61 + 12 \cdot 19$
6. $21 \cdot 14 + 26 \cdot 79 + 13 \cdot 82$
7. $34 \cdot 27 + 13 \cdot 75 + 14 \cdot 57$
8. $41 \cdot 05 + 16 \cdot 81 + 11 \cdot 56$
9. $51 \cdot 18 + 12 \cdot 55 + 19 \cdot 96$
10. $35 \cdot 47 + 10 \cdot 65 + 9 \cdot 83$

11. $8 \cdot 96 - 3 \cdot 79$	**12.** $7 \cdot 88 - 4 \cdot 67$	**13.** $15 \cdot 78 - 5 \cdot 84$	**14.** $13 \cdot 24 - 6 \cdot 31$
15. $27 \cdot 42 - 14 \cdot 18$	**16.** $32 \cdot 36 - 18 \cdot 29$	**17.** $35 \cdot 72 - 13 \cdot 69$	**18.** $44 \cdot 85 - 23 \cdot 91$
19. $48 \cdot 37 - 11 \cdot 16$	**20.** $52 \cdot 41 - 10 \cdot 09$		

21. $7 \cdot 13 \times 5 \cdot 96$	**22.** $9 \cdot 24 \times 3 \cdot 87$	**23.** $8 \cdot 32 \times 4 \cdot 69$	**24.** $12 \cdot 29 \times 2 \cdot 84$
25. $25 \cdot 36 \times 1 \cdot 91$	**26.** $14 \cdot 72 \times 4 \cdot 12$	**27.** $15 \cdot 59 \times 5 \cdot 37$	**28.** $19 \cdot 61 \times 3 \cdot 44$
29. $39 \cdot 56 \times 2 \cdot 07$	**30.** $17 \cdot 51 \times 2 \cdot 28$		

31. $12 \cdot 25 \div 1 \cdot 96$	**32.** $16 \cdot 08 \div 1 \cdot 92$	**33.** $9 \cdot 48 \div 0 \cdot 96$	**34.** $24 \cdot 32 \div 5 \cdot 12$
35. $10 \cdot 36 \div 1 \cdot 12$	**36.** $26 \cdot 28 \div 2 \cdot 88$	**37.** $13 \cdot 86 \div 1 \cdot 76$	**38.** $11 \cdot 97 \div 2 \cdot 5$
39. $5 \cdot 88 \div 2 \cdot 24$	**40.** $13 \cdot 77 \div 2 \cdot 16$		

Calculations involving several different operations should be carried out in stages. Write down the answer to each stage before beginning the next stage.

A calculation in brackets must be done before any other; and division and multiplication must be done before addition and subtraction.

Example 5

Calculate

(a) $6 \cdot 2 + 4 \cdot 3 \times 1 \cdot 5$

(b) $(6 \cdot 2 + 4 \cdot 3) \times 1 \cdot 5$

(c) $7 \times 3 \cdot 1 - 3 \cdot 69 \div 0 \cdot 82$

and check your answers with a rough estimate.

(a) [AC][4][.][3][×][1][.][5][=] 6·45 (Write down 6·45)
   [AC][6][.][2][+][6][.][4][5][=] 12·65

(b) [AC][6][.][2][+][4][.][3][=] 10·5 (Write down 10·5)
   [AC][1][0][.][5][×][1][.][5][=] 15·75

(c) [AC][7][×][3][.][1][=] 21·7 (Write down 21·7)
   [AC][3][.][6][9][÷][0][.][8][2] = 4·5 (Write down 4·5)
   [AC][2][1][.][7][−][4][.][5][=] 17·2

Check: write each number correct to 1 significant figure.

(a) $6 \cdot 2 + 4 \cdot 3 \times 1 \cdot 5 \approx 6 + 4 \times 2$
$$= 6 + 8 = 14 \checkmark$$

(b) $(6 \cdot 2 + 4 \cdot 3) \times 1 \cdot 5 \approx (6 + 4) \times 2$
$$= 10 \times 2 = 20 \checkmark$$

(c) $7 \times 3 \cdot 1 - 3 \cdot 69 \div 0 \cdot 82 \approx 7 \times 3 - 4 \div 0 \cdot 8$
$$= 21 - 5 = 16 \checkmark$$

Exercise 3.1e

Use your calculator to find each answer. Check your answers with a rough estimate.

1. $4 \cdot 8 + 6 \cdot 5 \times 3 \cdot 4$
2. $7 \cdot 2 \times 1 \cdot 5 - 3 \cdot 8$
3. $12 \cdot 4 + 7 \cdot 6 \times 8 \cdot 5$
4. $9 \cdot 4 \times 6 \cdot 2 - 7 \cdot 58$
5. $8 \cdot 7 - 0 \cdot 9 \times 5 \cdot 5$
6. $1 \cdot 87 + 3 \cdot 2 \times 0 \cdot 6$
7. $3 \cdot 1 + 0 \cdot 75 \times 6 \cdot 8$
8. $4 \cdot 2 \times 1 \cdot 6 - 3 \cdot 7$
9. $17 \cdot 7 - 4 \cdot 2 \times 3 \cdot 5$
10. $5 \cdot 4 - 0 \cdot 02 \times 12$
11. $9 \cdot 8 + 8 \cdot 96 - 2 \cdot 8$
12. $6 \cdot 3 - 11 \cdot 34 \div 5 \cdot 4$
13. $6 \cdot 65 + 8 \cdot 68 \div 2 \cdot 8$
14. $17 - 13 \cdot 91 \div 2 \cdot 14$
15. $60 \cdot 8 \div 6 \cdot 4 + 3 \cdot 7$
16. $38 \cdot 55 \div 5 \cdot 14 + 12 \cdot 62$
17. $59 \cdot 78 \div 6 \cdot 1 - 4 \cdot 46$
18. $1 \cdot 26 \div 0 \cdot 84 - 0 \cdot 76$
19. $3 \cdot 02 + 1 \cdot 44 \div 0 \cdot 64$
20. $1 \cdot 95 - 0 \cdot 21 \div 0 \cdot 25$
21. $4 \cdot 03 + 6 \cdot 2 \times 0 \cdot 75$
22. $0 \cdot 6 \times 0 \cdot 55 - 0 \cdot 29$
23. $3 \cdot 2 \times 0 \cdot 65 + 1 \cdot 93$
24. $14 + 104 \times 0 \cdot 06$
25. $(6 \cdot 1 + 2 \cdot 3) \times 3 \cdot 5$
26. $0 \cdot 94 \times (9 \cdot 91 + 2 \cdot 59)$
27. $(8 \cdot 48 - 0 \cdot 98) \times 0 \cdot 046$
28. $400 \times (12 \cdot 6 - 0 \cdot 06)$
29. $(6 \cdot 17 + 6 \cdot 23) \times 0 \cdot 95$
30. $(5 \cdot 36 - 1 \cdot 91) \times 120$
31. $6 \cdot 48 - 3 \cdot 8 \times 0 \cdot 4$
32. $7 \cdot 2 \times 1 \cdot 9 + 0 \cdot 04$
33. $(40 \cdot 8 + 22 \cdot 9) \div 6 \cdot 5$
34. $(4 \cdot 035 - 1 \cdot 09) \div 0 \cdot 31$
35. $3 \cdot 696 \div (3 \cdot 9 + 3 \cdot 8)$
36. $4 \cdot 08 \div (8 \cdot 5 - 3 \cdot 4)$
37. $(7 \cdot 12 + 8 \cdot 09) \div 0 \cdot 78$
38. $(27 \cdot 78 - 2 \cdot 32) \div 6 \cdot 7$
39. $16 \cdot 42 - 3 \cdot 6 \div 0 \cdot 48$
40. $6 \cdot 08 \div 3 \cdot 2 + 2 \cdot 3$
41. $(1 \cdot 09 + 0 \cdot 97) \times 3 \cdot 5$
42. $3 \cdot 2 \times (6 \cdot 1 - 1 \cdot 99)$
43. $0 \cdot 245 \div (2 \cdot 3 + 7 \cdot 5)$
44. $76 \cdot 22 \div (5 \cdot 56 - 1 \cdot 44)$
45. $(0 \cdot 059 + 0 \cdot 006) \times (29 + 35)$
46. $(3 \cdot 62 + 8 \cdot 13) \div (2 \cdot 3 - 1 \cdot 36)$
47. $(0 \cdot 81 - 0 \cdot 26) \times (0 \cdot 81 + 0 \cdot 43)$
48. $15 - 2(2 \cdot 58 + 3 \cdot 64)$
49. $6(3 \cdot 92 - 1 \cdot 08) - 9 \cdot 68$
50. $3 \cdot 91 + 3 \cdot 2(0 \cdot 9 + 0 \cdot 75)$

3.2 FRACTIONS AND DECIMALS

To change a fraction to a decimal, enter the numerator first then divide by the denominator.

Example 1

Use your calculator to change the following fractions into decimal fractions.

(a) $\frac{1}{2}$ (b) $\frac{7}{25}$ (c) $\frac{33}{125}$

(a) $\boxed{AC}\boxed{1}\boxed{\div}\boxed{2}\boxed{=}$ 0·5

(b) $\boxed{AC}\boxed{7}\boxed{\div}\boxed{2}\boxed{5}\boxed{=}$ 0·28

(c) $\boxed{AC}\boxed{3}\boxed{3}\boxed{\div}\boxed{1}\boxed{2}\boxed{5}\boxed{=}$ 0·264

Exercise 3.2a

Use your calculator to change the following common fractions to decimal fractions.

1. $\frac{1}{4}$
2. $\frac{3}{4}$
3. $\frac{1}{8}$
4. $\frac{5}{16}$
5. $\frac{3}{8}$
6. $\frac{2}{5}$
7. $\frac{7}{40}$
8. $\frac{3}{20}$
9. $\frac{4}{25}$
10. $\frac{3}{16}$

11. $\frac{7}{8}$ 12. $\frac{4}{5}$ 13. $\frac{9}{20}$ 14. $\frac{17}{40}$ 15. $\frac{9}{80}$

16. $\frac{15}{100}$ 17. $\frac{27}{50}$ 18. $\frac{9}{32}$ 19. $\frac{11}{80}$ 20. $\frac{13}{20}$

21. $\frac{5}{8}$ 22. $\frac{3}{40}$ 23. $\frac{27}{80}$ 24. $\frac{33}{120}$ 25. $\frac{7}{20}$

26. $\frac{53}{100}$ 27. $\frac{17}{50}$ 28. $\frac{15}{120}$ 29. $\frac{63}{2500}$ 30. $\frac{9}{500}$

31. $\frac{33}{125}$ 32. $\frac{19}{625}$ 33. $\frac{29}{40}$ 34. $\frac{7}{16}$ 35. $\frac{59}{80}$

36. $\frac{111}{625}$ 37. $\frac{13}{16}$ 38. $\frac{14}{25}$ 39. $\frac{21}{50}$ 40. $\frac{55}{400}$

41. $\frac{12}{125}$ 42. $\frac{124}{500}$ 43. $\frac{15}{16}$ 44. $\frac{16}{64}$ 45 $\frac{9}{160}$

46. $\frac{7}{250}$ 47. $\frac{57}{625}$ 48. $\frac{66}{3125}$ 49. $\frac{53}{1250}$ 50. $\frac{221}{6250}$

To change a mixed number to a decimal:
 (i) change the fraction to a decimal,
(ii) then add the whole number.

Example 2

Use your calculator to change the following mixed numbers into decimals:

(a) $1\frac{1}{2}$ (b) $12\frac{3}{25}$ (c) $8\frac{19}{40}$

(a) $\boxed{AC}\boxed{1}\boxed{\div}\boxed{2}\boxed{=}0{\cdot}5$ $\boxed{+}\boxed{1}\boxed{=}1{\cdot}5$

(b) $\boxed{AC}\boxed{3}\boxed{\div}\boxed{2}\boxed{5}\boxed{=}0{\cdot}12$ $\boxed{+}\boxed{1}\boxed{2}\boxed{=}12{\cdot}12$

(c) $\boxed{AC}\boxed{1}\boxed{9}\boxed{\div}\boxed{4}\boxed{0}\boxed{=}0{\cdot}475$ $\boxed{+}\boxed{8}\boxed{=}8{\cdot}475$

Exercise 3.2b

Use your calculator to change the following mixed numbers into decimals.

1. $1\frac{1}{4}$ 2. $2\frac{1}{4}$ 3. $3\frac{1}{2}$ 4. $4\frac{3}{8}$ 5. $2\frac{7}{16}$

6. $1\frac{5}{8}$ 7. $5\frac{2}{5}$ 8. $3\frac{4}{25}$ 9. $10\frac{1}{40}$ 10. $5\frac{3}{16}$

11. $8\frac{5}{20}$ 12. $2\frac{7}{8}$ 13. $4\frac{9}{40}$ 14. $11\frac{3}{50}$ 15. $6\frac{1}{8}$

16. $3\frac{19}{25}$ 17. $1\frac{11}{16}$ 18. $5\frac{3}{5}$ 19. $3\frac{17}{40}$ 20. $4\frac{21}{50}$

21. $6\frac{81}{100}$ 22. $12\frac{5}{32}$ 23. $9\frac{351}{2500}$ 24. $2\frac{98}{125}$ 25. $1\frac{3}{25}$

26. $7\frac{13}{125}$ 27. $88\frac{9}{16}$ 28. $21\frac{14}{625}$ 29. $11\frac{53}{400}$ 30. $13\frac{171}{1250}$

31. $6\frac{19}{40}$ 32. $9\frac{15}{16}$ 33. $2\frac{13}{16}$ 34. $1\frac{63}{3125}$ 35. $32\frac{81}{1250}$

36. $14\frac{173}{500}$ 37. $7\frac{43}{400}$ 38. $11\frac{21}{125}$ 39. $12\frac{69}{120}$ 40. $48\frac{17}{20}$

41. $182\frac{37}{125}$ 42. $17\frac{31}{40}$ 43. $9\frac{11}{40}$ 44. $4\frac{23}{3125}$ 45. $76\frac{132}{625}$

46. $16\frac{133}{250}$ 47. $3\frac{119}{1250}$ 48. $19\frac{27}{500}$ 49. $4\frac{191}{250}$ 50. $9\frac{31}{125}$

To find the value of a fraction containing more than one number in the numerator or denominator
 (i) Work out the values of the numerator and denominator separately, and write down their values,
 (ii) Divide the numerator by the denominator
(iii) Check your answers from the original numbers by means of a rough estimate.

Example 3

Calculate the following and check your answers.

(a) $\dfrac{19\cdot89}{5\cdot2 \times 0\cdot85}$ (b) $\dfrac{(24\cdot2 + 7\cdot92)}{(7\cdot9 - 3\cdot5)}$ (c) $62\frac{1}{2}\%$ of £12·88

(a)   (i)   The numerator is 19·89
Find the denominator:
[AC] [5] [.] [2] [×] [0] [.] [8] [5] [=] 4·42
Write down 4·42

     (ii)   Divide the numerator by the denominator:
[AC] [1] [9] [.] [8] [9] [÷] [4] [.] [4] [2] [=] 4·5

     (iii)   Check
$$\frac{19\cdot89}{5\cdot2 \times 0\cdot85} \approx \frac{20}{5 \times 0\cdot8} = \frac{20}{4} = 5 \checkmark$$
So the answer is 4·5

(b)   (i)   Find the numerator:
[AC] [2] [4] [.] [2] [+] [7] [.] [9] [2] [=] 32·12
Write down 32·12

Find the denominator:
[AC] [7] [.] [9] [−] [3] [.] [5] [=] 4·4
Write down 4·4

     (ii)   Divide the numerator by the denominator:
[AC] [3] [2] [.] [1] [2] [÷] [4] [.] [4] [=] 7·3

     (iii)   Check
$$\frac{(24\cdot2 + 7\cdot92)}{(7\cdot9 - 3\cdot5)} \approx \frac{20 + 8}{8 - 4} = \frac{28}{4} = 7 \checkmark$$
So the answer is 7·3

(c) $62\frac{1}{2}\%$ of £12·88 $= \dfrac{62\frac{1}{2} \times 12\cdot88}{100}$

$= \dfrac{125 \times 12\cdot88}{200}$

(i) Find the numerator:

[AC][1][2][5][×][1][2][.][8][8][=] 1610

Write down 1610

The denominator is 200

(ii) Divide the numerator by the denominator:

[AC][1][6][1][0][÷][2][0][0][=] 8·05

(iii) Check

$62\frac{1}{2}\%$ of £12·88 ≈ 60% of £13

$= \dfrac{60}{100} \times \dfrac{13}{1} = \dfrac{780}{100}$ = £7·80 ✓

So the answer is £8·05

Exercise 3.2c

Use your calculator to find each answer. Check your answers with a rough estimate.

1. $\dfrac{3}{2 \times 5}$ 2. $\dfrac{7}{4 \times 2}$ 3. $\dfrac{13}{8 \times 5}$

4. $\dfrac{6 \cdot 12}{6 \cdot 4 \times 4 \cdot 25}$ 5. $\dfrac{14 \cdot 2 \times 5 \cdot 8}{2 \cdot 84}$ 6. $\dfrac{18 \cdot 04 \times 4 \cdot 14}{8 \cdot 2}$

7. 75% of £8·24   8. 50% of £127   9. 85% of £6·40
10. 17% of £2500   11. 24% of £77·50   12. 90% of £34·60

13. $\dfrac{4 \cdot 8 \times 1 \cdot 5}{1 \cdot 44 \times 20}$ 14. $\dfrac{124 \times 62 \cdot 6}{9 \cdot 703 \times 320}$ 15. $\dfrac{16 \cdot 4 \times 3 \cdot 5}{205 \times 0 \cdot 14}$

16. $\dfrac{78 \cdot 8 \times 9 \cdot 6}{23 \cdot 64 \times 8}$ 17. $\dfrac{0 \cdot 075 \times 8 \cdot 4}{0 \cdot 126 \times 25}$ 18. $\dfrac{0 \cdot 43 \times 26}{559 \times 0 \cdot 16}$

19. $\frac{3}{5}$ of 800 tonnes   20. $\frac{3}{4}$ of 60 grams   21. $\frac{5}{8}$ of 4000 litres

22. $\frac{7}{16}$ of 36 metres   23. $\frac{3}{8}$ of 78 grams   24. $\frac{12}{25}$ of 170 tonnes

25. $\dfrac{5 \cdot 8 \times 5 \cdot 2}{3 \cdot 77}$ 26. $\dfrac{1 \cdot 443 \times 200}{50 \times 4 \cdot 81}$ 27. $\dfrac{52 \cdot 6 \times 36}{15 \cdot 1488}$

28. $\dfrac{11 \cdot 27}{196 \times 2 \cdot 3}$ 29. $\dfrac{0 \cdot 42 \times 1 \cdot 8}{0 \cdot 375 \times 4}$ 30. $\dfrac{98 \cdot 45 \times 0 \cdot 56}{1 \cdot 3783}$

31. $\dfrac{0 \cdot 63}{0 \cdot 28 \times 3 \cdot 6}$ 32. $\dfrac{3 \cdot 12}{6 \cdot 5 \times 2 \cdot 4}$ 33. $12\frac{1}{2}\%$ of 764

34. $6\frac{1}{4}\%$ of 396   35. $2\frac{3}{4}\%$ of 3860   36. $38\frac{1}{2}\%$ of 650

37. $3\frac{1}{8}\%$ of £25·60   38. $31\frac{1}{4}\%$ of £58·24   39. $9\frac{3}{4}\%$ of £264

40. $87\frac{1}{2}\%$ of £443·20   41. $57\frac{3}{4}\%$ of £4100   42. $27\frac{1}{4}\%$ of £980

43. $\dfrac{(4+6)}{5}$ 44. $\dfrac{(10-1)}{12}$ 45. $\dfrac{(3 \cdot 25 - 1 \cdot 05)}{20}$

46. $\dfrac{2 \cdot 9}{(23 \cdot 9 + 12 \cdot 35)}$ 47. $\dfrac{415}{(49 \cdot 6 + 54 \cdot 15)}$ 48. $\dfrac{(20 \cdot 3 - 4 \cdot 84)}{61 \cdot 84}$

49. $\dfrac{(1 \cdot 73 + 1 \cdot 92)}{(34 \cdot 9 + 23 \cdot 5)}$ 50. $\dfrac{(2 \cdot 954 + 1 \cdot 228)}{(1 \cdot 03 - 0 \cdot 21)}$

Usually not all the figures produced by a calculator are needed.
The answers should be given to a suitable number of decimal places or
significant figures.

Example 1

Find (a) $126 \cdot 53 \div 40$ (b) $5 \cdot 173 \times 0 \cdot 8$ to two decimal places
and check your answers with a rough estimate.

(a) $\boxed{AC}\boxed{1}\boxed{2}\boxed{6}\boxed{.}\boxed{5}\boxed{3}\boxed{\div}\boxed{4}\boxed{0}\boxed{=}$ $3 \cdot 16325$
$= 3 \cdot 16$ to two decimal places.

(b) $\boxed{AC}\boxed{5}\boxed{.}\boxed{1}\boxed{7}\boxed{3}\boxed{\times}\boxed{0}\boxed{.}\boxed{8}\boxed{=}$ $4 \cdot 1384$
$= 4 \cdot 14$ to two decimal places.

Check: write each number correct to 1 significant figure.

(a) $126 \cdot 53 \div 40 \approx 100 \div 40 = 2 \cdot 5 \checkmark$
(b) $5 \cdot 173 \times 0 \cdot 8 \approx 5 \times 0 \cdot 8 = 4 \checkmark$

Exercise 3.3a

Use your calculator to find each answer correct to two decimal places.
Check your answers with a rough estimate.

1. $35 \div 16$	**2.** $93 \div 48$	**3.** $138 \div 32$	**4.** $150 \div 96$
5. $73 \cdot 5 \div 24$	**6.** $57 \cdot 31 \div 40$	**7.** $60 \cdot 85 \div 40$	**8.** $51 \cdot 17 \div 40$
9. $197 \cdot 46 \div 80$	**10.** $251 \cdot 18 \div 80$	**11.** $187 \cdot 74 \div 80$	**12.** $170 \cdot 79 \div 120$
13. $387 \cdot 81 \div 120$	**14.** $284 \cdot 73 \div 120$	**15.** $250 \cdot 28 \div 160$	**16.** $708 \cdot 52 \div 160$
17. $594 \cdot 68 \div 160$	**18.** $301 \cdot 86 \div 240$	**19.** $575 \cdot 82 \div 240$	**20.** $762 \cdot 06 \div 240$
21. $0 \cdot 283 \times 4$	**22.** $0 \cdot 757 \times 3$	**23.** $0 \cdot 578 \times 6$	**24.** $0 \cdot 877 \times 7$
25. $0 \cdot 572 \times 8$	**26.** $4 \cdot 57 \times 0 \cdot 6$	**27.** $8 \cdot 91 \times 0 \cdot 4$	**28.** $1 \cdot 97 \times 0 \cdot 9$
29. $1 \cdot 61 \times 0 \cdot 7$	**30.** $7 \cdot 66 \times 0 \cdot 3$	**31.** $0 \cdot 3831 \times 4$	**32.** $0 \cdot 4022 \times 6$
33. $0 \cdot 3724 \times 9$	**34.** $0 \cdot 7231 \times 3$	**35.** $0 \cdot 5358 \times 8$	
36. $4 \cdot 934 \times 0 \cdot 7$	**37.** $6 \cdot 286 \times 0 \cdot 4$	**38.** $5 \cdot 413 \times 0 \cdot 6$	**39.** $5 \cdot 099 \times 0 \cdot 3$
40. $4 \cdot 984 \times 0 \cdot 9$			

Example 2

Given that 1 foot equals $0 \cdot 3048$ metres, change 6 feet to metres, giving
your answer correct to two decimal places (i.e. to the nearest centimetre)
Check your answer with a rough estimate.

$\boxed{AC}\boxed{6}\boxed{\times}\boxed{0}\boxed{.}\boxed{3}\boxed{0}\boxed{4}\boxed{8}\boxed{=}$ $1 \cdot 8288$
$= 1 \cdot 83$ to two decimal places
$= 1$ m 83 cm

Check: write each number correct to 1 significant figure.

$6 \times 0 \cdot 3048 \approx 6 \times 0 \cdot 3 = 1 \cdot 8$ m $\checkmark$

Exercise 3.3b

Given that 1 foot equals 0·3048 metres, use your calculator to change each of the following to metres, giving your answer correct to two decimal places (i.e. to the nearest centimetre) Check your answers with a rough estimate.

1. 7 feet   **2.** 9 feet   **3.** 3 feet   **4.** 4 feet   **5.** 8 feet
6. 12 feet   **7.** 2 feet   **8.** 5 feet   **9.** 15 feet   **10.** 10 feet
11. 35 feet   **12.** 45 feet

Example 3

(a) Given that 60 Belgian francs are equivalent to £1, change BF 123·75 to £'s giving the answer correct to two decimal places (i.e. to the nearest p.) Check your answer with a rough estimate.

$\boxed{AC}\boxed{1}\boxed{2}\boxed{3}\boxed{.}\boxed{7}\boxed{5}\boxed{\div}\boxed{6}\boxed{0}\boxed{=}$ 2·0625
$\qquad\qquad\qquad$ = 2·06 to two decimal places.
$\qquad\qquad\qquad$ = £2·06

Check: write each number correct to 1 significant figure.
123·75 ÷ 60 ≈ 100 ÷ 60 = 1·7 ✓

(b) Given that 3·84 German marks are equivalent to £1, change DM 10·32 to £'s, giving the answer correct to two decimal places (i.e. to the nearest p.) Check your answer with a rough estimate.

$\boxed{AC}\boxed{1}\boxed{0}\boxed{.}\boxed{3}\boxed{2}\boxed{\div}\boxed{3}\boxed{.}\boxed{8}\boxed{4}\boxed{=}$ 2·6875
$\qquad\qquad\qquad$ = 2·69 to two decimal places.
$\qquad\qquad\qquad$ = £2·69

Check: write each number correct to 1 significant figure.
10·32 ÷ 3·84 ≈ 10 ÷ 4 = 2·5 ✓

Exercise 3.3c

Use your calculator to change each of the following to £'s, given that BF 60 = £1 and that DM = 3·84 = £1. Express each answer to the nearest penny. (i.e. correct to two decimal places.) Check your answers with a rough estimate.

1. BF 93·75   **2.** BF 183·75   **3.** BF 131·25   **4.** BF 326·25
5. BF 247·50   **6.** DM 7·56   **7.** DM 17·88   **8.** DM 13·20
9. DM 10·92   **10.** DM 13·56   **11.** DM 11·88   **12.** DM 17·52

Example 4

Find (a) 78·765 ÷ 60 (b) 248·5 ÷ 160 to three decimal places and check your answers with a rough estimate.

(a) $\boxed{AC}\boxed{7}\boxed{8}\boxed{.}\boxed{7}\boxed{6}\boxed{5}\boxed{\div}\boxed{6}\boxed{0}\boxed{=}$ 1·31275 = 1·313 (to 3 D.P.)
(b) $\boxed{AC}\boxed{2}\boxed{4}\boxed{8}\boxed{.}\boxed{5}\boxed{\div}\boxed{1}\boxed{6}\boxed{0}\boxed{=}$ 1·553125 = 1·553 (to 3 D.P.)

Check: write each number correct to 1 significant figure.

(a) $78{\cdot}765 \div 60 \approx 80 \div 60 = 1{\cdot}3$ ✓
(b) $248{\cdot}5 \div 160 \approx 200 \div 200 = 1$ ✓

Exercise 3.3d

Using your calculator, find each of the following to three decimal places. Check your answers with a rough estimate.

Exercise 3.3d

1. $105 \div 32$
2. $665 \div 160$
3. $237 \div 96$
4. $690 \div 192$
5. $315 \div 224$
6. $283{\cdot}302 \div 80$
7. $349{\cdot}146 \div 80$
8. $502{\cdot}254 \div 80$
9. $578{\cdot}876 \div 160$
10. $840{\cdot}756 \div 160$
11. $699{\cdot}652 \div 160$
12. $859{\cdot}47 \div 400$
13. $773{\cdot}69 \div 400$
14. $853{\cdot}79 \div 400$
15. $529{\cdot}998 \div 240$
16. $859{\cdot}854 \div 240$
17. $323{\cdot}178 \div 240$
18. $326{\cdot}379 \div 120$
19. $642{\cdot}297 \div 120$
20. $737{\cdot}799 \div 120$

21. $0{\cdot}3783 \times 4$
22. $0{\cdot}4759 \times 7$
23. $0{\cdot}4307 \times 3$
24. $0{\cdot}8262 \times 6$
25. $0{\cdot}3973 \times 8$
26. $0{\cdot}7052 \times 9$
27. $7{\cdot}613 \times 0{\cdot}3$
28. $2{\cdot}221 \times 0{\cdot}7$
29. $8{\cdot}795 \times 0{\cdot}5$
30. $5{\cdot}233 \times 0{\cdot}6$
31. $7{\cdot}383 \times 0{\cdot}4$
32. $5{\cdot}747 \times 0{\cdot}3$
33. $7{\cdot}376 \times 0{\cdot}9$
34. $18{\cdot}39 \times 0{\cdot}07$
35. $77{\cdot}69 \times 0{\cdot}06$
36. $48{\cdot}43 \times 0{\cdot}09$
37. $93{\cdot}47 \times 0{\cdot}07$
38. $82{\cdot}09 \times 0{\cdot}04$
39. $42{\cdot}38 \times 0{\cdot}03$
40. $92{\cdot}67 \times 0{\cdot}06$

Example 5

Given that 1 mile = $1{\cdot}6093$ km, change 32 miles to kilometres, giving the answers correct to three decimal places (i.e. to the nearest metre.) Check your answers with a rough estimate.

$$\boxed{AC}\;\boxed{3}\;\boxed{2}\;\boxed{\times}\;\boxed{1}\;\boxed{.}\;\boxed{6}\;\boxed{0}\;\boxed{9}\;\boxed{3}\;\boxed{=}\; 51{\cdot}4976$$
$$= 51{\cdot}498 \text{ (to 3 D.P.)}$$
$$= 51 \text{ km } 498 \text{ m}$$

Check: write each number correct to 1 significant figure.

$32 \times 1{\cdot}6093 \approx 30 \times 2 = 60$ ✓

Exercise 3.3e

Given that 1 mile = $1{\cdot}6093$ km, use your calculator to change each of the following to km, giving your answers correct to three decimal places (i.e. to the nearest m.) Check your answers with a rough estimate.

1. 4 miles
2. 7 miles
3. 2 miles
4. 6 miles
5. 9 miles
6. 8 miles
7. 24 miles
8. 21 miles
9. 12 miles
10. 36 miles
11. 5 miles
12. 15 miles
13. 27 miles
14. 18 miles
15. 28 miles

Example 6

Given that 1 pound (lb) = 0·4536 kg, change 4 lb to kg, giving the answer correct to 3 decimal places (i.e. to the nearest g.) Check your answer with a rough estimate.

 1·8144
= 1·814 (to 3 D.P.)
= 1 kg 814 g

Check: write each number correct to 1 significant figure.
4 × 0·4536 ≈ 4 × 0·5 = 2 ✓

Exercise 3.3f

Given that 1 lb = 0·4536 kg, use your calculator to change each of the following into kg, giving each answer correct to 3 decimal places (i.e. to the nearest g.) Check your answers with a rough estimate.

1. 18 lb   **2.** 21 lb   **3.** 7 lb   **4.** 12 lb   **5.** 2 lb
6. 9 lb   **7.** 14 lb   **8.** 6 lb   **9.** 8 lb   **10.** 16 lb
11. 3 lb   **12.** 28 lb

Example 7

Given that 1 gallon = 4·5448 litres, change 14 gallons to litres, giving your answer correct to 3 decimal places (i.e. to the nearest m*l*.) Check your answers with a rough estimate.

 63·6272
= 63·627 (to 3 D.P.)
= 63 litres 627 m*l*

Check: write each number correct to 1 significant figure.
14 × 4·5448 ≈ 10 × 5 = 50 ✓

Exercise 3.3g

Given that 1 gallon = 4·5448 litres and that 1 pint = 0·5681 litres, use your calculator to change each of the following into litres, giving each answer correct to 3 decimal places (i.e. to the nearest m*l*.) Check your answers with a rough estimate.

1. 3 gallons   **2.** 4 gallons   **3.** 7 gallons   **4.** 16 gallons
5. 12 gallons   **6.** 8 gallons   **7.** 18 gallons   **8.** 6 gallons
9. 2 gallons   **10.** 21 gallons   **11.** 2 pints   **12.** 3 pints
13. 8 pints   **14.** 6 pints   **15.** 5 pints

Example 8

Find each of the following correct to 3 significant figures. Check your answers with a rough estimate.

(a) $45 \div 32$ (b) $965 \cdot 1 \div 40$
(c) $8653 \cdot 5 \div 20$ (d) $20514 \div 8$

(a) [AC][4][5][÷][3][2][=] 1·40625
                  = 1·41 (to 3 S.F.)

      [AC][9][6][5][.][1][÷][4][0][=] 24·1275
                       = 24·1 (to 3 S.F.)

(c) [AC][8][6][5][3][.][5][÷][2][0][=] 432·675
                       = 433 (to 3 S.F.)

(d) [AC][2][0][5][1][4][÷][8][=] 2564·25
                     = 2560 (to 3 S.F.)

Check: write each number correct to 1 significant figure.

(a) $45 \div 32 \approx 40 \div 30 = 1 \cdot 3$ ✓
(b) $965 \cdot 1 \div 40 \approx 1000 \div 40 = 25$ ✓
(c) $8653 \cdot 5 \div 20 \approx 9000 \div 20 = 450$ ✓
(d) $20514 \div 8 \approx 20\,000 \div 8 = 2500$ ✓

Exercise 3.3h

Use your calculator to find each of the following to three significant figures. Check your answers with a rough estimate.

1. $37 \div 16$	**2.** $285 \div 80$	**3.** $195 \div 48$
4. $498 \div 96$	**5.** $385 \div 112$	**6.** $86 \cdot 75 \div 40$
7. $55 \cdot 17 \div 40$	**8.** $141 \cdot 49 \div 40$	**9.** $273 \cdot 02 \div 80$
10. $338 \cdot 54 \div 80$	**11.** $494 \cdot 66 \div 80$	**12.** $285 \cdot 39 \div 120$
13. $185 \cdot 61 \div 120$	**14.** $442 \cdot 23 \div 120$	**15.** $795 \cdot 83 \div 280$
16. $440 \cdot 93 \div 280$	**17.** $978 \cdot 25 \div 280$	**18.** $556 \cdot 62 \div 240$
19. $995 \cdot 22 \div 240$	**20.** $431 \cdot 58 \div 240$	**21.** $510 \div 16$
22. $4350 \div 80$	**23.** $630 \div 48$	**24.** $2460 \div 96$
25. $6790 \div 112$	**26.** $617 \cdot 3 \div 40$	**27.** $1986 \cdot 2 \div 80$
28. $2255 \cdot 7 \div 120$	**29.** $3601 \cdot 5 \div 280$	**30.** $3451 \cdot 8 \div 240$
31. $8338 \cdot 5 \div 60$	**32.** $30327 \div 120$	**33.** $28106 \div 80$
34. $19756 \div 160$	**35.** $22407 \div 40$	**36.** $17251 \div 4$
37. $20186 \div 8$	**38.** $26019 \div 12$	**39.** $21596 \div 16$
40. $72642 \div 24$		

Example 9

Change (a) $\frac{6}{19}$ (b) $\frac{8}{15}$ to a decimal and express your answer correct to 3 decimal places. Check your answers with a rough estimate.

(a) $\boxed{\text{AC}}\boxed{6}\boxed{\div}\boxed{1}\boxed{9}\boxed{=}$ 0·3157894
$= 0·316$ (to 3 D.P.)

(b) $\boxed{\text{AC}}\boxed{8}\boxed{\div}\boxed{1}\boxed{5}\boxed{=}$ 0·5333333
$= 0·533$ (to 3 D.P.)

Check: write each number correct to 1 significant figure.
(a) $6 \div 19 \approx 6 \div 20 = 3 \div 10 = 0·3$ ✓
(b) $8 \div 15 \approx 8 \div 20 = 4 \div 10 = 0·4$ ✓

Exercise 3.3i

Use your calculator to change each of the following fractions to decimals and express each answer correct to three decimal places. Check your answers with a rough estimate.

1. $\frac{2}{7}$ 2. $\frac{5}{13}$ 3. $\frac{3}{19}$ 4. $\frac{5}{19}$ 5 $\frac{8}{13}$ 6. $\frac{4}{7}$

7. $\frac{5}{12}$ 8. $\frac{2}{3}$ 9. $\frac{8}{9}$ 10. $\frac{7}{12}$ 11. $\frac{1}{3}$ 12. $\frac{5}{18}$

13. $\frac{5}{9}$ 14. $\frac{2}{9}$ 15. $\frac{1}{12}$ 16. $\frac{6}{17}$ 17. $\frac{4}{21}$ 18. $\frac{7}{9}$

Example 10

Change (a) $\frac{4}{39}$ (b) $\frac{7}{127}$ to a decimal and express your answer correct to 3 significant figures. Check your answers with a rough estimate.

(a) $\boxed{\text{AC}}\boxed{4}\boxed{\div}\boxed{3}\boxed{9}$ $= 0·10256$
$= 0·103$ (to 3 S.F.)

(b) $\boxed{\text{AC}}\boxed{7}\boxed{\div}\boxed{1}\boxed{2}\boxed{7}\boxed{=}$ $= 0·055118$
$= 0·0551$ (to 3 S.F.)

(a) $4 \div 39 \approx 4 \div 40 \; = 1 \div 10 = 0·1$ ✓
(b) $7 \div 127 \approx 7 \div 100 = 0·07$ ✓

Exercise 3.3j

Use your calculator to change each of the following fractions to decimals and express each answer correct to three significant figures. Check your answers with a rough estimate.

1. $\frac{9}{53}$ 2. $\frac{14}{27}$ 3. $\frac{3}{61}$ 4. $\frac{6}{19}$ 5. $\frac{4}{121}$ 6. $\frac{8}{23}$

7. $\frac{7}{33}$ 8. $\frac{11}{131}$ 9. $\frac{1}{67}$ 10. $\frac{7}{97}$ 11. $\frac{13}{81}$ 12. $\frac{15}{47}$

13. $\frac{8}{123}$ 14. $\frac{13}{99}$ 15. $\frac{5}{136}$ 16. $\frac{9}{111}$ 17. $\frac{4}{77}$ 18. $\frac{6}{83}$

A *discount* is a sum of money taken off an account. It is usually written as a percentage.

Example 1

Find (a) the discount of 5% on a bill of £6·50,
(b) the amount actually paid.

(a)   Discount = 5% of £6·50

$$= \frac{5}{100} \times \frac{650}{1} \text{ pence}$$

$$= \frac{65}{2} = 32\frac{1}{2} \text{ pence}$$

(b)   Amount to be paid = £6·50 − £0·32$\frac{1}{2}$
$$= £6·17\frac{1}{2}$$

Exercise 4.1a

For the following, find
(a) the discount
(b) the amount actually paid

1. A car is priced in a catalogue at £2400, but the dealer offers a 20% discount to the purchaser.
2. A three-piece suite which has a catalogue price of £600, but is sold with a 15% discount during a sale.
3. A colour television, with a marked price of £400, which is sold with a 9% discount to a man who pays immediate cash.
4. A contractor makes an estimate of £250 for a cavity-wall insulation, but then offers an 8% discount to the house owner for immediate payment.
5. A man is to have his windows double-glazed and the firm's representative offers him a 12% discount on an estimate of £350.
6. A black-and-white television, with a marked price of £150 which is offered for sale at a 16% discount rate because the woodwork is slightly defaced.
7. An electric cooker of catalogue price £125, which is sold with a 12% discount to a man who is willing to collect it.
8. A record player of marked price £80, which is sold to a purchaser with a 15% discount in exchange for handing in his old record player to the dealer.

9. A washing machine of catalogue price £180 which is sold at a discount rate of 25% to a man who is willing to do his own delivering and installing.
10. A refrigerator of marked price £85, which is offered for sale with a 20% discount because the dealer is over-stocked.
11. A spin drier of marked price £60, which is sold with an 8% discount for immediate cash payment.
12. A man is to have his loft insulated. A contractor makes an estimate of £54, and offers a 5% discount for prompt payment on completion of the contract.
13. A bicycle of catalogue price £63 which is sold with a 6% discount because the paintwork is slightly scratched.
14. A radio of catalogue price £22 which is sold to a man with a 5% discount in exchange for his being able to do his own repairs and therefore forfeit the guarantee.
15. A steam iron is priced at £15, but is sold at 4% discount because it was part of a bulk purchase.

A discount may be stated as a number of pence in the £.
Multiply the number of pence discount in £1 by the total cost (in £).

Example 2

Find (a) the discount on a bill of £8·50 at 12p in the £,
(b) the amount actually paid.

(a) Discount = 12p × 8·50
$$= 102p = £1·02$$
(b) Amount to be paid = £8·50 − £1·02
$$= £7·48$$

Exercise 4.1b

For the following, find
(a) the discount
(b) the amount actually paid

1. A book club offers a discount of 5p in the £, on a bill of £15.
2. A woman receives a bill of £12·50 for photographs, but finds she will be given a discount of 10p in the £, for prompt payment.
3. A carpet shop advertises a discount of 6p in the £, when a customer buys a carpet costing £72.
4. A grocer gives a pensioner a discount of 12p in the £, on a bill of £7·25.

5. A local council sends a householder a rates bill of £742, but offers 4p in the £ discount if the bill is paid within 14 days.

6. A model railway club has a subscription of £6·50 per member, but offers 2p in the £ discount for payment before the 1st of January.

7. A boutique owner offers a discount of 8p in the £ when selling sweat shirts slightly damaged in a flood. The sweat shirts normally cost £4·25,

8. A market stall owner offers electronic watches at £58, less 7p in the £ discount.

9. A motorcycle, normally priced at £3240, is sold at a discount of $7\frac{1}{2}$p in the £, because it is last year's model.

10. In a stocktaking sale a furniture dealer gives 9p in the £ discount on an armchair which normally costs £84.

11. A plumber buys a shower unit priced at £52·50, and is given a trade discount of 24p in the £.

12. A newsagent is sent a bill for £96, but receives $12\frac{1}{2}$p in the £ discount for paying cash.

13. A china shop offers mugs normally costing £1·25 at 16p in the £ discount, because they are "seconds".

14. A paint manufacturer offers 5 litre tins of gloss paint, which normally cost £7·25, at a discount of 8p in the £, to anyone buying direct from the factory.

15. A model shop offers a discount of 13p in the £ to anyone spending more than £50 on balsa wood. Two men club together and buy balsa worth £63·50.

A *surcharge* is a sum of money added to an account. It is usually written as a percentage.

Example 3

Find (a) the surcharge of 10% on a bill of £12·50.
　　(b) the amount actually paid.

(a) Surcharge = 10% of £12·50

$$= \frac{10}{100} \times \frac{1250}{1} \text{ pence}$$

$$= 125 \text{ pence} = £1·25$$

(b) Amount to be paid = £12·50 + £1·25
$$= £13·75$$

Exercise 4.1c

For the following, find
(a) the surcharge
(b) the amount actually paid.

1. A young man is told he must pay a 25% surcharge on insurance which normally costs £80, because he drives a sports car.

2. A grocer adds a surcharge of 5% to a bill of £6·40 because he delivers the groceries.

3. A woman is two weeks late in paying the bill of £150 for a stereo tape recorder, and is asked to pay a surcharge of 6%.

4. A restaurant applies a surcharge of 4% to a meal costing £8·25, because it is served after 11 p.m.

5. A blacksmith must pay a surcharge of 2% when he buys a new van costing £4850 because it is to be painted in a non-standard colour.

6. A businessman agrees to a surcharge of 3% for rustproofing when he buys a new car costing £3800.

7. A man buys a microcomputer costing £520, and is offered an extra year's guarantee on payment of an 8% surcharge.

8. A surcharge of 16% is made on a bill of £250 from a joiner, because of extra work carried out after his estimate had been accepted.

9. An ironmonger makes a surcharge of 18% on a box of bolts which normally cost £6·50, because they are chrome plated.

10. An amateur radio operator buys a new transmitter costing £840, and agrees to a surcharge of 1% to cover special delivery costs.

11. A travel agent finds he must make a surcharge of 5% on an air fare of £260 because of rising fuel costs.

12. A sawmill applies a 15% surcharge to a builder's account of £480, because he does not pay within 30 days.

13. A girl buys a new dress costing £56, and pays a surcharge of 25% to have lace trimmings sewn on.

14. Mail order goods worth £84 are subject to a $2\frac{1}{2}$% surcharge to cover postage and packing.

15. A company selling chauffeurs' uniforms for £120 makes a surcharge of $6\frac{1}{4}$% when gold braid is sewn on the shoulders and cap.

A *commission* is a sum of money received by an agent on the value of goods he sells. It is usually written as a percentage.

Example 1

The table shows the commission charged by an estate agent for selling a house.

Sales price	Commission
first £5000	2%
all over £5000	$1\frac{1}{4}\%$

What is the total commission on a house sold for £12 500?

$$\text{Commission on £5000} = 2\% \text{ of £5000}$$

$$= \frac{2}{\underset{1}{\cancel{100}}} \times \frac{\overset{50}{\cancel{5000}}}{1}$$

$$= £100$$

$$\text{Remainder} = £12\,500 - £5000 = £7500$$

$$\text{Commission on remainder} = 1\frac{1}{4}\% \text{ of £7500}$$

$$= \frac{1\frac{1}{4}}{100} \times \frac{7500}{1}$$

$$= \frac{5}{\underset{4}{\cancel{400}}} \times \frac{\overset{75}{\cancel{7500}}}{1}$$

$$= £\frac{375}{4}$$

$$= £93\frac{3}{4} = £93 \cdot 75$$

So total commission $= £100 + £93 \cdot 75$

$$= £193 \cdot 75$$

Exercise 4.2

1. A man who sells newspapers receives for his wages 15% of the money he takes. If he sells 800 newspapers at 5p a copy, how much does he receive?
2. A man who sells refreshments at a kiosk receives a 5% commission on the value of his sales. If his takings are £180 on a certain day, how much commission does he receive?

3. A stock auctioneer receives a 4% commission on his sales. His takings for five days in a certain week are given below. Find the commission he receives for each day.
Monday £1250   Wednesday £925   Friday £1025.
Tuesday £1100   Thursday £775
4. An ice cream seller receives for his wages 16% of the money he takes. Over the three days of the Summer Bank Holiday his takings were: Saturday £150; Sunday £105; Monday £120. Find how much he earns on each day.
5. A door-to-door sales agent is paid commission on his sales at a rate of 2%. Find how much commission he receives for the sale of each of the following:
   (a) Carpets costing £450
   (b) A lounge suite costing £720
   (c) A dining-room suite costing £635.
6. The commission that an estate agent charges on house sales is $2\frac{1}{2}\%$ up to £10 000 and 2% above that figure. Find the commission he receives for the sale of each property below.
   (a) Selling price = £7200
   (b) Selling price = £9400
   (c) Selling price = £12 750
   (d) Selling price = £15 650
7. A man buys a franchise for a coin-operated launderette. He has to pay 10% of his earnings up to £5000 and 5% over that amount. Find how much he has to pay if his earnings are
   (a) £4500 (b) £7500 (c) £11 450 (d) £5275
8. A cinema manager pays a proportion of his ticket receipts in return for showing a film. He pays 15% of the first £1500, and 4% of sales after that. Find the amount he must pay if his receipts are
   (a) £780 (b) £2300 (c) £3150 (d) £3260
9. An agency finds work in TV commercials for actors. Its fee is $2\frac{1}{2}\%$ of the first £1000 or less of the actor's wage, and 6% of anything more than this. Find the fee paid by actors who earn:
   (a) £2400 (b) £10 000 (c) £850 (d) £12 450
10. The manager of a rock band charges commission on the profits from each concert. He charges $12\frac{1}{2}\%$ of the first £5000, and 18% above that amount. Find the commission charged on profits of:
   (a) £4560 (b) £7200 (c) £8950 (d) £12 656

Interest is the amount earned on money which is invested or the sum charged on money which is borrowed. It is usually written as a percentage of the money lent or borrowed for each year.

Example 1

(a)   £200 is invested at 4% interest. How much interest is earned in each year?

$$4\% \text{ of } £200 = \frac{4}{\underset{1}{\cancel{100}}} \times \frac{\overset{2}{\cancel{200}}}{1}$$

$$= £8 \text{ yearly.}$$

(b)   £480 is borrowed for 3 years at $7\frac{1}{2}\%$ interest. What is the total interest paid?

The interest is $7\frac{1}{2}\%$ of £480 each year.
The total interest is $3 \times (7\frac{1}{2}\% \text{ of } £480)$

$$= \frac{3}{1} \times \frac{7\frac{1}{2}}{100} \times \frac{480}{1}$$

$$= \frac{3}{1} \times \frac{\overset{3}{\cancel{15}}}{\underset{1}{\cancel{200}}} \times \frac{\overset{12}{\cancel{480}}}{1}$$

$$= £108$$

Exercise 4.3a

For the following find
(a)   the yearly interest,
(b)   the total interest chargeable after the time stated,
(c)   the debt (total interest + loan).

	Loan	Interest rate	Time
1.	£150	10%	4 years
2.	£120	15%	4 years
3.	£250	12%	3 years
4.	£350	16%	2 years
5.	£120	$12\frac{1}{2}\%$	4 years
6.	£240	$18\frac{3}{4}\%$	3 years
7.	£160	$21\frac{1}{4}\%$	3 years
8.	£210	15%	4 years
9.	£315	12%	2 years
10.	£125	9%	2 years

For the following find
(a)   the yearly interest,
(b)   the total interest payable after the time stated,
(c)   the value of the investor's savings (total interest + investment).

	Investment	Interest rate	Time
11.	£350	6%	4 years
12.	£280	5%	4 years
13.	£425	8%	3 years
14.	£360	$7\frac{1}{2}\%$	3 years
15.	£112	$6\frac{1}{4}\%$	5 years
16.	£256	$3\frac{1}{8}\%$	4 years
17.	£160	$9\frac{3}{8}\%$	5 years
18.	£450	9%	4 years
19.	£135	8%	2 years
20.	£360	4%	2 years

Simple interest is the interest calculated at a fixed yearly rate on the sum of money first borrowed or invested.

Example 2

Find the simple interest on £420 at 12% for 2 years 9 months.
Interest for 1 year = 12% of £420
Interest for 2 years 9 months (i.e. $2\frac{3}{4}$ years)

$$= 2\frac{3}{4} \times (12\% \text{ of } £420)$$

$$= \frac{11}{\underset{1}{\cancel{4}}} \times \frac{\overset{3}{\cancel{12}}}{\underset{5}{\cancel{100}}} \times \frac{\overset{21}{\cancel{420}}}{1}$$

$$= £\frac{693}{5} = £138\frac{3}{5} = £138 \cdot 60$$

Exercise 4.3b

For the following find
(a) the yearly simple interest,
(b) the total interest chargeable after the time stated,
(c) the debt (total interest + loan).

	Loan	Interest rate	Time
1.	£250	12%	1 year 6 months
2.	£320	15%	1 year 3 months
3.	£450	16%	2 years 3 months
4.	£150	12%	1 year 4 months

For the following find
(a) the yearly simple interest,
(b) the total interest payable after the time stated,
(c) the value of the investor's savings (total interest + investment).

	Investment	Interest rate	Time
5.	£150	8%	2 years 6 months
6.	£320	10%	1 year 9 months
7.	£120	5%	3 years 6 months
8.	£750	8%	3 years 4 months
9.	£250	6%	2 years 8 months
10.	£720	5%	1 year 10 months
11.	£450	8%	3 years 2 months

The annual rate of interest is found by making the interest earned in one year a percentage of the amount invested.

Example 3

An investment of £400 earns £20 interest in 6 months. What is the annual rate of interest?

$$\text{Interest in 1 year} = £20 \times 2$$
$$= £40$$

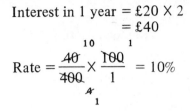

$$\text{Rate} = \frac{40}{400} \times \frac{100}{1} = 10\%$$

Exercise 4.3c

Find the yearly rate of interest for the following:

	Investment	Interest	Time
1.	£300	£15	1 year
2.	£500	£12·50	6 months
3.	£1200	£9	3 months
4.	£750	£7·50	4 months
5.	£2500	£150	2 years

4.4 COMPOUND INTEREST AND DEPRECIATION

Compound interest is the interest calculated on the amount of money at the beginning of each year, where the amount is
    original sum + interest earned.

The original sum invested is often called the principal.

Example 1

Find the compound interest on £200 for 3 years at 10%.

original sum   = £200
1st year interest = £ 20 (10% of £200)

amount = £220 (£200 + £20)
2nd year interest = £ 22 (10% of £220)

amount = £242 (£220 + £22)
3rd year interest = £ 24·20 (10% of £242)

amount   = £266·20
original sum = £200

∴ interest = £ 66·20

Exercise 4.4a

For the following find (for the time stated):
(a) the total debt,
(b) the compound interest chargeable.

	Loan	Compound interest rate	Time
1.	£300	20%	3 years
2.	£400	15%	2 years
3.	£600	12%	2 years
4.	£500	10%	3 years
5.	£500	16%	2 years
6.	£200	18%	2 years
7.	£400	5%	3 years
8.	£1000	6%	2 years
9.	£200	9%	2 years
10.	£500	4%	2 years

Usually interest is calculated only on complete pounds.
e.g.

£134·20 becomes £134
£134·96 becomes £134
(i.e. it is not rounded up to £135)

Example 2

Find the compound interest on £450 for 3 years at 8%.

principal = £450
1st year interest = £ 36 (8% of £450)

amount = £486 (£450 + £36)
2nd year interest = £ 38·88 (8% of £486)

amount = £524·88 (£486 + £38·88)
3rd year interest = £ 41·92 (8% of £524)
Note: interest = 8% of £524; not 8% of £524·88 or 8% of £525

amount = £566·80
(£524·88 + £41·92)
principal = £450

interest = £116·80

Exercise 4.4b

For the following find
(a) the total value of the investor's savings,
(b) the compound interest payable.

	Principal	Compound interest rate	Time
1.	£600	8%	2 years
2.	£250	10%	3 years
3.	£1200	15%	3 years
4.	£750	6%	2 years
5.	£1000	12%	2 years
6.	£900	16%	2 years
7.	£1600	25%	4 years
8.	£480	20%	3 years
9.	£700	3%	2 years
10.	£5600	30%	4 years

Depreciation is a compound decrease in the value of an item, from year to year.
The value at the end of a year is
(value at the beginning of the year) − (depreciation)

Example 3

Find the cost of depreciation over 3 years at 10% on an item costing £1400.

original value = £1400
depreciation during
1st year = £140
(10% of £1400)
value after 1 year = £1260
(£1400 − £140)
depreciation during
2nd year = £126
(10% of £1260)
value after 2 years = £1134
(£1260 − £126)
depreciation during
3rd year = £113·40
(10% of £1134)
value after 3 years = £1020·60
(£1134 − £113·40)

depreciation over
3 years = £379·40
(£1400 − £1020·60)

Exercise 4.4c

For the following find the value after the time stated.

	Initial value	Rate of depreciation	Time
1.	£500	10%	2 years
2.	£1000	20%	3 years
3.	£200	5%	2 years
4.	£4000	25%	3 years
5.	£1500	2%	2 years
6.	£800	50%	5 years
7.	£500	6%	2 years
8.	£5000	4%	3 years
9.	£8400	18%	2 years
10.	£400 000	15%	5 years

When goods are bought on hire purchase a deposit must be paid, followed by a number of equal payments called instalments.

Total hire purchase price = deposit + instalments

Example 1

A car costs £2500 if payment is made by cash.
The hire purchase terms are: a deposit of £700 and 24 monthly instalments of £85.
Find (a) the total hire purchase price,
and (b) the extra cost of hire purchase.

(a) Deposit = £ 700
24 instalments of £85 = £85 × 24 = £2040
Total hire purchase price = £2740 (£700 + £2040)

(b) Extra cost of hire purchase = hire purchase price − cash price
= £2740 − £2500
= £240

Exercise 4.5a

For the following find:
(a) the hire purchase price,
(b) the difference between the hire purchase price and the cash price.

Article	Cash price	Deposit	Time to pay	Monthly amount
1. bicycle	£108	£24	12 months	£8
2. office chair	£94	£45	18 months	£4
3. digital watch	£72	£20	12 months	£5
4. diamond ring	£220	£82	12 months	£14·50
5. carpet	£355	£44	15 months	£24
6. typewriter	£300	£250	18 months	£3·30
7. golf clubs	£285	£73	24 months	£10·60
8. greenhouse	£88	£12	30 months	£3
9. enlarger	£330	£78	20 months	£15·10
10. microwave oven	£158	£43	6 months	£22·85

The deposit is often stated as a percentage of the cash price.

Example 2

The cash price of a bicycle is £64. It can be bought on hire purchase with a deposit of 25% and 18 monthly payments of £3. What is the difference between the hire purchase price and the cash price?

$$\text{Deposit} = 25\% \text{ of } £64 = \frac{25}{100} \times \frac{64}{1} = £16$$

Total monthly payments = 18 × 3 = £54

∴ total hire purchase price = £70   i.e. £54 + £16

and difference = £70 − £64 = £6

Exercise 4.5b

For the following find:
(a) the hire purchase price,
(b) the difference between the hire purchase price and the cash price.

Article	Cash price	Deposit	Time to pay	Monthly amount
1. cooker	£90	20%	1 year	£7
2. spin dryer	£60	25%	2 years	£2
3. record player	£75	12%	1 year	£6
4. television	£135	20%	2 years	£5
5. washing machine	£240	15%	2 years	£9
6. kitchen suite	£450	16%	1 year 6 months	£24
7. colour television	£350	12%	2 years	£15
8. car	£2500	15%	2 years	£95
9. motor cycle	£900	25%	1 year 6 months	£40
10. van	£1800	16%	2 years	£70

Example 3

A stereo tape recorder can be bought on hire purchase at a total cost of £185.
The deposit required is £80, and the remainder must be paid in 15 equal monthly instalments.
Find the cost of each instalment.

Cost of 15 instalments $=$ total cost $-$ deposit
$$= £185 - £80 = £105$$

Cost of 1 instalment $= \dfrac{£105}{15} = £7$

Exercise 4.5c

For the following find the cost of each monthly instalment.

Article	Total hire purchase price	Deposit	Number of monthly instalments
1. clock	£110	£26	12
2. suit	£86	£32	18
3. stereo centre	£352	£64	36
4. camera	£192	£48	24
5. encyclopedia	£62	£14	24
6. computer	£883	£19	12
7. dinghy	£425	£83	18
8. estate car	£3390	£950	16
9. tape recorder	£216	£30	12
10. motorcycle	£825	£230	25

Cost price (wholesale price) is the amount a shopkeeper *pays* for his goods.
Selling price (retail price) is the amount a shopkeeper *receives* for his goods.

profit = (selling price) − (cost price)

A loss is made when the selling price is *less* than the cost price.

loss = (cost price) − (selling price)

Example 1

Find the profit or loss when
(a) eggs are bought at 5p each and sold for 65p a dozen.
(b) a tonne of potatoes is bought for £60 and sold for 5p a kg.

(a)　Cost price of eggs = 12 × 5p = 60p a dozen
　　　Selling price of eggs = 65p a dozen
　　　∴ profit = (65 − 60)p = 5p a dozen

(b)　Cost price of potatoes = £60 a tonne
　　　Selling price of potatoes = 1000 × 5p = £50 a tonne
　　　∴ loss = £60 − £50 = £10 a tonne

Exercise 4.6a

For each question find the profit or loss on each single item.

Commodity	Retail price	Wholesale price
1. bars of chocolate	12p each	£9 per 100 bars
2. pencils	10p each	£40 per 500
3. post cards	9p each	£32 per 400
4. oranges	5p each	£12 per 200
5. grapefruits	8p each	£9 per 150
6. cigarettes	52p for twenty	£125 for 5000
7. sheets of paper	40p for eighty	£9 for 2000
8. tomatoes	60p per kg	£540 per tonne
9. apples	50p per kg	£400 per tonne
10. bananas	30p per kg	£360 per tonne
11. sugar	28p per kg	£15 for 50 kg
12. cheese	£1·30 per kg	£30 for 20 kg
13. butter	£1·20 per kg	£54 for 50 kg
14. wire	20p per m	£120 for $\frac{1}{2}$ km
15. carpet	£5·40 per m²	£240 for 50 m²
16. linoleum	£1·20 per m²	£75 for 50 m²
17. paraffin	12p per litre	£4·00 for 50 *l*
18. beer	60p per litre	£21·60 for 40 *l*
19. lemonade	18p per can	£3·60 for 24 cans
20. milk	10p per $\frac{1}{4}$ *l* cup	£2·40 for 15 *l*

Profit (or loss) is usually written as a percentage of the cost price (wholesale price).

Example 2

Find the percentage profit or loss when
(a) pens are bought at 6p each and sold at 75p a dozen
(b) cheese is bought at £1·25 per kg and sold at 10p per 100g

(a)   Cost price = 72p per dozen
        Selling price = 75p per dozen
              $\therefore$ profit = 3p per dozen

Profit as a fraction of cost price $= \dfrac{3}{72}$

$\therefore$ percentage profit $= \dfrac{\overset{1}{\cancel{3}}}{\cancel{72}} \times \dfrac{\overset{25}{\cancel{100}}}{1} = \dfrac{25}{6} = 4\tfrac{1}{6}\%$

(b)   Cost price = £1·25 per kg
        Selling price = £1·00 per kg
              $\therefore$ loss = 25p per kg

Loss as a fraction of cost price $= \dfrac{25}{125}$

$\therefore$ percentage loss $= \dfrac{\overset{1}{\cancel{25}}}{\cancel{125}} \times \dfrac{\overset{20}{\cancel{100}}}{1} = 20\%$

Exercise 4.6b

Find the percentage profit (or loss) for each question

Commodity	Retail price	Wholesale price
1. bars of chocolate	12p each	£9 per 100 bars
2. pencils	10p each	£40 per 500
3. post cards	9p each	£32 per 400
4. oranges	5p each	£12 per 200
5. grapefruits	8p each	£9 per 150
6. cigarettes	52p for twenty	£125 for 5000
7. sheets of paper	40p for eighty	£9 for 2000
8. tomatoes	60p per kg	£540 per tonne
9. apples	50p per kg	£400 per tonne
10. bananas	30p per kg	£360 per tonne
11. sugar	28p per kg	£15 for 50 kg
12. cheese	£1·30 per kg	£30 for 20 kg
13. butter	£1·20 per kg	£54 for 50 kg
14. wire	20p per m	£120 for $\frac{1}{2}$ km
15. carpet	£5·40 per m²	£240 for 50 m²
16. linoleum	£1·20 per m²	£75 for 50 m²
17. paraffin	12p per litre	£4·00 for 50 l
18. beer	60p per litre	£21·60 for 40 l
19. lemonade	18p per can	£3·60 for 24 cans
20. milk	10p per $\frac{1}{4}$ l cup	£2·40 for 15 l

Profit (or loss) may be written as a percentage of the selling price (retail price).

Example 3

Express the profit or loss as a percentage of the selling price when
(a) pencils are bought at 5p each and sold at £1·25 for 20
(b) a 10m coil of rope is bought for £4·20 and sold at 36p per metre

(a)   Cost price = £1 for 20
       Selling price = £1·25 for 20
              profit = 25p on 20

$$\text{Profit as a fraction of selling price} = \frac{25}{125}$$

$$\text{percentage profit} = \frac{\overset{1}{25}}{\underset{\underset{1}{5}}{125}} \times \frac{\overset{20}{100}}{1}$$

$$= 20\%$$

(b)   Cost price = £4·20 for 10m
       Selling price = £3·60 for 10m
              loss = £0·60 on 10m

$$\text{Loss as a fraction of selling price} = \frac{60}{360}$$

$$\text{percentage loss} = \frac{\overset{1}{60}}{\underset{\underset{3}{6}}{360}} \times \frac{\overset{50}{100}}{1}$$

$$= \frac{50}{3} = 16\tfrac{2}{3}\%$$

Exercise 4.6c

Find the profit (or loss) as a percentage of the selling price, in each question.

Commodity	Retail price	Wholesale price
1. television	£100 each	£75 each
2. radio	£30 each	£24 each
3. food mixer	£80 each	£88 each
4. fridge	£200 each	£190 each
5. golf clubs	£120 each	£138 each
6. shoes	£20 each	£17·60 each
7. gloves	£3 each	£2·28 each
8. hammer	£2·50 each	£5 each
9. vice	£16 each	£14 each
10. tent	£135 each	£126·90 each

If an article is sold at a profit
    selling price = cost price + profit
If an article is sold at a loss
    selling price = cost price − loss

Example 4

Find the selling price when
(a) a washing machine is bought for £85 and sold at a profit of £28
(b) a bicycle is bought for £62 and sold at a loss of £14
(c) sweets are bought at £1·20 per kg, and sold in 250 g bags, at a profit
    of 60% of the cost price

(a)   Cost price = £85
                Profit = £28
    ∴ selling price = £85 + £28 = £113

(b)   Cost price = £62
                 Loss = £14
    ∴ selling price = £62 − £14 = £48

(c)   Cost price = 120 p per kg

$$\text{Profit} = \frac{\cancel{60}^{12}}{\cancel{100}} \times \frac{\cancel{120}^{6}}{1} = 72 \text{ p}$$

$$\therefore \text{selling price} = 120 \text{ p} + 72 \text{ p} = 192 \text{ p per kg}$$

$$= \frac{192}{4} = 48 \text{ p per 250 g bag}$$

Exercise 4.6d

Find the selling price for each single item.

Commodity	*Wholesale price*	*Profit or loss*
1. calculator	£6 each	£2 profit
2. bag	£11.50	£5·25 loss
3. rug	£26.30	£7·84 profit
4. cassettes	£84 for 100	£7 loss
5. radios	£170 for 5	£13 profit
6. loaves	£2 for 10	10% profit
7. lemonade	£45 for 100 cans	20% profit
8. towing hooks	£120 for 20	25% loss
9. dish-towels	£26·40 for 12	5% profit
10. nuts	£7·50 for 50	50% profit
11. golf balls	£12·60 for 15	$33\frac{1}{3}$% profit
12. cans of paint	£68·75 for 25	8% loss
13. sheets of card	£96 for 400	$12\frac{1}{2}$% loss
14. ball bearings	£42·50 for 25	150% profit
15. petrol	£240 for 1000 *l*	$6\frac{1}{4}$% profit

Find the cost of each item in a bill, and add, to find the total cost.

'@' means 'at'

Example 1

Mr. Jackson buys

> 5 m² plywood @ £2·51 per m²
> 4 kg nails @ 62p per kg
> 250 screws @ £2·50 for 100
> 4 litres of paint @ £2·30 per litre
> 16 rolls wallpaper @ £1·82 per roll

(a) Find the total cost of these items.
(b) If VAT at 15% is added to the total cost, how much must **Mr. Jackson** finally pay?

(a)

5 m² plywood @ £2·51 per m²	= £12·55 (£2·51 × 5)
4 kg nails @ 62p per kg	= £ 2·48 (£0·62 × 4)
250 screws @ £2·50 for 100	= £ 6·25 (£2·50 × 2.5)
4 litres of paint @ £2·30 per litre	= £ 9·20 (£2·30 × 4)
16 rolls wallpaper @ £1·82 per roll	= £29·12 (£1·82 × 16)
total	= £59·60

(b)

$$\text{VAT} = 15\% \text{ of } £59\cdot60 = \frac{\overset{3}{\cancel{15}}}{\underset{\underset{1}{20}}{\cancel{100}}} \times \frac{\overset{298}{\cancel{5960}}}{1}$$

$$= 3 \times 298 = 894p$$

$$= £8\cdot94$$

Final cost = £59·60 + £8·94 = £68·54

Exercise 5.1

Find the cost of each bill.

1. 4 fish suppers @ 66p each
   2 pies @ 32p each
   3 single fish @ 42p each
   6 pickled onions @ $2\frac{1}{2}$p each
   3 cans of lemonade @ 17p each

2. 12 apples @ 8p each
   4 kg nuts @ £1·98 per kg
   $\frac{1}{4}$ kg margarine @ 80p per kg
   4 l milk @ 32p per l
   24 eggs @ 82p per dozen

3. 4 kg steak @ £3·45 per kg
   5 kg potatoes @ 15p per kg
   $\frac{1}{2}$ kg butter @ £1·84 per kg
   one $1\frac{1}{2}$ kg chicken @ £1·76 per kg
   9 cans of lemonade @ 63p for 3

4. 16 m² cloth @ £2·34 per m²
   3 needles @ 25p for 5
   5 m thread @ 16p for 20 m
   $\frac{1}{4}$ kg wool @ £17·92 for 2 kg
   1 pattern @ 93p
   VAT at 15% is added to the **total cost.**

5. 250 ml developer @ £2·60 per *l*
   21 rolls film @ £1·73 per roll
   7 lightbulbs @ £1·04 for 2
   3 pairs of film clips @ 98p per pair
   100 ml wetting agent @ £4·40 per *l*
   VAT at 15% is added.

Gas and electricity bills are paid quarterly, i.e. every 13 weeks.

The gas bill is based on the number of therms used each quarter.

The electricity bill is based on the number of units used each quarter. These quantities are found by reading the meter.

Example 1

An electricity meter was read as follows:

Jan. 1st	16943
April 1st	18491
July 1st	19631

Find the number of units used in each quarter.

Units used in quarter ending
April 1st = 18491 − 16943
= 1548 units

Units used in quarter ending
July 1st = 19631 − 18491
= 1140 units

Exercise 5.2a

Find the number of electricity units used in each quarter from the following meter readings.

1.	Jan. 1st	13562	2.	Jan. 1st	17722
	April 1st	15825		April 1st	19587
	July 1st	16499		July 1st	20492
3.	July 1st	12133	4.	July 1st	18576
	Oct. 1st	12588		Oct. 1st	19197
	Jan. 1st	13792		Jan. 1st	20739
5.	July 1st	25361	6.	April 1st	18932
	Oct. 1st	25895		July 1st	19559
	Jan. 1st	27531		Oct. 1st	20075
7.	April 1st	25324	8.	Jan. 1st	17532
	July 1st	26037		April 1st	19853
	Oct. 1st	26466		July 1st	20615
9.	July 1st	48382		Oct. 1st	21163
	Oct. 1st	48915	10.	Jan. 1st	29234
	Jan. 1st	50662		April 1st	31127
	April 1st	53117		July 1st	31942
				Oct. 1st	32691
				Jan. 1st	35777

Electricity bills are usually calculated on a 'two-tier' basis, as shown.

Example 2

A householder uses 1548 units of electricity in a certain quarter. Calculate his bill if the first 72 units cost 5·5p per unit and all remaining units cost 2·25p each.

72 @ 5·5p each cost £3·96
Remaining units = 1548 − 72 = 1476
1476 @ 2·25 each cost £33·21
∴ total bill = £3·96 + £33·21 = £37·17

Exercise 5.2b

Calculate the bill for each quarter below if the first 72 units cost 5·5p each and all remaining units cost 2·25p each. Questions 9 to 12 give the meter readings.

1. Jan. 1st to April 1st, 1712 units
2. Jan. 1st to April 1st, 1800 units
3. July 1st to Oct. 1st, 492 units
4. July 1st to Oct. 1st, 720 units
5. April 1st to July 1st, 1032 units
6. April 1st to July 1st, 936 units
7. Oct. 1st to Jan. 1st, 1352 units
8. Oct. 1st to Jan. 1st, 1224 units
9. Jan. 1st: 16362, April 1st: 18194
10. Jan. 1st: 12143, April 1st: 14519
11. April 1st: 14534, July 1st: 15726
12. April 1st: 27781, July 1st: 28637

Gas bills usually consist of a Standing Charge to be paid each quarter in addition to the cost of the gas.

Example 3

In the quarter ending 31 December, a man used 174 therms of gas. In the quarter ending 31 March he used 206 therms of gas. Calculate the total amount he paid for the gas if the gas cost 22p per therm and the Standing Charge was £3·50 per quarter.

Total number of therms used = 174 + 206
$$= 380$$
Cost of gas = 22p × 380 = £83·60
Standing Charge (2 quarters) = £3·50 × 2
$$= £7·00$$
∴ total bill = £83·60 + £7·00
$$= £90·60$$

Exercise 5.2c

Calculate the total bill for each quarter if the
Standing Charge for each quarter is £3·50 and
the price of gas is 22p per therm. Questions
9 to 15 give the meter readings.

1. Dec. 31st to March 31st, 190 therms
2. March 31st to June 30th, 43 therms
3. June 30th to Sept. 30th, 36 therms
4. Sept. 30th to Dec. 31st, 165 therms
5. Dec. 31st to March 31st, 154 therms
   and March 31st to June 30th, 32 therms
6. March 31st to June 30th, 41 therms and
   June 30th to Sept. 30th, 25 therms
7. Sept. 30th to Dec. 31st, 135 therms and
   Dec. 31st to March 31st, 142 therms
8. Dec. 31st to March 31st, 123 therms and
   March 31st to June 30th, 37 therms

9. Dec. 31st: 234
   March 31st: 386
   June 30th: 429
10. March 31st: 154
    June 30th: 188
    Sept. 30th: 215
11. June 30th: 386
    Sept. 30th: 447
    Dec. 31st: 573
12. Sept. 30th: 058
    Dec. 31st: 195
    March 31st: 406
13. Dec. 31st: 135
    March 31st: 339
    June 30th: 392
    Sept. 30th: 430
    Dec. 31st: 611
14. Dec. 31st: 257
    March 31st: 426
    June 30th: 474
    Sept. 30th: 518
    Dec. 31st: 653
15. Dec. 31st: 008
    March 31st: 202
    June 30th: 251
    Sept. 30th: 290
    Dec. 31st: 448

5.3 RATES

Rates are paid on the rateable value (R.V.)
of a property.

Example 1

Find the rates paid on a shop with a
rateable value of £725 if the rate levied is
72p in the £.
R.V. = £725; rate = 72p in the £
∴ rates paid = 725 × 72p
$$= £725 \times \frac{72}{100}$$
$$= £522$$

The rateable value is sometimes called the
assessed rental.

Exercise 5.3a

Find the rates that are payable by the owners of
the following properties.

	Rateable value	Rate levied
1.	£200	50p in the £
2.	£300	60p in the £
3.	£250	80p in the £
4.	£180	70p in the £
5.	£360	75p in the £
6.	£400	84p in the £
7.	£350	72p in the £
8.	£450	66p in the £
9.	£480	75p in the £
10.	£320	55p in the £
11.	£121	150p in the £
12.	£162	110p in the £
13.	£508	102p in the £
14.	£335	75p in the £
15.	£495	107p in the £

Example 2

Find the rateable value of a property
where the owner paid rates of £153 when
the rate levied was 90p in the £.
Rates paid = £153; rate = 90p in the £

$$\therefore \text{R.V.} = \frac{£153}{90p} = £\frac{\overset{17}{\cancel{153}} \times \overset{10}{\cancel{100}}}{\underset{\underset{1}{9}}{\cancel{90}}}$$

$$= £170$$

Exercise 5.3b

Find the rateable value of the following properties.

Rates paid	Rate levied
1. £180	50p in the £
2. £360	80p in the £
3. £144	60p in the £
4. £320	64p in the £
5. £140	56p in the £
6. £484	88p in the £
7. £315	75p in the £
8. £221	65p in the £
9. £187·50	50p in the £
10. £171·60	60p in the £
11. £409·60	80p in the £
12. £302·40	90p in the £
13. £177·80	70p in the £
14. £211.20	110p in the £
15. £472.80	120p in the £
16. £569.10	105p in the £
17. £296.40	104p in the £
18. £331.50	102p in the £

Example 3

A small town has a total rateable value of
£250 000. If the town council wishes to
raise £180 000 in rates, what rate in the £
has it to levy?

Rateable value = £250 000
Amount to be raised in rates = £180 000

$$\therefore \text{rate levied} = \frac{£180\ 000}{£250\ 000} = £0·72$$

so rate is 72p in the £.

Exercise 5.3c

Calculate the rate that each of the following
towns or villages levy on the inhabitants from
the details given below.

Town or village	Amount to be raised	Rateable value
1. Douglas	£75 000	£100 000
2. Lochwinnoch	£160 000	£200 000
3. Buckie	£240 000	£400 000
4. Wick	£350 000	£500 000
5. Monifieth	£270 000	£300 000
6. Albury	£216 000	£216 000
7. Shere	£341 000	£310 000
8. Gomshall	£234 000	£225 000
9. Peaslake	£231 000	£220 000
10. Cranleigh	£561 000	£550 000

If the calculated rate per £ is not a whole
number of pence, it must always be rounded
up to the next whole number of pence.
This means that a small surplus may arise when
all the rates are collected.
Surplus = (amount actually raised) − (amount
to be raised)

Example 4

A town has a total rateable value of £1 250 000.
(a) If the amount to be raised from the rates is
    £1 405 000, what rate in the £ must be
    levied?
(b) What surplus will result from this rate?

(a) Rateable value = £1 250 000
    Amount to be raised in rates = £1 405 000

$$\text{rate levied} = \frac{1\ 405\ 000}{1\ 250\ 000} = £1·124$$

so the rate must be £1·13 in the £
(b) R.V. = £1 250 000; rate = £1·13 in the £
    rates paid = £1·13 × 1 250 000
              = £1 412 500
    Surplus = £1 412 500 − £1 405 000
            = £7500

Exercise 5.3d

For each question find
(a) the rate which must be levied (correct to the nearest necessary 1p)
(b) the surplus

Town or village	Amount to be raised	Rateable value
1. Irvine	£859 000	£1 000 000
2. Airdrie	£1 928 000	£2 000 000
3. Lanark	£250 000	£400 000
4. Dunfermline	£1 746 000	£3 000 000
5. Kilsyth	£437 000	£500 000
6. Ayr	£2 560 000	£2 500 000
7. Motherwell	£3 647 000	£3 500 000
8. Ipswich	£5 859 000	£5 400 000
9. Ledbury	£133 000	£125 000
10. Perth	£2 484 000	£2 250 000

Exercise 5.3e

1. A town council wishes to raise £492 000 in rates, from a town with a rateable value of £480 000.
   (a) Calculate the rate they must charge.
   (b) Calculate the surplus, if any.
   (c) Find the rates paid on property in this town with rateable value £1250.
2. A woman moves from a house with rateable value £642 to a house with rateable value £480. If the rate is 84p in the £, by how much is her rates bill reduced?
3. A man owns a house with rateable value £450, and pays £441 in rates.
   (a) Find the rate levied.
   (b) If the same rate applies to a hotel of rateable value £276, how much must the owner pay in rates?

5.4 WAGES AND SALARIES

A wage is the amount earned at a fixed rate per hour for a given number of hours per week. Any extra hours worked are called *overtime*.

Example 1

Find the wage of a man who earns £1·17 per hour for a 40 hour week.

wage = £1·17 × 40 = £46·80

Exercise 5.4a

In each case find the weekly wage for the information given.

1. £1·20 per hour, 40 hours
2. £1·35 per hour, 40 hours
3. £1·05 per hour, 40 hours
4. £1·24 per hour, 40 hours
5. £1·48 per hour, 40 hours
6. £1·50 per hour, 42 hours
7. £1·10 per hour, 42 hours
8. £1·40 per hour, 42 hours
9. £1·15 per hour, 42 hours
10. £1·32 per hour, 42 hours
11. £1·25 per hour, 36 hours
12. £1·30 per hour, 36 hours
13. £1·60 per hour, 36 hours
14. £1·45 per hour, 36 hours
15. £1·04 per hour, 36 hours
16. £1·20 per hour, 44 hours

Overtime is paid at a higher rate per hour, e.g. 'Time and a quarter', 'time and a half', etc.

'Time and a half' means that for each hour worked, you are paid for $1\frac{1}{2}$ hours work.

Example 2

A man earns £1·20 per hour. What is his overtime pay per hour
(a) at 'time and a quarter'?
(b) 'at double time'?

(a) Time and a quarter = £1·20 × $1\frac{1}{4}$
                                      = £1·50 per hour
(b) Double time   = £1·20 × 2
                                      = £2·40 per hour.

Exercise 5.4b

In each case find the pay per hour for
(a) overtime rate of 'time and a half'
(b) rest-day rate of 'double time'.

1. £1·10	2. £1·24	3. £1·70
4. £1·32	5. £1·40	6. £1·36

In each case find the pay per hour for
(a) overtime rate of 'time and a quarter'
(b) rest-day rate of 'double time and a quarter'

7. £1·16	8. £1·60	9. £1·44
10. £1·28		

Example 3

An apprentice earns 95p per hour. If his basic week is 40 hours, find his total wage in the week in which he works 47 hours when all overtime is at 'time and a half'.

Basic wage = 95p × 40 = £38
Overtime pay
$$= 95 × 1\tfrac{1}{2}$$
$$= £1·42\tfrac{1}{2} \text{ per hour}$$
No. of hours overtime = 47 − 40 = 7 hours
Total overtime pay
$$= £1·42\tfrac{1}{2} × 7$$
$$= £9·97\tfrac{1}{2}$$
Total wage = £38 + £9·97$\tfrac{1}{2}$ = £47·97$\tfrac{1}{2}$

Exercise 5.4c

Find the total wage if the overtime rate is 'time and a half'.

	Standard rate per hour	Hours in basic week	Actual hours worked
1.	£1·20	40	48
2.	£1·30	40	45
3.	£1·10	40	50
4.	£1·28	42	48
5.	£1·36	42	50

Find the total wage if the overtime rate is 'time and a quarter'.

6.	£1·24	40	44
7.	£1·32	40	42
8.	£1·20	42	50
9.	£1·40	44	46
10.	£1·16	44	54

A salary is the amount earned in a year. It is paid in 12 equal monthly instalments.

Example 4

Find the salary of a person who earns £240 per month.

Salary = £240 × 12 = £2880.

Exercise 5.4d

Find the annual salary for each monthly payment below.

1. £250		2. £320
3. £355		4. £275
5. £316		6. £344
7. £418		8. £293
9. £320·50		10. £284·50
11. £310·25		12. £275·25
13. £386·25		14. £290·75
15. £372·75		

Example 5

A man's salary is £3684 per year. How much does he earn per month?

Monthly pay
$$= £3684 ÷ 12$$
$$= £307 \text{ per month.}$$

Exercise 5.4e

Find the monthly earnings from each annual salary below.

1. £3600	2. £4500	3. £3204
4. £3912	5. £3828	6. £3984
7. £3336	8. £3408	9. £4083
10. £4371	11. £3489	12. £4521
13. £2886	14. £4146	15. £4038

Example 6

A man has a choice of two jobs
(a). £320·50 per month
(b) £74 per week

Which is the better rate of pay and by how much a year?

(a) £320·50 per month = £320·50 × 12
$$= £3846 \text{ per year}$$
(b) £74 per week = £74 × 52
$$= £3848 \text{ per year}$$
∴ £74 per week is better by £2 per year.

Exercise 5.4f

In each case find which job gives the higher pay and by how much per year.

1. £64 per week or £277 per month
2. £55 per week or £239 per month
3. £68 per week or £294·50 per month
4. £80 per week or £346·75 per month
5. £56·50 per week or £244·50 per month
6. £72·50 per week or £313·75 per month
7. £75 per week or £326 per month
8. £66 per week or £286·25 per month

5.5 INCOME TAX

Income tax is paid at a given rate on income after various allowances are deducted.

Allowances = (Tax Code Number) × 10
Taxable income
= (total income) − (allowances)

Example 1

A man earns £5000 per year. If his Tax Code Number is 192, calculate his taxable income.

Total allowances = 192 × 10 = £1920
Taxable income = £5000 − £1920 = £3080

Exercise 5.5a

In each case find the taxable income from the details given.

	Annual earnings	Tax Code Number
1.	£3000	150
2.	£4000	175
3.	£3800	125
4.	£3600	115
5.	£3200	95
6.	£4500	132
7.	£3750	114
8.	£4150	96
9.	£2850	112
10.	£3360	128

Individual allowances may be found from a list.

Allowances:

lower personal (single person)	£1375
higher personal (married man)	£2145
wife's earned income	£1375
additional personal	£ 770
widow's bereavement	£ 770
dependent relative	£ 100
housekeeper	£ 100
blind person	£ 180
age (single person)	£1820
age (married man)	£2895
son's or daughter's services	£ 55

Everyone is entitled to either the lower personal allowance, or the higher personal allowance. Other allowances depend on individual circumstances.
Add individual allowances to find the total allowances.

Example 2

A married man earns £8450 per year, and his wife earns £1500. The man is entitled to a dependent relative allowance.
Find (a) the man's total allowances, and
(b) his taxable income.

(a) Allowances:

higher personal	£2145
wife's earned income	£1375
dependent relative	£ 100
total =	£3620

(b) total income = £8450 + £1500
= £9950
total allowances = £3610
taxable income = £9950 − £3620
= £6330

Exercise 5.5b

For each question find
(a) the total allowances, and
(b) the taxable income.

1. Mr. Jones is single and earns £4800 per year.
2. Jason earns £3250 per year and his wife Diane earns £1800.
3. Mr. MacIntosh is married and earns £7600 per year. His wife does not work, but he receives an additional personal allowance for a son living abroad.
4. Mr. & Mrs. Gibson are both artists and earn £4750 per year each.
5. Victor earns £4200 per year as a singer. He is single and receives a blind person's allowance.
6. Ann McTaggart is unmarried and earns £2860 per year as a waitress. She also earns tips amounting to £382 which must be added to her income before deduction of allowances.
7. Miss Rennie earns £4520 per year and receives a dependent relative allowance of £145 because she cares for her mother.
8. Mr. Belcher is a widower and earns £8420 per year as a tax consultant. In addition to a lower personal allowance he receives a dependent relative allowance, an additional personal allowance, and an allowance of £685 for business equipment.
9. Agnes is a retired widow. Because she was born before 6th April 1915, Agnes receives a single person's age allowance instead of the lower personal allowance. She also receives a daughter's services allowance. Agnes' income is a pension of £36·50 per week.
10. Mr. Torrance is a photographer earning £638 per month. He is divorced and receives a single person's allowance. He also has an allowance of 30% of £2100 alimony paid to his former wife.

The next exercise assumes that income tax is charged on taxable income at the following rates:

30% on the first	£11250
40% on the next	£ 2000
45% on the next	£ 3500
50% on the next	£ 5500
55% on the next	£ 5500
60% on the remainder	

Example 3

A man earns £14400 per year and receives allowances of £2895 per year.
Find (a) the yearly tax he pays
      (b) the monthly tax he pays
      (c) his monthly salary, after deduction of tax.

(a) Taxable income = £14400 − £2895
                   = £11505
Tax at 30% of the first £11250

$$= \frac{\overset{3}{\cancel{30}}}{\underset{10}{\cancel{100}}} \times \frac{\overset{1125}{\cancel{11250}}}{1}$$

$$= 3 \times 1125 = £3375$$

Income still to be taxed
                   = £11505 − £11250
                   = £255

Tax at 40% of £255 $= \dfrac{\overset{2}{\cancel{40}}}{\underset{5}{\cancel{100}}} \times \dfrac{\overset{51}{\cancel{255}}}{1}$

$$= 2 \times 51 = £102$$
Total yearly income tax = £3375 + £102
                       = £3477

(b) Monthly tax $= \dfrac{£3477}{12} = £289·75$

(c) Annual salary after deduction of income
    tax   = £14400 − £3477 = £10923
    Monthly salary after deduction of income
    tax
    $= \dfrac{£10923}{12} = £910·25$

Exercise 5.5c

For each question find
(a) the yearly tax
(b) the monthly tax
(c) the monthly salary after deduction of tax.

	Annual salary	Allowances
1.	£2904	£1224
2.	£3624	£1544
3.	£2616	£1256
4.	£2184	£1304
5.	£5148	£2188
6.	£4152	£1592
7.	£4548	£1868

8.	£2952	£1512
9.	£2436	£1356
10.	£7056	£2816
11.	£11484	£1964
12.	£13248	£8568
13.	£12492	£2772
14.	£5262	£2222
15.	£7740	£2480

For the following find the annual tax payable.

16.	£14200	£2750
17.	£17400	£3330
18.	£16800	£2910

5.6 INSURANCE

An *annual premium* is an amount paid each year to maintain an insurance policy.
Endowment life assurance premiums are paid for a fixed term e.g. 20 years.
Whole life assurance premiums are paid until the policy holder dies.

Example 1

A young woman takes out an endowment life assurance policy in which the sum assured is £4000. The policy is payable in 20 years, and the annual premium is £6·20 per £100 of sum assured.
(a) Find the annual premium.
(b) If income tax relief is given, at 30% of the annual premium, find the net cost of the policy over the 20 year period.

(a) £100 assurance costs £6·20,

so £4000 assurance costs $\dfrac{\overset{40}{\cancel{4000}}}{\underset{1}{\cancel{100}}} \times \dfrac{620}{1}$ p

$$= 40 \times 620 = 24800 \text{ p}$$
$$= £248·00$$

(b) Tax relief at 30% $= \dfrac{\overset{3}{\cancel{30}}}{\underset{\underset{5}{10}}{\cancel{100}}} \times \dfrac{\overset{124}{\cancel{248}}}{1}$

$$= \frac{372}{5} = £74·40$$

Net annual premium $= £248·00 - £74·40$
$$= £173·60$$
Net cost over 20 years $= £173·60 \times 20$
$$= £3472·00$$

Exercise 5.6a

1. Mr. Goldstein buys an endowment policy in which the sum assured is £2000. The annual premium is charged at the rate of £5·20 per £100 of sum assured. Find the annual premium.
2. Mrs. Johnstone assures her husband's life for £15 000 at the rate of £4·10 per £100. Find the annual premium.
3. Jason takes out a whole life assurance policy for £2500. The premium is calculated at the rate of £5·80 per £100. Find the annual premium.
4. Ann Taylor assures her life for £40 000 at the rate of £6·30 per £100. Find the annual premium.
5. Dick assures his partner's life for £25 000 at the rate of £7·85 per £100. Find the annual premium.

6. Jill assures her life for £15 000 at the rate of £5·60 per £100. If she pays one premium every year for 20 years, find
   (a) her annual premium, and
   (b) the total cost of the policy over the 20 year period.
7. Mr. Cooper assures his life for £35 000 at the rate of £4·35 per £100, over a 25 year period. Find
   (a) his annual premium, and
   (b) the total cost of the assurance.
8. Terence takes out an endowment policy worth £8500. If the annual premium rate is £3·86 per £100 and the policy runs for 30 years, find
   (a) his annual premium, and
   (b) the total cost of the assurance.
9. Miss Paterson assures her life for £75 000, at £4·95 per £100, for a period of 25 years. Find
   (a) her annual premium, and
   (b) the total cost of the policy.
10. Mr. Higgins pays £6·63 per £100 for a 20 year endowment policy worth £3500. Find
    (a) his annual premium, and
    (b) the total cost of this policy.
11. Dennis assures his wife's life for £38 000 at the rate of £5·42 per £100, over a 25 year period.
    (a) Find the annual premium.
    (b) If tax relief is given on the premium, at 30%, find the net annual premium.
12. Helen assures her life for £4200 at £4·25 per £100, over a 30 year period.
    ·(a) Find her annual premium.
    (b) If tax relief is given, at 40% of the premium, find her net annual premium.
13. Mrs. Dempster buys a whole life policy valued at £82 000 paying a premium of £2·82 per £100.
    (a) Find her annual premium.
    (b) If tax relief is given, at 55% of the premium, find the net annual premium.
14. Mrs. Gable assures her husband's life for £5500 by paying a premium of £4·32 per £100.
    (a) Find the annual premium.
    (b) If tax relief is given at 45%, find the net premium.
    (c) If the premiums are paid for 25 years, find the net cost of the policy.
15. Terry assures his life with an endowment policy of £6500. He pays a premium of £4·28 per £100. After 5 years Terry dies, and his wife collects the sum assured.
    (a) Find the annual premium.
    (b) Find the total amount paid in premiums until Terry's death.
    (c) Find the loss made by the assurance company, on this policy.

Example 2

A man owns a house valued at £18 000, and its contents are worth £7200. The annual insurance rates are: £0·22 per cent for buildings, and £0·30 per cent for contents.
(a) Find the total annual premium required to insure the house and contents.
(b) Find the total monthly premium.

(a) Premium required for house

$$= \frac{\overset{180}{\cancel{18\,000}}}{\underset{1}{\cancel{100}}} \times \frac{22}{1}\,\text{p}$$

$$= 3960\,\text{p} = £39·60$$

Premium required for contents

$$= \frac{\overset{72}{\cancel{7200}}}{\underset{1}{\cancel{100}}} \times \frac{30}{1}\,\text{p}$$

$$= 2160\,\text{p} = £21·60$$

Total annual premium $= £39·60 + £21·60$
$$= £61·20$$

(b) Total monthly premium

$$= \frac{6120}{12}\,\text{p} = 510\,\text{p}$$

$$= £5·10$$

Exercise 5.6b

Find (a) the total annual premium to insure house and contents, and (b) the monthly premium.

	Value		Rate per £100	
	house	contents	house	contents
1.	£6000	£2500	10p	12p
2.	£8000	£2400	15p	25p
3.	£12 000	£4000	8p	15p
4.	£15 000	£7500	12p	16p
5.	£24 000	£30 000	9p	24p
6.	£8200	£2500	25p	38p
7.	£70 000	£35 000	20p	26p
8.	£60 000	£12 000	13p	20p
9.	£16 800	£6400	11p	18p
10.	£27 000	£7500	7p	22p
11.	£30 000	£16 200	28p	36p
12.	£50 000	£4560	24p	30p

13.	£6240	£1500	30p	40p
14.	£15 600	£3760	32p	30p
15.	£120 000	£40 000	16p	21p
16.	£20 000	£7800	21p	28p
17.	£3750	£1800	26p	28½p
18.	£5400	£7200	14p	18½p
19.	£34 800	£12 600	12½p	17p
20.	£25 200	£14 160	22½p	27½p

The basic premium for motor insurance is often found from a table. A no-claims discount (or no-claims bonus) is given to drivers who have had no accidents in previous years.

Area classification	Insurance group		
	2	4	6
Low	£54	£84	£120
Medium	£72	£90	£150
High	£95	£126	£185

Number of years accident-free	Discount
1	25%
2	33⅓%
3	50%
4 or more	60%

Example 3

A girl aged 19 lives in a medium-rated area, and her car is classified as group 4. She has had 2 years accident-free driving.

Find her insurance premium if she pays a 10% surcharge on the net premium because of her age.

Basic premium $= £90$

No-claims discount $= \dfrac{33\frac{1}{3}}{100} \times £90$

$= \dfrac{\overset{1}{\cancel{100}}}{\underset{3}{\cancel{300}}} \times \dfrac{\overset{30}{\cancel{90}}}{1} = £30$

Net premium $= £90 - £30 = £60$
Surcharge $= 10\%$ of £60

$= \dfrac{\overset{1}{\cancel{10}}}{\underset{10}{\cancel{100}}} \times \dfrac{\overset{6}{\cancel{60}}}{1} = £6$

Final premium $= £60 + £6 = £66$

Exercise 5.6c

Find the annual premium, in each question.

	Area rating	Group	Number of years accident-free
1.	low	4	0
2.	low	6	1
3.	medium	2	1
4.	low	2	2
5.	high	4	3
6.	medium	4	0
7.	medium	2	2
8.	low	6	5
9.	high	6	4
10.	high	4	2

11. Trevor owns a group 4 car and lives in a low-rated area. He has had 2 years accident-free driving, but must pay a surcharge of 50% on the net premium because of his age.

12. Martha lives in a high-rated area and owns a group 2 car. She has had 5 years accident-free driving. Because her car is not kept in a garage, she must pay a surcharge of 15% of the annual premium before deduction of her no-claims bonus.

13. Hector owns a group 6 car and lives in a medium rated area. He has had 3 years accident-free driving. Hector is also allowed a discount of 10% of the premium after deduction of no-claims discount because he is a Civil Servant.

14. Mr. Allan has had 13 years accident-free driving. He lives in a low-rated area and owns a group 4 car. He must pay a surcharge of 25% of the basic premium because his car is a foreign make. Mr. Allan's no-claim discount is calculated on the premium after adding the surcharge.

15. Carlos drives a group 6 car in a high-rated area. He must pay the following surcharges, all calculated on the basic premium; 25% for under age 25, 20% for a foreign car, 10% for car not garaged. He has had 6 years accident-free driving, and his no-claims discount is calculated on the total premium after addition of all the surcharges.

In Britain the pound (£) is the basic unit of currency.
Because £1 = 100 pence, all sums of money can be written as
decimals. Most other countries use a similar system.
    e.g. Germany, 1 mark = 100 pfennigs
In the following exercises, the rates of exchange to be used are shown
in the table below.

Country	Unit of currency	Exchange rate
Belgium	1 franc = 100 centimes	BF 60 = £1
France	1 franc = 100 centimes	Fr. 9·10 = £1
Germany	1 mark = 100 pfennigs	DM 3·84 = £1
Spain	1 peseta = 100 cents	Ptas. 156 = £1
Switzerland	1 franc = 100 centimes	SF 3·60 = £1
U.S.A.	1 dollar = 100 cents	$1·96 = £1

Example 1

How many dollars would a man receive
for £50?
    number of dollars = 1·96 × 50
                      = $98

Exercise 5.7a

Find how much each amount of British money
is worth in the foreign currency.

1. £200 (dollars)   **2.** £150 (dollars)
3. £25 (dollars)   **4.** £40 (French francs)
5. £300 (French francs) **6.** £140 (French francs)
7. £75 (marks)   **8.** £200 (marks)
9. £250 (marks)   **10.** £45 (Swiss francs)
11. £105 (Swiss francs) **12.** £16 (pesetas)
13. £21 (pesetas)   **14.** £125 (Belgian francs)
15. £72 (Belgian francs)

Example 2

How many £'s would a man receive for
728 French francs?

    number of £'s = 728 ÷ 9·10
                  = 7280 ÷ 91 = £80

Exercise 5.7b

Find how much each amount of foreign currency
is worth in £'s.

1. $588   **2.** $490
3. $147   **4.** Fr. 546

5. Fr. 3640   **6.** Fr. 1456
7. DM 192   **8.** DM 1344
9. DM 480   **10.** SF 54
11. SF 126   **12.** Ptas. 2808
13. Ptas. 3744   **14.** BF 6900
15. BF 3840

Example 3

Petrol in England costs 18p per litre. In
Spain it costs Ptas. 39 per litre. In which
country is petrol cheaper, and by how
much?

    Ptas. 39 = £ (39 ÷ 156)
             = £0·25 = 25p
∴ petrol is cheaper in England by 7p per
litre.

Exercise 5.7c

1. A typewriter costs £100 in Britain and $245 in
   the United States. Find in terms of British money
   the country where the selling price is the cheaper,
   and by how much.
2. A motor cycle costs £900 in Britain and Fr. 7826
   in France. Find in terms of British money the
   country where the selling price is the cheaper,
   and by how much.
3. A kilogram of flour costs 25p in Britain, whereas
   in Switzerland the same kind of flour costs 72
   centimes per kilogram. In which country is it the
   cheaper and by how many pence per kilogram?

4. A kilogram of sugar costs 30p in Britain, whereas in Belgium the same kind of sugar costs 24 francs per kilogram. In which country is it the cheaper and by how many pence per kilogram?

5. In Britain, Spanish oranges cost £1·00 for ten, but in Spain itself the price is Ptas. 117 for the same ten oranges. In which country do the ten oranges cost less and by how many pence?

6. A bottle of whisky costs £4·50 in Britain and $9·80 in the United States. In which country is it cheaper and by how many pence?

7. A bottle of wine costs £1·50 in Britain and DM 4·80 in Germany. In which country is it cheaper and by how many pence?

8. A certain type of car is sold in several countries at the prices given below.

Britain	£2400
U.S.A.	$4900
Switzerland	SF 8820
Spain	Ptas. 343 200
Germany	DM 8640
France	Fr. 24 570
Belgium	BF 156 000

Find in terms of British money the price in each other country, then write a list of the prices in order, beginning with the cheapest.

9. A certain brand of cigarettes is sold in several countries and the price for a packet of twenty is given below.

Britain	60p
Switzerland	SF 2·34
Spain	Ptas. 98·28
France	Fr. 4·55
Belgium	BF 34·20

Find in terms of British money the price in each country, then write a list of the prices in order, beginning with the cheapest.

10. An average rail fare for a journey of 200 kilometres is worked out for several countries and the results are given below.

Britain	£5·00
U.S.A.	$9·31
Switzerland	SF 17·28
Spain	Ptas. 702
Germany	DM 20·16
France	Fr. 47·32
Belgium	BF 294

Find the average fare for each country in terms of British money, then write a list of the average fares in order, beginning with the cheapest.

11. A businessman spent the day in Belgium. He changed £65 into francs, spent 2070 francs, and changed the remaining francs into £. How much British currency did he receive?

12. A man sends an order for car spares to the U.S.A. The value of his order is £247·25, but he is given a discount of $55·86.
   (a) What is the value of his order, in dollars?
   (b) What is the value of his discount, in pounds?

13. The director of a travel firm changes Ptas. 40 014 into pounds, and then into marks. How many marks does he receive?

14. A sailor changes £1280 into dollars at $1·96 = £1. Some time later he changes the same amount back into pounds, at $2·45 = £1. How much money has he lost?

15. A tourist changes £114·70 into French francs, and pays Fr. 34·58 to travel to his hotel. He buys goods costing Fr. 598·78 and pays a hotel bill of Fr. 222·04.
   His return journey to the airport costs Fr. 29·12 and he gives the taxi driver a tip of Fr. 2·73. On his return to Britain he changes his remaining francs to pounds.
   (a) How much does he receive?
   (b) If his air fare was £38·65, how much has his holiday cost altogether?
   (c) How much was his hotel bill in pounds, and how much was the taxi driver's tip?

When time is written in a.m. and p.m. notation, a.m. means *before noon,* and p.m.
means *after noon.* So, 4.30 a.m. means 4.30
in the morning, and 4.30 p.m. means 4.30 in
the afternoon.

$$1 \text{ hour} = 60 \text{ minutes}$$
$$1 \text{ day } = 24 \text{ hours}$$

To find the difference between one time and
another, subtract the earlier time from the
later time.

Example 1

A bus leaves a depot at 6.40 a.m. and returns
at 8.30 a.m. It remains in the depot for $\frac{3}{4}$ of
an hour before leaving again.
(a) How long does the first journey take?
(b) When does the bus leave the depot again?

(a)   7.90
       8̶.̶3̶0̶
     − 6.40
     ─────────
       1.50

the first journey takes 1 hour 50 minutes.

(b) $\frac{3}{4}$ of an hour $= \frac{3}{\overset{}{\underset{1}{4}}} \times \frac{\overset{15}{\cancel{60}}}{1} = 45$ minutes

       8.30
     + 0.45
     ─────────
       9.15

the bus leaves again at 9.15 a.m.

Exercise 5.8a

Find the difference between the times in each
question.

1. 4 a.m. and 10 a.m.
2. 6 p.m. and 11 p.m.
3. 1 a.m. and 11 a.m.
4. 4.30 a.m. and 10.45 a.m.
5. 7.15 p.m. and 11.35 p.m.
6. 10.50 p.m. and 11.55 p.m.
7. 6.25 a.m. and 11.37 a.m.
8. 9.04 a.m. and 10.42 a.m.
9. 11.32 a.m. and 11.50 a.m.
10. 8.14 p.m. and 10.20 p.m.
11. 9.16 a.m. and 10.15 a.m.

12. 6.40 p.m. and 9.25 p.m.
13. 7.15 p.m. and 10.02 p.m.
14. 2.05 a.m. and 8 a.m.
15. 6.44 a.m. and 10.30 a.m.
16. 9.55 p.m. and 11.17 p.m.
17. 3.13 a.m. and 9 a.m.
18. 11.14 p.m. and 11.31 p.m.
19. 1.21 a.m. and 11.10 a.m.
20. 5.05 p.m. and 10.02 p.m.

If a journey starts before midday and ends
after midday,
(i) find the time taken until midday,
(ii) add the later time.

If a journey is continued into the next day,
(i) find the time taken until midnight,
(ii) add the time taken after midnight.

Example 2

The sun rises one day at 4.30 a.m. and sets
at 6.15 p.m. It rises on the following day at
4.25 a.m.
(a) How long does daylight last on the first
    day?
(b) How long from sunset to sunrise?

(a)   11.60   7.30
       1̶2̶.̶0̶0̶   + 6.15
     − 4.30   ─────────
     ─────────   13.45
       7.30

∴ daylight lasts for 13 hours 45 minutes.

(b)   11.60   5.45
       1̶2̶.̶0̶0̶   + 4.25
     − 6.15   ─────────
     ─────────   10.10
       5.45

∴ from sunset to sunrise is 10 hours 10
  minutes.

Exercise 5.8b

Find the time taken for each journey.

1. A car leaves Glasgow at 5 a.m. and arrives in
   London at 3 p.m.
2. Two lorries leave Edinburgh at 10 a.m. and arrive
   in Ayr at 1.30 p.m.
3. A bus leaves Birmingham at 5 p.m. on Wednesday
   and arrives in Glasgow at 7 a.m. on Thursday.

4. A yacht leaves Troon Marina at 9.15 p.m. and arrives in Kip Marina at 11.45 a.m. the next day.
5. An aircraft leaves Gatwick at 3.40 a.m. and arrives at Abbotsinch at 1.05 p.m.
6. A cyclist leaves Dumfries at 11.05 a.m. and arrives in Carlisle at 3 p.m.
7. A motorcyclist leaves Thurso at 2.15 p.m. one day and arrives in Oxford at 6.05 a.m. the next day.
8. A scout leaves Newcastle at 10 a.m. on Friday and arrives in Manchester at 5 p.m. on Saturday.
9. A man leaves Stranraer at 8.45 a.m. and arrives in Dublin at 10.05 p.m.
10. A girl leaves Monkton at 10.54 a.m. and arrives in Straiton at 1.06 p.m.
11. Four men leave Dover at 5.15 a.m. one Thursday and arrive in Edinburgh at 6.52 p.m. the following day.
12. A boat leaves St. Ives at 6.15 p.m. and arrives in Falmouth at 11.40 p.m. the following day.
13. A mountaineer leaves camp at 4.40 a.m. on Wednesday and returns at 5.12 p.m. on Thursday.
14. A train leaves York at 11.15 p.m. on Friday and arrives in Glasgow at midday on Saturday.
15. Two women leave a village at 10.20 a.m. on Tuesday and arrive home at 11.15 p.m. on Friday.

16. 10.00 p.m.	17. 10.24 p.m.	18. 11.45 p.m.
19. 11.05 p.m.	20. 8.00 a.m.	21. 6.00 a.m.
22. 9.15 a.m.	23. 8.42 a.m.	24. 7.35 a.m.
25. 6.08 a.m.	26. 9.03 a.m.	27. 10.00 a.m.
28. 10.15 a.m.	29. 11.55 a.m.	30. 11.32 a.m.

Example 4

Write these times as 12-hour times
(a) 01.00 h (b) 16.27 h (c) 23.59 h
(a) 01.00 h = 1 a.m.
(b) 16.27 h = 4.27 p.m.
(c) 23.59 h = 11.59 p.m.

Exercise 5.8d

Write the following as 12-hour times

1. 13.00 h	2. 16.00 h	3. 19.00 h
4. 13.30 h	5. 13.50 h	6. 19.35 h
7. 19.18 h	8. 16.25 h	9. 16.05 h
10. 18.40 h	11. 18.12 h	12. 15.55 h
13. 14.36 h	14. 21.00 h	15. 21.45 h
16. 23.00 h	17. 23.30 h	18. 22.45 h
19. 22.06 h	20. 09.00 h	21. 07.00 h
22. 08.45 h	23. 09.53 h	24. 08.56 h
25. 07.02 h	26. 08.09 h	27. 11.00 h
28. 11.25 h	29. 10.50 h	30. 10.36 h

Timetables generally use the 24-hour clock system instead of a.m. and p.m. In this system, all times are written using four figures. The first two figures indicate hours and the last two indicate minutes past the hour. All p.m. times have 12 hours added to them.

Example 3

Write these times as 24-hour clock times:
(a) 8 a.m. (b) 10.27 a.m. (c) 1.05 p.m.
(a) 8 a.m. = 08.00 h
(b) 10.27 a.m. = 10.27 h
(c) 1.05 p.m. = 13.05 h

Exercise 5.8c

Write the following as 24-hour clock times.

1. 2.00 p.m.	2. 5.00 p.m.	3. 6.00 p.m.
4. 2.15 p.m.	5. 2.40 p.m.	6. 5.20 p.m.
7. 5.32 p.m.	8. 6.30 p.m.	9. 6.48 p.m.
10. 7.50 p.m.	11. 7.05 p.m.	12. 4.35 p.m.
13. 8.00 p.m.	14. 8.30 p.m.	15. 9.15 p.m.

If time is written using the 24-hour clock system, then the difference between two times is easily found by subtraction.

Example 5

A bus leaves Dunoon at 19.35 h and arrives in Inveraray at 21.15 h. How long does the journey take?

$$\begin{array}{r} 20.75 \\ \cancel{21.15} \\ - 19.35 \\ \hline 1.40 \end{array}$$

∴ time taken is 1 h 40 min.

Exercise 5.8e

Find the time of each journey for questions 1 to 10.

1. A train leaves Crianlarich at 08.10 h and arrives in Roy Bridge at 09.55 h.
2. A bus leaves Hull at 11.05 h and arrives in Bridlington at 12.40 h.

3. A bus leaves Aberdeen at 14.15 h and arrives in Aboyne at 15.30 h.

4. A train leaves Liverpool at 16.20 h and arrives in Crewe at 16.54 h.

5. A bus leaves Elgin at 11.30 h and arrives in Aviemore at 13.55 h.

6. A train leaves Glasgow at 12.25 h and arrives in Aberdeen at 15.55 h.

7. A train leaves Inverness at 10.30 h and arrives in Achnasheen at 12.20 h.

8. A bus leaves Birmingham at 15.30 h and arrives in Worcester at 17.15 h.

9. A bus leaves Fraserburgh at 08.40 h and arrives in Maud at 09.25 h.

10. A train leaves London (King's Cross) at 11.20 h and arrives in Leeds at 14.05 h.

11. Find the time taken by the train from London to each station.

London (Waterloo)	(depart)	09.05 h
Southampton	(arrive)	10.17 h
Bournemouth	(arrive)	10.50 h
Poole	(arrive)	11.05 h
Weymouth	(arrive)	11.45 h

12. Find the time taken by the train from Perth to each station.

Perth	(depart)	01.10 h
Pitlochry	(arrive)	01.53 h
Kingussie	(arrive)	03.24 h
Aviemore	(arrive)	03.39 h
Inverness	(arrive)	04.45 h

13. Find the time taken by the train from London to each station.

London (Paddington)	(depart)	12.30 h
Taunton	(arrive)	14.42 h
Exeter	(arrive)	15.10 h
Plymouth	(arrive)	16.15 h
Penzance	(arrive)	18.30 h

14. Bus timetable

Torranhill–Stanebrig				
From Torranhill				
Torranhill	08.10	11.45	13.35	17.40
Barlannie	08.24	11.59	13.49	17.54
Leebank	08.28	12.03	13.53	17.58
Dunwhin	08.36	12.11	14.01	18.06
Stanebrig	08.50	12.25	14.15	18.20
From Stanebrig				
Stanebrig	07.20	10.35	14.30	18.25
Dunwhin	07.34	10.49	14.44	18.39
Leebank	07.42	10.57	14.52	18.47
Barlannie	07.46	11.01	14.56	18.51
Torranhill	08.00	11.15	15.10	19.05

(a) When (in 12-hour time) does the first bus after 12 noon leave Torranhill for Stanebrig? When does it arrive in Stanebrig? How long does the journey take?

(b) How long does it take to go from Barlannie to Stanebrig?

(c) A man who lives at Leebank works in Stanebrig between 9.30 a.m. and 5.30 p.m.
   (i) At what time does he catch the bus in the morning?
   (ii) At what time does he arrive in Stanebrig?
   (iii) How long does his journey take?
   (iv) At what time does he catch the bus after work?
   (v) How long after he finishes work does the bus depart?

(d) I arrive in Stanebrig 35 minutes before the last bus leaves for Torranhill. When do I arrive in Stanebrig?

(e) If I arrive in Barlannie at 11.15 h, by how many minutes have I missed a bus to Torranhill? How long is it before the next one?

15. Bus timetable

Stranraer–Drummore			
From Stranraer			
Stranraer	07.15	12.25	17.30
Stoneykirk	07.27	12.37	17.42
Sandhead	07.32	12.42	17.47
Ardwell	07.40	12.50	17.55
Port Logan	07.48	12.58	18.03
Drummore	08.00	13.10	18.15
From Drummore			
Drummore	08.00	13.15	18.20
Port Logan	08.12	13.27	18.32
Ardwell	08.20	13.35	18.40
Sandhead	08.28	13.43	18.48
Stoneykirk	08.33	13.48	18.53
Stranraer	08.45	14.00	19.05

(a) I travel on the first bus from Stranraer to Drummore, but I alight at Ardwell to visit a friend. I travel on to Drummore on the next bus and finally return to Stranraer on the last bus of the day.
   (i) How long do I have to visit my friend?
   (ii) How long after leaving Stranraer do I eventually reach Drummore?
   (iii) How long do I spend in Drummore before finally returning to Stranraer?
   (iv) At what time (in 12-hour time) do I arrive back in Stranraer?

(b) A lady who lives in Sandhead has a dentist's appointment in Stranraer from 11.15 to 11.45 a.m.
  (i) At what time does she catch the bus to Stranraer?
  (ii) At what time does she arrive there?
  (iii) How long does her journey take?
  (iv) How long does she have in Stranraer before her appointment?
  (v) At what time does she catch the bus home?
  (vi) How long after 11.45 a.m. is she in Stranraer before catching the bus home?
  (vii) At what time, on the 12-hour clock, does she arrive back at Sandhead?

(c) I travel from Stranraer on the last bus as far as Port Logan.
  (i) How long does the journey take?
  (ii) At what time, on the 12-hour clock, do I arrive at Port Logan?

16. Bus timetable

Brigg–Worlaby–Barton					
From Brigg					
Brigg	07.15	08.10	10.30	13.55	21.45
Wrawby	07.20	08.15	10.35	14.00	21.50
Worlaby	07.30	08.25	10.45	14.10	22.00
Ferriby	07.43	08.38	10.58	14.23	22.13
Barton	07.50	08.45	11.05	14.30	22.20
From Barton					
Barton	08.00	09.00	12.15	16.20	22.30
Ferriby	08.07	09.07	12.22	16.27	22.37
Worlaby	08.20	09.20	12.35	16.40	22.50
Wrawby	08.30	09.30	12.45	16.50	23.00
Brigg	08.35	09.35	12.50	16.55	23.05

(a) When (in 12-hour time) does the last bus leave Brigg for Barton? How long does the journey take?
(b) How long does it take to go from Brigg to Worlaby?
(c) A man lives in Wrawby and travels to and from work in Barton by bus.
  (i) If he starts work at 9 a.m. at what time does he catch the bus?
  (ii) At what time does he arrive in Barton?
  (iii) How long does the journey take?
  (iv) If he finishes work at 4.15 p.m. at what time does he catch the bus home?
  (v) At what time does he arrive back in Wrawby?

(d) If the 13.55 bus is 5 minutes late leaving Brigg, at what time will it reach Worlaby? If owing to road works at Ferriby the bus does not arrive in Barton until 14.40, how long has the journey from Worlaby taken?
(e) The 12.15 bus from Barton arrives in Brigg 15 minutes late. At what time does it reach Brigg? How long has a passenger on this bus to wait in Brigg to catch the next train to London which leaves at twenty to five?

Thirty days hath September,
April, June and November.
All the rest have thirty-one,
except February alone
which has twenty-eight days clear,
and twenty-nine in each leap year.

1 year = 52 weeks
1 year = 365 days
1 leap year (every fourth year) = 366 days

When subtracting one date from another within the same month, one of the days (first or last) is not included.
e.g. Jan 4th to Jan 7th = 7 − 4 = 3 days
    Including both dates; Jan 4th to Jan 7th
                  = 3 + 1 = 4 days.

Example 6

Find the number of days from 3rd December 1968 to 4th March 1969
(a) excluding both given days
(b) including both given days.

(a) 3rd Dec to 31st Dec
  (excluding 3rd Dec) = 31 − 3 = 28 days
                 January = 31 days
              February = 28 days
  1st March to 4th March
  (excluding 4th March) = 4 − 1 =   3 days
                 total = 90 days

There are 90 days between 3rd Dec 1968 and 4th March 1969 (excluding both given days).

(b) Number of days, excluding those given

$$= 90 \text{ days}$$
$$\text{3rd Dec} = 1 \text{ day}$$
$$\text{4th March} = 1 \text{ day}$$
$$\text{total} = \overline{92} \text{ days}$$

There are 92 days from 3rd Dec 1968 to 4th March 1969, inclusive of both days.

Exercise 5.8f

In each question, find the number of days from one date to the other
(a) excluding both given days
(b) including both given days.

1. Jan 5th 1967 and Feb 8th 1967
2. April 4th 1950 and May 7th 1950
3. June 17th 1970 and Aug 3rd 1970
4. Sept 15th 1932 and Nov 19th 1932
5. Feb 12th 1966 and May 3rd 1966
6. July 21st 1958 and Sept 30th 1958
7. Aug 19th 1970 and Oct 23rd 1970
8. June 21st 1980 and July 30th 1980
9. Feb 18th 1979 and April 5th 1979
10. Aug 4th 1923 and Nov 17th 1923
11. Oct 5th 1969 and Jan 16th 1970
12. March 17th 1972 and Feb 3rd 1973
13. Oct 11th 1982 and March 7th 1983
14. July 15th 1981 and Aug 10th 1982
15. Feb 18th 1977 and May 19th 1978
16. Oct 1st 1973 and Dec 12th 1974
17. May 5th 1972 and July 22nd 1973
18. Oct 23rd 1979 and Jan 7th 1980
19. Dec 2nd 1990 and Jan 30th 1992
20. Sept 3rd 1997 and Feb 1st 2000

Except for century years which end in 00, every year which is divisible by 4 is a leap year.
Century years which are divisible by 400 are leap years.

Example 7

Which of the following are leap years?
(a) 1980   (b) 2000   (c) 1642

(a) $1980 \div 4 = 495$, so 1980 is divisible by 4.
    1980 is a leap year.

(b) $2000 \div 400 = 5$, so 2000 is a century year divisible by 400.
    2000 is a leap year.

(c) 1642 is not divisible by 4
    1642 is not a leap year.

Exercise 5.8g

State which of the following are leap years.

1. 2020	2. 1979	3. 1984	4. 1920
5. 1812	6. 1930	7. 1600	8. 1740
9. 1880	10. 1706	11. 1830	12. 1723
13. 1612	14. 2700	15. 1902	16. 1696
17. 3800	18. 2800	19. 1952	20. 2400

5.9 SPEED, TIME AND DISTANCE

Average speed is the ratio of the distance travelled to the time taken. It is usually calculated in kilometres per hour (km/h).

Example 1

A car travels 144 km in $2\frac{1}{4}$ h. What is the average speed?

$$\text{Average speed} = \frac{\text{distance}}{\text{time}}$$

$$= \frac{144}{2\frac{1}{4}}$$

$$= \frac{\overset{16}{\cancel{144}} \times 4}{\underset{1}{\cancel{9}}} = 64 \text{ km/h}.$$

Exercise 5.9a

Find the average speed for the following air flights:

1. London to Gibraltar, 1896 km in 3 h.
2. London to Rome, 1464 km in 3 h.
3. London to Montreal, 5040 km in 7 h.
4. London to New York, 5184 km in 6 h.
5. London to Havana, 6952 km in 11 h.

Find the average speed for the following sea voyages:

6. Hull to Hamburg, 600 km in 15 h.
7. Hull to Rotterdam, 336 km in 14 h.
8. Liverpool to New York, 4800 km in 5 days.
9. Liverpool to Monte Video, 10 080 km in 10 days.
10. Southampton to Lisbon, 2112 km in 2 days.

Find the average speed for the following journeys by road:

11. Edinburgh to Dundee, 96 km in $1\frac{1}{2}$ h.
12. Glasgow to Dunblane, 60 km in $1\frac{1}{4}$ h.
13. Inverness to Dalwhinnie, 91 km in $1\frac{3}{4}$ h.
14. Glasgow to Stranraer, 150 km in $2\frac{1}{2}$ h.
15. Edinburgh to Aberdeen, 182 km in $3\frac{1}{2}$ h.
16. Kirkcaldy to Carlisle, 190 km in 2 h 30 min.
17. Lincoln to Northampton, 153 km in 2 h 15 min.
18. Glasgow to Kirkcaldy, 84 km in 1 h 20 min.
19. London to Stranraer, 646 km in 9 h 30 min.
20. Doncaster to Inverness, 625 km in 8 h 20 min.

Example 2

A lorry leaves Glasgow at noon and arrives at Ayr harbour at 1.18 p.m. If the distance is 52 km, what is the average speed of the lorry?

Distance = 52 km
Time = 1 h 18 min = $1\frac{18}{60}$ h

Average speed $= \dfrac{\text{distance}}{\text{time}} = \dfrac{52}{1\frac{18}{60}}$

$= 52 \div 1\frac{18}{60}$

$= 52 \div \frac{78}{60}$

$= \dfrac{\overset{4}{\cancel{52}}}{1} \times \dfrac{\overset{10}{\cancel{60}}}{\underset{1}{\cancel{\underset{6}{78}}}} = 40$ km/h

Exercise 5.9b

1. A train leaves St. Andrews at 9.00 a.m. and arrives in Aberdeen at 10.20 a.m. If the distance is 120 km, find the average speed.
2. A bus leaves Glasgow at 10.00 a.m. and arrives in Kilmarnock at 11.10 a.m. If the distance is 42 km, find the average speed.
3. A lorry leaves a Glasgow warehouse at 8.30 a.m. and arrives at a factory in Ayr at 9.36 a.m. If the distance is 55 km, find the average speed.
4. A car leaves Oban at 2.00 p.m. and arrives at Fort William, 80 km away, at 3.40 p.m. Find the average speed.
5. A train leaves Stirling at 8.00 a.m. and travels 231 km to Aberdeen by 10.20 a.m. Find the average speed.
6. A train leaves London at 5.00 p.m. and arrives in Liverpool at 7.40 p.m. If the distance is 312 km, find the average speed.

7. A boat leaves Stranraer at 6.30 p.m. and reaches Larne at 8.35 p.m. having sailed a distance of 50 km. Find the average speed.
8. A bus leaves Newcastle at 7.00 a.m. and travels 27 km to Durham, arriving there at 7.54 a.m. Find the average speed.
9. A car leaves Belmullet at 6.00 p.m. and arrives in Galway at 7.48 p.m. If the car travels 153 km between the two towns find its average speed.
10. A bus leaves Fraserburgh at 9.05 a.m. and arrives in Aberdeen at 10.55 a.m. If the distance is 66 km, find the average speed.

The time taken for a journey is the total distance divided by the average speed.

Example 3

A van travels 363 km at an average speed of 66 km/h. How long does the journey take?

Time taken $= \dfrac{\text{distance}}{\text{speed}} = \dfrac{\overset{11}{\cancel{363}}}{\underset{2}{\cancel{66}}}$

$= \dfrac{11}{2}$ hours $= 5\frac{1}{2}$ hours

$= 5$ hours 30 minutes.

Exercise 5.9c

Find the time taken for the following journeys:

1. Aberdeen to Glasgow, 250 km at 50 km/h.
2. Aberdeen to Oxford, 760 km at 40 km/h.
3. Aberdeen to Braemar, 96 km at 6 km/h.
4. Dover to Barcelona, 3650 km at 25 km/h.
5. Dumfries to Lochguilphead, 256 km at 16 km/h.
6. Perth to Elgin, 186 km at 12 km/h.
7. Exeter to London, 273 km at 42 km/h.
8. St. Andrews to Inverness, 225 km at 18 km/h.
9. Thurso to Aberdeen, 260 km at 80 km/h.
10. Ayr to Kirkcaldy, 132 km at 48 km/h.
11. Lima to Santiago, 2400 km at 450 km/h.
12. Mallaig to Braemar, 256 km at 80 km/h.
13. Rotterdam to Strasbourg, 660 km at 99 km/h.
14. Inverness to Dover, 972 km at 60 km/h.
15. Edinburgh to Newcastle, 169 km at 78 km/h.
16. Campbeltown to Carlisle, 368 km at 48 km/h.
17. 240 cm at 75 cm/s.
18. 189 m at 84 m/s.
19. 533 mm at 6·5 mm/min.
20. 4·5 mm at 3·6 mm/s.

Distance travelled is found by multiplying the average speed by the time taken.

Example 4

An aircraft flies at a speed of 654 km/h for 2 h 10 min. How far does it fly?

Distance = speed × time = $654 \times 2\frac{10}{60}$

$$= \frac{\overset{109}{\cancel{654}}}{1} \times \frac{\overset{13}{\cancel{130}}}{\underset{1}{\cancel{60}}}$$

$$= 109 \times 13 = 1417 \text{ km.}$$

Exercise 5.9d

Find the distance travelled during each journey:

1. Fraserburgh to Dornoch, 4 h at 60 km/h.
2. London to Campbeltown, 12 h at 71 km/h.
3. London to Braemar, 12 h at 65 km/h.
4. London to Lochguilphead, 14 h at 55 km/h.
5. London to Perth, 7 h at 96 km/h.
6. Stirling to Elgin, 3 h 30 min at 70 km/h.
7. Cork to Larne, 5 h 15 min at 84 km/h.
8. Singapore to Hong Kong, 10 h 45 min at 240 km/h.
9. Dover to Colwyn Bay, 6 h 20 min at 75 km/h.
10. Zeebrugge to Vienna 3 h 10 min at 654 km/h.
11. Hong Kong to Wake Island, 11 h 12 min at 550 km/h.
12. Ayr to Galashiels, 1 h 40 min at 87 km/h.
13. Leeds to London, 6 h 48 min at 45 km/h.
14. Athens to Khartoum, 4 h 3 min at 880 km/h.
15. Edinburgh to Southampton, 2 h 1 min at 360 km/h.
16. 8 h 50 min at 426 m/h.
17. 5 h 18 min at 20 cm/h.
18. 9 h 55 min at 972 m/h.
19. 3 h 30 min at 3.6 km/h.
20. 14 h 5 min at 144 m/h.

5.10 FLOW CHARTS

A flow chart contains instructions which must be carried out in order.

Exercise 5.10

1. A shopkeeper uses the flow chart below to calculate the total cost of items in a sale.
   Find the total cost of
   (a) a television : basic price £80
   (b) a coffee table : basic price £50
   (c) a carpet : basic price £120.

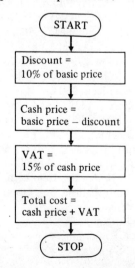

2. A road haulage firm uses the flow chart below to calculate the carriage charge for each item they transport.
   Find the charge for packages weighing
   (a) 4 kg
   (b) 16 kg
   (c) 12·5 kg

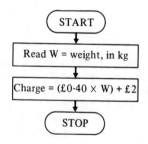

3. This flow chart allows a sales manager to calculate the weekly wage of his salesmen. Find the wage earned by a salesman who makes sales of
(a) £900
(b) £1200
(c) £2500
(d) £1000

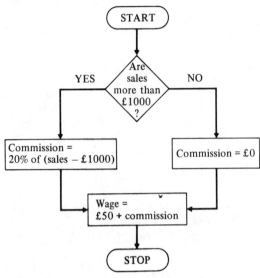

4. The flowchart below shows how to calculate the insurance premium for photographic equipment. Find the premium for equipment of value
(a) £250
(b) £550
(c) £560
(d) £760

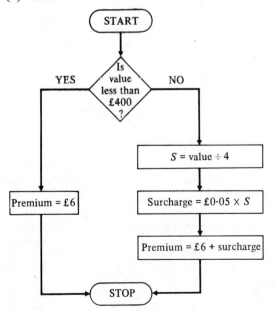

5. The flowchart shows one way of finding whether a number is divisible by 3. Use the flowchart to find whether the following numbers are divisible by 3.
(a) 144
(b) 275
(c) 342
(d) 2151

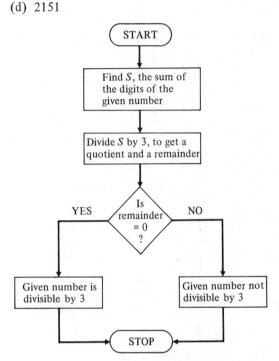

6. A personnel officer calculates labourers' weekly wages, using the following flowchart.
Find the weekly wage of a labourer aged
(a) 17 (b) 24 (c) 42 (d) 25 (e) 18

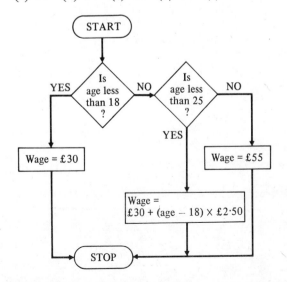

7. A milk marketing board uses a flowchart to
   calculate the allowance made to each farm, for
   new packing machinery.
   Find the allowance made to a farm with a weekly
   output of
   (a) 600 *l* (b) 1000 *l* (c) 1850 *l* (d) 2000 *l*
   (e) 2700 *l*

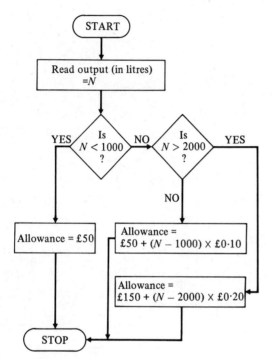

8. Use the flowchart to convert the following base
   ten numbers to base two.
   (a) 7 (b) 14 (c) 27 (d) 32 (e) 109

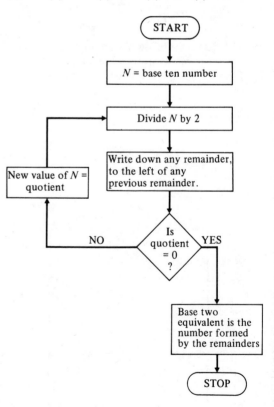

9. The flowchart shows how to estimate the population of a region.
   Estimate the population in the following cases.

	(a)	(b)	(c)
Current population	10 000	200 000	4 000 000
Number of years growth	2	3	4

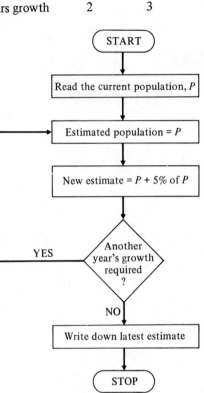

10. This flowchart shows how to calculate the bus fare in a city.
    Find the fare required from passengers making journeys of
    (a) 4 fare stages, at 5.00 am
    (b) 3 fare stages, at 4.30 pm
    (c) 9 fare stages, at 7.15 pm
    (d) 2 fare stages, at 1.30 pm.

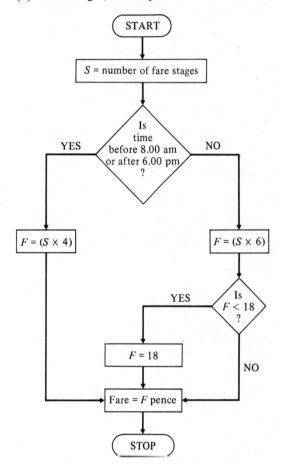

The square of $3 = 3^2 = 3 \times 3 = 9$

The square of $40 = 40^2 = 40 \times 40 = 1600$

The square of $9000 = 9000^2 = 9000 \times 9000$
$$= 81\ 000\ 000$$

The square of $0.05 = 0.05^2 = 0.05 \times 0.05$
$$= 0.0025$$

The square of $2\frac{1}{4} = (2\frac{1}{4})^2 = 2\frac{1}{4} \times 2\frac{1}{4}$
$$= \frac{9}{4} \times \frac{9}{4} = \frac{81}{16} = 5\frac{1}{16}$$

Exercise 6.1a

Find the squares of the following.

1. 2	**2.** 5	**3.** 9	**4.** 20
5. 60	**6.** 70	**7.** 400	**8.** 800
9. 100	**10.** 2000	**11.** 5000	**12.** 6000
13. 0·5	**14.** 0·8	**15.** 0·1	**16.** 0·03
17. 0·02	**18.** 0·06	**19.** $\frac{1}{4}$	**20.** $\frac{1}{7}$
21. $\frac{1}{12}$	**22.** $\frac{3}{4}$	**23.** $\frac{2}{5}$	**24.** $\frac{4}{11}$
25. $1\frac{1}{5}$	**26.** $1\frac{1}{3}$	**27.** $1\frac{1}{4}$	**28.** $1\frac{2}{3}$
29. $2\frac{1}{5}$	**30.** $3\frac{1}{3}$		

The square root of $9 = \sqrt{9} = 3$
$$\text{because } 3 \times 3 = 9$$

The square root of $25 = \sqrt{25} = 5$
$$\text{because } 5 \times 5 = 25$$

The square root of $2\frac{1}{4} = \sqrt{2\frac{1}{4}}$
$$= \sqrt{\frac{9}{4}}$$
$$= \frac{\sqrt{9}}{\sqrt{4}} = \frac{3}{2} = 1\frac{1}{2}$$
$$\text{because } 1\frac{1}{2} \times 1\frac{1}{2} = 2\frac{1}{4}$$

Exercise 6.1b

Find the square roots of the following.

1. 4	**2.** 16	**3.** 49	**4.** 64
5. 36	**6.** 100	**7.** 81	**8.** 144
9. 121	**10.** 1	**11.** $\frac{1}{9}$	**12.** $\frac{1}{64}$
13. $\frac{1}{121}$	**14.** $\frac{9}{25}$	**15.** $\frac{49}{64}$	**16.** $\frac{25}{144}$
17. $6\frac{1}{4}$	**18.** $7\frac{1}{9}$	**19.** $3\frac{1}{16}$	**20.** $1\frac{24}{25}$

The square roots of much larger or smaller numbers can be found by factorisation. Write the given number as either a product or a fraction using numbers whose square roots are known.

Example 1

(a) $\sqrt{400} = \sqrt{4 \times 100} = \sqrt{4} \times \sqrt{100}$
$$= 2 \times 10 = 20$$

(b) $\sqrt{360\ 000} = \sqrt{36 \times 100 \times 100}$
$$= \sqrt{36} \times \sqrt{100} \times \sqrt{100}$$
$$= 6 \times 10 \times 10$$
$$= 600$$

(c) $\sqrt{0.04} = \sqrt{\dfrac{4}{100}} = \dfrac{\sqrt{4}}{\sqrt{100}}$
$$= \frac{2}{10} = 0.2$$

(d) $\sqrt{0.0081} = \dfrac{\sqrt{81}}{\sqrt{10000}} = \dfrac{\sqrt{81}}{\sqrt{100 \times 100}}$
$$= \frac{9}{10 \times 10} = \frac{9}{100}$$
$$= 0.09$$

(e) $\sqrt{225} = \sqrt{25 \times 9}$
$$= \sqrt{25} \times \sqrt{9}$$
$$= 5 \times 3 = 15$$

Exercise 6.1c

Find the square roots of the following by using factorisation.

1. 900	**2.** 2500
3. 3600	**4.** 810 000
5. 490 000	**6.** 250 000
7. 12 100	**8.** 1 000 000
9. 40 000	**10.** 640 000
11. 0·09	**12.** 0·04
13. 0·36	**14.** 0·16
15. 0·64	**16.** 0·49
17. 0·25	**18.** 0·81
19. 0·0025	**20.** 0·0064
21. 0·0016	**22.** 0·0001
23. 0·0144	**24.** 196
25. 256	**26.** 324
27. 441	**28.** 729

A *perfect square* is the square of a whole number.

e.g.

4 is a perfect square, because $4 = 2^2$

81 is a perfect square, because $81 = 9^2$

Every number lies between two consecutive perfect squares.

e.g.

8 lies between 4 ($= 2^2$) and 9 ($= 3^2$)

i.e.   4, 8, 9

1·7 lies between 1 ($= 1^2$) and 4 ($= 2^2$)

i.e   1, 1·7, 4

Example 2

Find two consecutive perfect squares between which the following lie.

(a) 24   (b) 39   (c) 7·3

(a) $4^2 = 16$, and $5^2 = 25$,

so 24 lies between 16 and 25

i.e.   16, 24, 25

(b) $6^2 = 36$, and $7^2 = 49$,

so 39 lies between 36 and 49

i.e.   36, 39, 49

(c) $2^2 = 4$, and $3^2 = 9$,

so 7·3 lies between 4 and 9

i.e.   4, 7·3, 9

Exercise 6.1d

Find two consecutive perfect squares between which the following lie.

1. 3	2. 5	3. 7
4. 2	5. 10	6. 14
7. 17	8. 23	9. 12
10. 29	11. 18	12. 39
13. 42	14. 51	15. 37
16. 63	17. 70	18. 21
19. 2·2	20. 3·1	21. 9·4
22. 11·6	23. 17·3	24. 12·9
25. 21·8	26. 42·2	27. 48·9
28. 63·7	29. 26·2	30. 80·4

To estimate the square root of a number, first find the nearest perfect square.

Example 3

Estimate (a) $\sqrt{24}$ (b) $\sqrt{39}$ (c) $\sqrt{6·2}$

(a) 24 lies between the perfect squares 16 and 25.

Since 24 is nearest to 25,

$\sqrt{24} \approx 5$ (because $\sqrt{25} = 5$)

(b) 39 lies between the perfect squares 36 and 49.

Since 39 is nearest to 36,

$\sqrt{39} \approx 6$ (because $\sqrt{36} = 6$)

(c) 6·2 lies between the perfect squares 4 and 9.

Since 6·2 is nearest to 4,

$\sqrt{6·2} \approx 2$ (because $\sqrt{4} = 2$)

Exercise 6.1e

Estimate the square root of each of the following.

1. 3	2. 10	3. 15
4. 8	5. 14	6. 18
7. 23	8. 19	9. 26
10. 35	11. 28	12. 38
13. 48	14. 50	15. 47
16. 52	17. 62	18. 66
19. 3·5	20. 1·2	21. 8·1
22 9·3	23. 7·9	24. 10·2
25. 15·4	26. 16·3	27. 23·4
28. 25·9	29. 35·7	30. 66·6

To estimate the square roots of much larger or smaller numbers, use factorisation. Convenient factors may be 100 or 10 000.

Example 4

Estimate (a) $\sqrt{928}$ (b) $\sqrt{352\,000}$
(c) $\sqrt{0·15}$ (d) $\sqrt{0·0082}$

(a) $\sqrt{928} = \sqrt{9·28 \times 100} = \sqrt{9·28} \times \sqrt{100}$

Replace 9·28 by the nearest perfect square

$\therefore \sqrt{900} = \sqrt{9} \times \sqrt{100} = 3 \times 10$

$= 30$

So, $\sqrt{928} \approx 30$

(b) $\sqrt{352\,000} = \sqrt{35{\cdot}2 \times 10\,000}$
$= \sqrt{35{\cdot}2 \times 100 \times 100}$
$= \sqrt{35{\cdot}2} \times \sqrt{100} \times \sqrt{100}$

Replace 35·2 by the nearest perfect square

i.e. $\sqrt{360\,000} = \sqrt{36} \times \sqrt{100} \times \sqrt{100}$
$= 6 \times 10 \times 10 = 600$

so, $\sqrt{352\,000} \approx 600$

(c) $\sqrt{0{\cdot}15} = \sqrt{\dfrac{15}{100}} = \dfrac{\sqrt{15}}{\sqrt{100}}$

Replace 15 by the nearest perfect square

i.e. $\sqrt{0{\cdot}16} = \dfrac{\sqrt{16}}{\sqrt{100}} = \dfrac{4}{10} = 0{\cdot}4$

so, $\sqrt{0{\cdot}15} \approx 0{\cdot}4$

(d) $\sqrt{0{\cdot}0082} = \sqrt{\dfrac{82}{10\,000}} = \dfrac{\sqrt{82}}{\sqrt{100} \times \sqrt{100}}$

Replace 82 by the nearest perfect square

i.e. $\sqrt{0{\cdot}0081} = \dfrac{\sqrt{81}}{\sqrt{100 \times 100}} = \dfrac{9}{10 \times 10}$
$= \dfrac{9}{100} = 0{\cdot}09$

so, $\sqrt{0{\cdot}0082} \approx 0{\cdot}09$

Exercise 6.1f

Estimate the square root of each of the following.

1. 420	2. 940	3. 3700
4. 2400	5. 1500	6. 6300
7. 5000	8. 3628	9. 8094
10. 41 000	11. 920 000	12. 110 000
13. 260 000	14. 340 000	15. 500 000
16. 660 000	17. 171 000	18. 143 700
19. 0·08	20. 0·17	21. 0·67
22. 0·33	23. 0·51	24. 0·821
25. 0·0017	26. 0·0008	27. 0·0035
28. 0·0023	29. 0·0062	30. 0·00158

6.2 CUBES AND CUBE ROOTS

The cube of $2 = 2^3 = 2 \times 2 \times 2 = 8$
The cube of $40 = 40^3 = 40 \times 40 \times 40$
$= 64\,000$
The cube of $0{\cdot}3 = 0{\cdot}3^3 = 0{\cdot}3 \times 0{\cdot}3 \times 0{\cdot}3$
$= 0{\cdot}027$
The cube of $1\frac{1}{4} = (1\frac{1}{4})^3 = 1\frac{1}{4} \times 1\frac{1}{4} \times 1\frac{1}{4}$
$= \frac{5}{4} \times \frac{5}{4} \times \frac{5}{4} = \frac{125}{64} = 1\frac{61}{64}$

Exercise 6.2a

Find the cubes of the following.

1. 3	2. 4	3. 5	4. 7
5. 9	6. 10	7. 12	8. 20
9. 100	10. 300	11. 0·1	12. 0·2
13. 0·5	14. 0·6	15. 0·9	16. 0·003
17. $\frac{2}{3}$	18. $\frac{3}{5}$	19. $\frac{6}{7}$	20. $\frac{9}{10}$
21. $\frac{4}{11}$	22. $\frac{3}{4}$	23. $1\frac{1}{2}$	24. $2\frac{1}{4}$

The cube root of $27 = \sqrt[3]{27} = 3$
because $3 \times 3 \times 3 = 27$
The cube root of $125 = \sqrt[3]{125} = 5$
because $5 \times 5 \times 5 = 125$

The cube root of $3\frac{3}{8} = \sqrt[3]{3\frac{3}{8}} = \sqrt[3]{\frac{27}{8}}$
$= \dfrac{\sqrt[3]{27}}{\sqrt[3]{8}} = \frac{3}{2} = 1\frac{1}{2}$

because $1\frac{1}{2} \times 1\frac{1}{2} \times 1\frac{1}{2} = 3\frac{3}{8}$

Exercise 6.2b

Find the cube roots of the following.

1. 8	2. 1	3. 64	4. 1000
2. 216	6. 343	7. $\frac{1}{8}$	8. $\frac{1}{64}$
9. $\frac{1}{27}$	10. $\frac{1}{216}$	11. $\frac{1}{125}$	12. $\frac{8}{27}$
13 $\frac{27}{64}$	14. $\frac{64}{125}$	15. $\frac{125}{216}$	16. $2\frac{10}{27}$
17. $4\frac{17}{27}$	18. $1\frac{61}{64}$	19. $15\frac{5}{8}$	20. $1\frac{91}{125}$

The cube roots of much larger or smaller numbers can be found by factorisation. Write the given number either as a product or a fraction, using numbers whose cube roots are know.

Example 1

(a) $\sqrt[3]{27\,000} = \sqrt[3]{27 \times 1000} = \sqrt[3]{27} \times \sqrt[3]{1000}$
$\qquad = 3 \times 10 = 30$

(b) $\sqrt[3]{8\,000\,000} = \sqrt[3]{8 \times 1000 \times 1000}$
$\qquad = \sqrt[3]{8} \times \sqrt[3]{1000} \times \sqrt[3]{1000}$
$\qquad = 2 \times 10 \times 10 = 200$

(c) $\sqrt[3]{0 \cdot 027} = \sqrt[3]{\dfrac{27}{1000}} = \dfrac{\sqrt[3]{27}}{\sqrt[3]{1000}}$

$\qquad = \frac{3}{10} = 0 \cdot 3$

(d) $\sqrt[3]{0 \cdot 000064} = \sqrt[3]{\dfrac{64}{1\,000\,000}}$

$\qquad = \dfrac{\sqrt[3]{64}}{\sqrt[3]{1000} \times \sqrt[3]{1000}}$

$\qquad = \dfrac{4}{10 \times 10}$

$\qquad = \dfrac{4}{100} = 0 \cdot 04$

(e) $\sqrt[3]{1728} = \sqrt[3]{8 \times 216} = \sqrt[3]{8 \times 8 \times 27}$

$\qquad = \sqrt[3]{8} \times \sqrt[3]{8} \times \sqrt[3]{27}$

$\qquad = 2 \times 2 \times 3 = 12$

Exercise 6.2c

Find the cube roots of the following by using factorisation.

1. 8 000	2. 64 000
3. 125 000	4. 27 000 000
5. 1 000 000 000	6. 343 000
7. 8 000 000	8. 216 000 000
9. 0·008	10. 0·064
11. 0·125	12. 0·216
13. 0·001	14. 0·343
15. 0·000 008	16. 0·000 001

17. 0·000 125	18. 0·000 027
19. 0·000 216	20. 0·000 343
21. 512	22. 2744
23. 5832	24. 4096
25. 13824	

A *perfect cube* is the cube of a whole number. e.g.

8 is a perfect cube, because $8 = 2^3$
125 is a perfect cube, because $125 = 5^3$

Every number lies between two consecutive perfect cubes. e.g.

17 lies between 8 $(= 2^3)$ and 27 $(= 3^3)$
i.e.   8, 17, 27
4·9 lies between 1 $(= 1^3)$ and 8 $(= 2^3)$
i.e.   1, 4·9, 8

Example 2

Find two consecutive perfect cubes between which the following lie.

(a) 11   (b) 32   (c) 9·6

(a) $2^3 = 8$, and $3^3 = 27$,
    so 11 lies between 8 and 27
    i.e.   8, 11, 27

(b) $3^3 = 27$, and $4^3 = 64$,
    so 32 lies between 27 and 64
    i.e.   27, 32, 64

(c) $2^3 = 8$, and $3^3 = 27$,
    so 9·6 lies between 8 and 27
    i.e.   8, 9·6, 27

Exercise 6.2d

Find two consecutive perfect cubes between which the following lie.

1. 3	2. 7	3. 10
4. 16	5. 6	6. 23
7. 32	8. 47	9. 70
10. 98	11. 62	12. 120
13. 145	14. 85	15. 58
16. 19	17. 38	18. 67
19. 1·4	20. 5·6	21. 9·2
22. 7·8	23. 10·6	24. 21·4
25. 26·7	26. 27·8	27. 36·5
28. 57·6	29. 82·1	30. 130·8

To estimate the cube root of a number, first find the nearest perfect cube.

Example 3

Estimate (a) $\sqrt[3]{12}$ (b) $\sqrt[3]{75}$ (c) $\sqrt[3]{7\cdot3}$

(a) 12 lies between the perfect cubes 8 and 27.
Since 12 is nearest to 8,
$\sqrt[3]{12} \approx 2$ (because $\sqrt[3]{8} = 2$)

(b) 75 lies between the perfect cubes 64 and 125.
Since 75 is nearest to 64,
$\sqrt[3]{75} \approx 4$ (because $\sqrt[3]{64} = 4$)

(c) 7.3 lies between the perfect cubes 1 and 8.
Since 7.3 is nearest to 8,
$\sqrt[3]{7\cdot3} \approx 2$ (because $\sqrt[3]{8} = 2$)

Exercise 6.2e

Estimate the cube root of each of the following.

1. 2	2. 7	3. 9
4. 25	5. 10	6. 28
7. 62	8. 34	9. 59
10. 66	11. 122	12. 68
13. 120	14. 129	15. 65
16. 26	17. 60	18. 130
19. 1·1	20. 7·9	21. 2·3
22. 8·2	23. 26·8	24. 11·2
25. 30·1	26. 63·9	27. 29·7
28. 67·2	29. 6·6	30. 0·8

Estimate the cube roots of much larger or smaller numbers by using factorisation. Convenient factors may be 1000 or 1 000 000.

Example 4

Estimate (a) $\sqrt[3]{28\,000}$ (b) $\sqrt[3]{7\,900\,000}$
         (c) $\sqrt[3]{0.062}$ (d) $\sqrt[3]{0.000129}$

(a) $\sqrt[3]{28\,000} = \sqrt[3]{28 \times 1000} = \sqrt[3]{28} \times \sqrt[3]{1000}$

Replace 28 by the nearest perfect cube

i.e. $\sqrt[3]{27\,000} = \sqrt[3]{27} \times \sqrt[3]{1000} = 3 \times 10$

$= 30$

So, $\sqrt[3]{28\,000} \approx 30$

(b) $\sqrt[3]{7\,900\,000} = \sqrt[3]{7\cdot9 \times 1\,000\,000}$

$= \sqrt[3]{7\cdot9} \times \sqrt[3]{1000} \times \sqrt[3]{1000}$

Replace 7·9 by the nearest perfect cube

i.e. $\sqrt{8\,000\,000} = \sqrt[3]{8} \times \sqrt[3]{1000} \times \sqrt[3]{1000}$

$= 2 \times 10 \times 10 = 200$

so, $\sqrt[3]{7\,900\,000} \approx 200$

(c) $\sqrt[3]{0\cdot062} = \sqrt[3]{\dfrac{62}{1000}} = \dfrac{\sqrt[3]{62}}{\sqrt[3]{1000}}$

Replace 62 by the nearest perfect cube

i.e $\sqrt[3]{0\cdot064} = \dfrac{\sqrt[3]{64}}{\sqrt[3]{1000}} = \dfrac{4}{10} = 0\cdot4$

so, $\sqrt[3]{0\cdot062} \approx 0\cdot4$

(d) $\sqrt[3]{0\cdot000\,129} = \sqrt[3]{\dfrac{129}{1\,000\,000}}$

$= \dfrac{\sqrt[3]{129}}{\sqrt[3]{1000} \times \sqrt[3]{1000}}$

Replace 129 by the nearest perfect cube

i.e. $\sqrt[3]{0\cdot000\,125} = \dfrac{\sqrt[3]{125}}{\sqrt[3]{1000} \times \sqrt[3]{1000}}$

$= \dfrac{5}{10 \times 10} = \dfrac{5}{100} = 0\cdot05$

so, $\sqrt[3]{0\cdot000\,129} \approx 0\cdot05$

Exercise 6.2f

Estimate the cube root of each of the following.

1. 26 000	2. 9000	3. 65 000
4. 217 000	5. 7800	6. 25 800
7. 9200	8. 63 800	9. 1100
10. 7 000 000	11. 29 000 000	12. 9 000 000
13. 25 000 000	14. 8 200 000	15. 1 300 000
16. 62 700 000	17. 121 900 000	18. 67 100 000
19. 0·009	20. 0·011	21. 0·029
22. 0·065	23. 0·023	24. 0·13
25. 0·000 009	26. 0·000 012	25. 0·000 059
28. 0·000 026	29. 0·000 127	30. 0·000 06

Numbers may be written in standard form thus: $a \times 10^n$ where a is a number between 1 and 10 and n is a whole number.

Example 1

Write in standard form:
(a) 186 000;   (b) 32·76

(a) 186 000 $= 1·86 \times 100\ 000$
$= 1·86 \times 10 \times 10 \times 10$
$\times 10 \times 10$
$= 1·86 \times 10^5$

(b) $32·76 = 3·276 \times 10$
$= 3·276 \times 10^1$

Exercise 6.3a

Write the following numbers in standard form:

1. 800 000	2. 360 000	3. 548 000
4. 50 000	5. 35 000	6. 22 400
7. 9000	8. 7500	9. 2840
10. 1563	11. 700	12. 290
13. 342	14. 186·3	15. 80
16. 36	17. 29·8	18. 4
19. 8·2	20. 7·36	21. 1·23
22. 427·4	23. 1169·2	24. 4000·6
25. 1111 111	26. 2020 206	27. 10 000 001
28. 40 060 007	29. 186 000	30. 10·0001

Example 2

Write as ordinary numbers:

(a) $2·34 \times 10^3$;   (b) $4·00 \times 10^6$

(a) $2·34 \times 10^3 = 2·34 \times 10 \times 10 \times 10$
$= 2·34 \times 1000$
$= 2340$

(b) $4·00 \times 10^6$
$= 4·00 \times 10 \times 10 \times 10 \times 10 \times 10 \times 10$
$= 4·00 \times 1\ 000\ 000$
$= 4\ 000\ 000$

Exercise 6.3b

Write the following as ordinary numbers.

1. 7×10^5	2. $8·25 \times 10^5$
3. $9·663 \times 10^5$	4. $5·001 \times 10^5$
5. 8×10^4	6. $6·3 \times 10^4$

7. $7·506 \times 10^4$	8. 6×10^3
9. $3·91 \times 10^3$	10. $5·376 \times 10^3$
11. $4·5 \times 10^2$	12. $9·38 \times 10^2$
13. $8·837 \times 10^2$	14. $6·206 \times 10^2$
15. 5×10^1	16. $7·28 \times 10^1$
17. $1·532 \times 10^1$	18. $5·086 \times 10^1$
19. $9·51 \times 10^0$	20. $4·537 \times 10^0$
21. $1·6 \times 10^0$	22. $1·06 \times 10^1$
23. $4·82 \times 10^2$	24. $4·007 \times 10^5$
25. $2·7 \times 10^6$	26. $9·91 \times 10^5$
27. $1·02 \times 10^7$	28. $9·436 \times 10^7$
29. $8·42 \times 10^8$	30. $9·63 \times 10^{10}$

Example 3

Write in standard form:
(a) 0·482   (b) 0·000 53

(a) $0·482 = 4·82 \div 10$
$= 4·82 \times 10^{-1}$

(b) $0·000\ 53 = 5·3 \div 10\ 000$
$= 5·3 \div (10 \times 10 \times 10 \times 10)$
$= 5·3 \times 10^{-4}$

Exercise 6.3c

Write the following numbers in standard form:

1. 0·56	2. 0·819
3. 0·704	4. 0·02
5. 0·035	6. 0·0601
7. 0·0004	8. 0·000 57
9. 0·000 08	10. 0·000 0965
11. 0·000 794	12. 0·000 0007
13. 0·000 000 821	14. 0·36
15. 0·091	16. 0·000 423
17. 0·6241	18. 0·0079
19. 0·000 805	20. 0·07
21. 0·62	22. 0·0909
23. 0·6040	24. 0·0003
25. 0·069	26. 0·000 0792
27. 0·043	28. 0·000 037
29. 0·0084	30. 0·68

Example 4

Write as ordinary numbers:
(a) $6·21 \times 10^{-3}$ (b) $9·08 \times 10^{-5}$

(a) $6·21 \times 10^{-3} = 6·21 \div (10 \times 10 \times 10)$
$= 6·21 \div 1000$
$= 0·006\ 21$

(b) $9 \cdot 08 \times 10^{-5} =$
$$9 \cdot 08 \div (10 \times 10 \times 10 \times 10 \times 10)$$
$$= 9 \cdot 08 \div 100\ 000$$
$$= 0 \cdot 000\ 0908$$

Exercise 6.3d

Write the following as ordinary numbers.

1. 4×10^{-3}
2. $6 \cdot 2 \times 10^{-2}$
3. $7 \cdot 1 \times 10^{-5}$
4. $4 \cdot 03 \times 10^{-7}$
5. $6 \cdot 35 \times 10^{-4}$
6. $7 \cdot 4 \times 10^{-1}$
7. $8 \cdot 9 \times 10^{-6}$
8. $1 \cdot 98 \times 10^{-5}$
9. $1 \cdot 24 \times 10^{-2}$
10. $2 \cdot 2 \times 10^{-8}$
11. 3×10^{-6}
12. $5 \cdot 0 \times 10^{-4}$
13. $9 \cdot 091 \times 10^{-3}$
14. $1 \cdot 01 \times 10^{-7}$
15. $1 \cdot 9 \times 10^{-5}$
16. $7 \cdot 97 \times 10^{-3}$
17. $1 \cdot 41 \times 10^{-1}$
18. $6 \cdot 24 \times 10^{-4}$
19. $8 \cdot 182 \times 10^{-2}$
20. $4 \cdot 73 \times 10^{-10}$
21. $5 \cdot 6 \times 10^{0}$
22. $7 \cdot 34 \times 10^{-4}$
23. $6 \cdot 49 \times 10^{-2}$
24. $8 \cdot 91 \times 10^{-3}$
25. $4 \cdot 32 \times 10^{-4}$
26. $7 \cdot 05 \times 10^{-1}$
27. $6 \cdot 6 \times 10^{-5}$
28. $4 \cdot 083 \times 10^{-3}$
29. $1 \cdot 73 \times 10^{-2}$
30. $3 \cdot 89 \times 10^{-5}$

Exercise 6.3e

Write the following numbers in standard form:

1. 642
2. 0·0035
3. 17·6
4. 0·91
5. 374 000
6. 0·0045
7. 0·0079
8. 139·4
9. 2·76
10. 0·840
11. 0·000 76
12. 9700
13. 100 040
14. 0·1019
15. 1·4

Write the following as ordinary numbers:

16. 5×10^{3}
17. 4×10^{-2}
18. $6 \cdot 3 \times 10^{5}$
19. $7 \cdot 91 \times 10^{0}$
20. $4 \cdot 81 \times 10^{-4}$
21. 6×10^{3}
22. $7 \cdot 15 \times 10^{-6}$
23. $8 \cdot 93 \times 10^{4}$
24. $5 \cdot 6 \times 10^{3}$
25. $7 \cdot 4 \times 10^{-5}$
26. 8×10^{-3}
27. $1 \cdot 51 \times 10^{2}$
28. $4 \cdot 28 \times 10^{-1}$
29. $3 \cdot 2 \times 10^{1}$
30. $5 \cdot 23 \times 10^{-4}$

6.4 NUMBER BASES

In any number base N the column headings are

$$\ldots N^4; N^3; N^2; N; 1; \ldots$$

Example 1

Write the number 4322_{ten} in column form.

Power	10^3	10^2	10	1
Column value	1000's	100's	10's	units
$4322_{ten} =$	4	3	2	2

Example 2

Write the number 110101_{two} in column form.

Power	2^5	2^4	2^3	2^2	2	1
Column value	32's	16's	8's	4's	2's	units
$110101_{two} =$	1	1	0	1	0	1

Exercise 6.4a

Write the following numbers in column form.

1. 3654_{ten}
2. 4037_{ten}
3. 905_{ten}
4. 2431_{five}
5. 4032_{five}
6. 340_{five}
7. 34_{five}
8. 101101_{two}
9. 10110_{two}
10. 1011_{two}
11. 111_{two}
12. 20121_{three}
13. 1202_{three}
14. 210_{three}
15. 3231_{four}
16. 2303_{four}
17. 320_{four}
18. 2547_{eight}
19. 650_{eight}
20. 4351_{six}

By writing a number in column form, it is easy to change a number written in any base to its equivalent in base ten.

Example 3

Change 4026_{eight} into base ten.

Power	8^3	8^2	8	1
Column value	512's	64's	8's	units
4026_{eight} =	4	0	2	6
Value	2048	0	16	6

Hence

$$4026_{eight} = (4 \times 512) + (0 \times 64) + \\ (2 \times 8) + (6 \times 1)$$
$$= 2048 + 0 + 16 + 6$$
$$= 2070_{ten}$$

Example 4

Find the 'odd value out' in base ten of 32_{eight}; 51_{five}; 221_{three}.

$$32_{eight} = (3 \times 8) + (2 \times 1)$$
$$= 24 + 2 = 26$$

$$51_{five} = (5 \times 5) + (1 \times 1) = 25 + 1 = 26$$

$$221_{three} = (2 \times 9) + (2 \times 3) + (1 \times 1)$$
$$= 18 + 6 + 1 = 25$$

So 221_{three} is the 'odd value out' because its base-ten value is 25.

Exercise 6.4b

In questions **1** to **10** change each number to base ten.

1. 1241_{five} 2. 2230_{five} 3. 312_{five}

4. 100101_{two} 5. 11010_{two} 6. 1110_{two}

7. 1231_{four} 8. 1042_{six} 9. 272_{eight}

10. 2021_{three}

In questions **11** to **20** change each number to base 10 and find the 'odd value out'.

11. (a) 10100_{two} (b) 102_{four} (c) 33_{five}

12. (a) 11101_{two} (b) 131_{four} (c) 102_{five}

13. (a) 110100_{two} (b) 302_{four} (c) 200_{five}

14. (a) 1102_{three} (b) 100111_{two} (c) 47_{eight}

15. (a) 1200_{three} (b) 134_{five} (c) 55_{eight}

16. (a) 2002_{three} (b) 321_{four} (c) 71_{eight}

17. (a) 52_{six} (b) 200_{four} (c) 11110_{two}

18. (a) 115_{six} (b) 101110_{two} (c) 142_{five}

19. (a) 140_{six} (b) 2022_{three} (c) 330_{four}

20. (a) 244_{five} (b) 1023_{four} (c) 2210_{three}

A number written in base ten is changed into another base by repeated division by the number of the required base.

Example 5

Change 222_{ten} into base eight.

```
8) 222
  8) 27  +  6 units (1)
    8) 3  + 3 eights (8)
       0  + 3 sixty-fours (8)²
```

The remainders are read upwards to give the answer.

Hence $222_{ten} = 336_{eight}$

Exercise 6.4c

Change each base ten number into the required base.

1. 328 to base five   2. 556 to base five
3. 435 to base five   4. 270 to base five
5. 46 to base two   6. 39 to base two
7. 25 to base two   8. 19 to base two
9. 318 to base four   10. 431 to base four
11. 192 to base four   12. 725 to base six
13. 438 to base six   14. 157 to base six
15. 104 to base three   16. 51 to base three
17. 96 to base three   18. 624 to base eight
19. 336 to base eight   20. 211 to base eight

When changing a base ten number into base twelve, use T for 10 and E for 11.

21. 146 to base twelve   22. 432 to base twelve
23. 213 to base twelve   24. 132 to base twelve
25. 1728 to base twelve   26. 730 to base twelve.

To change a number in one base to another base, first change the number into base ten and then into the new base.

Example 6

Change 1003_{six} into base five.

1. 1003_{six} into base ten.

Power	6^3	6^2	6	1
Column value	216's	36's	6's	units
1003_{six} =	1	0	0	3

$$\therefore 1003_{six} = (1 \times 216) + (3 \times 1)$$
$$= 216 + 3 = 219_{ten}$$

2. Base ten to base five.

$5\underline{)219}$
$\quad 5\underline{)43}$ + 4 units (1)
$\quad\quad 5\underline{)8}$ + 3 fives (5)
$\quad\quad\quad 5\underline{)1}$ + 3 twenty-fives $(5)^2$
$\quad\quad\quad\quad 0$ + 1 one-hundred-and-twenty-fives
$\quad\quad\quad\quad\quad\quad (5)^3$

Reading upwards:
$$219_{ten} = 1334_{five}$$

Hence $1003_{six} = 1334_{five}$

Exercise 6.4d

1. Change 144_{five} to base two.
2. Change 132_{five} to base two.
3. Change 202_{five} to base two.
4. Change 111101_{two} to base five.
5. Change 100111_{two} to base five.
6. Change 101001_{two} to base five.
7. Change 2033_{four} to base six.
8. Change 2201_{four} to base six.
9. Change 1332_{four} to base six.
10. Change 2130_{four} to base eight.
11. Change 3002_{four} to base eight.
12. Change 332_{four} to base eight.
13. Change 312_{five} to base three.
14. Change 420_{six} to base four.
15. Change 121_{three} to base two.
16. Change 131_{seven} to base eight.
17. Change 214_{five} to base nine.
18. Change 313_{four} to base six.
19. Change 120_{six} to base seven.
20. Change 35_{seven} to base five.

When adding numbers in any base N, carry 1 into the next column when the total in any column reaches N.

Example 7

(a) $\quad 1011_{two}$
$\quad + \;\; 111_{two}$
$\quad \overline{10010_{two}}$

From the right:
column 1: $1 + 1 = 2_{ten} = 10_{two}$;
$\quad\quad\quad$ write 0, carry 1
column 2: $1 + 1 + 1 = 3_{ten} = 11_{two}$;
$\quad\quad\quad$ write 1, carry 1
column 3: $1 + 1 + 0 = 2_{ten} = 10_{two}$;
$\quad\quad\quad$ write 0, carry 1
column 4: $1 + 1 = 2_{ten} = 10_{two}$;
$\quad\quad\quad$ write 0, write 1

(b) $\quad 341_{five}$
$\quad + \;\; 34_{five}$
$\quad \overline{430_{five}}$

From the right:
Column 1: $1 + 4 = 5_{ten} = 10_{five}$;
$\quad\quad\quad$ write 0, carry 1
column 2: $1 + 3 + 4 = 8_{ten} = 13_{five}$;
$\quad\quad\quad$ write 3, carry 1
column 3: $1 + 3 = 4_{ten} = 4_{five}$;
$\quad\quad\quad$ write 4

Exercise 6.4e

1. $101_{two} + 10_{two}$
2. $110_{two} + 11_{two}$
3. $111_{two} + 11_{two}$
4. $210_{three} + 12_{three}$
5. $202_{three} + 12_{three}$
6. $112_{three} + 22_{three}$
7. $222_{five} + 14_{five}$
8. $322_{five} + 33_{five}$
9. $424_{five} + 33_{five}$
10. $52_{eight} + 64_{eight}$
11. $64_{eight} + 37_{eight}$
12. $36_{eight} + 43_{eight}$
13. $541_{six} + 23_{six}$
14. $450_{six} + 122_{six}$
15. $44_{six} + 55_{six}$
16. $1011_{two} + 111_{two}$
17. $111_{two} + 11_{two} + 11_{two}$
18. $212_{three} + 20_{three} + 12_{three}$
19. $243_{five} + 13_{five} + 32_{five}$
20. $486_{ten} + 398_{ten} + 71_{ten}$

Taking 1 from the next column during subtraction means adding N to the present column.

Example 8

(a)
$$\begin{array}{r} 101_{two} \\ - \quad 11_{two} \\ \hline 10_{two} \end{array}$$

From the right:
In *column 2*, $0_{two} - 1_{two}$ becomes
$$10_{two} - 1_{two} = 1_{two}$$
(i.e. $2_{ten} - 1_{ten} = 1_{ten}$)

(b)
$$\begin{array}{r} 341_{five} \\ - \quad 33_{five} \\ \hline 303_{five} \end{array}$$

From the right:
In *column 1*, $1_{five} - 3_{five}$ becomes
$$11_{five} - 3_{five} = 3_{five}$$
(i.e. $6_{ten} - 3_{ten} = 3_{ten}$)

Exercise 6.4f

1. $111_{two} - 10_{two}$
2. $101_{two} - 11_{two}$
3. $1101_{two} - 111_{two}$
4. $221_{three} - 11_{three}$
5. $212_{three} - 22_{three}$
6. $110_{three} - 22_{three}$
7. $312_{four} - 11_{four}$
8. $212_{four} - 23_{four}$
9. $302_{four} - 33_{four}$
10. $433_{five} - 123_{five}$
11. $401_{five} - 21_{five}$
12. $311_{five} - 43_{five}$
13. $264_{seven} - 133_{seven}$
14. $203_{seven} - 156_{seven}$
15. $634_{eight} - 223_{eight}$
16. $701_{eight} - 63_{eight}$
17. $1010_{two} - 111_{two}$
18. $1000_{two} - 111_{two}$
19. $384_{nine} - 187_{nine}$
20. $600_{ten} - 299_{ten}$

Example 9

(a)
$$\begin{array}{r} 101_{two} \\ \times \quad 10_{two} \\ \hline 000 \\ 101- \\ \hline 1010_{two} \end{array}$$
$000 \quad (101_{two} \times 0_{two})$
$101- \quad (101_{two} \times 1_{two})$

(b)
$$\begin{array}{r} 321_{five} \\ \times \quad 43_{five} \\ \hline 2013 \\ 2334- \\ \hline 30403_{five} \end{array}$$
$2013 \quad (321_{five} \times 3_{five})$
$2334- \quad (321_{five} \times 4_{five})$

Exercise 6.4g

1. $100_{two} \times 10_{two}$
2. $101_{two} \times 11_{two}$
3. $111_{two} \times 101_{two}$
4. $21_{three} \times 11_{three}$
5. $22_{three} \times 12_{three}$
6. $212_{three} \times 21_{three}$
7. $312_{five} \times 12_{five}$
8. $232_{five} \times 23_{five}$
9. $402_{five} \times 32_{five}$
10. $304_{six} \times 41_{six}$
11. $245_{six} \times 43_{six}$
12. $1101_{two} \times 111_{two}$

The length of the above line is 6·9 cm.

This statement means that the measurement can be accurately made to within 0·1 cm.

The *real* length of the line may be between 6·85 cm and 6·95 cm: the difference between the greatest and the least possible values is 0·1 cm.

Hence the line must measure 6·9 cm ± 0·05 cm
i.e. between 6·95 cm (or 6·9 + 0·05)
    and 6·85 cm (or 6·9 − 0·05)

The maximum error possible is called the *absolute error*; this is half of the smallest unit used during the measurement.

Example 1

Find the greatest and the least possible values for the following.

(a) 6 cm   (b) 6·6 m   (c) 4·20 litres   (d) 20 kg

(a) 6 cm
 Smallest unit used is 1 cm
 The maximum error is 1 ÷ 2 = 0·5 cm
 ∴ the measurement is 6 ± 0·5 cm
 or between 6·5 cm and 5·5 cm.

(b) 6·6 m
 Smallest unit used is 0·1 m
 Maximum error is 0·1 ÷ 2 = 0·05 m
 ∴ the measurement is 6·6 ± 0·05 m
 or between 6·65 and 6·55 m.

(c) 4·20 litres
 Smallest unit used is 0·01 litre
 Maximum error used is 0·01 ÷ 2 = 0·005 l
 ∴ the measurement is 4·20 ± 0·005 l
 or between 4·205 litres and 4·195 litres

(d) 20 kg
 Smallest unit used is 1 kg
 Maximum error is 1 ÷ 2 = 0·5 kg
 ∴ the measurement is 20 ± 0·5 kg
 or between 20·5 kg and 19·5 kg.

Exercise 6.5a

Find the greatest and least acceptable values of the following measurements.

1. (5 ± 0·5) m
2. (3 ± 0·5) cm
3. (14 ± 0·5) g
4. (25 ± 0·5) kg
5. (8 ± 0·1) cm
6. (17 ± 0·1) g
7. (16 ± 0·2) cm
8. (12 ± 0·2) m
9. (30 ± 0·5) mg
10. (26 ± 0·5) cm
11. (6·5 ± 0·05) l
12. (9·8 ± 0·05) g
13. (25·5 ± 0·05) cm
14. (42·6 ± 0·05) mm
15. (37·6 ± 0·01) kg
16. (14·7 ± 0·01) l
17. (48·2 ± 0·01) kg
18. (173·5 ± 0·01) mg
19. (63·4 ± 0·02) N
20. (82·0 ± 0·02) l

Exercise 6.5b

Find the greatest and least possible values of the following measurements.

1. 2 mm
2. 14 cm
3. 8 ml
4. 16 cm
5. 7 mm
6. 9 mg
7. 5·4 m
8. 6·2 cm
9. 1·8 l
10. 5·0 m
11. 1·2 g
12. 2·8 cm
13. 6·92 m
14. 0·76 kg
15. 4·64 g
16. 13·58 l
17. 21·50 mg
18. 7·80 ml
19. 30 m
20. 20 cm
21. 100 g
22. 4·0 m
23. 7·0 g
24. 5·30 mg
25. 4·862 l
26. 8·704 mg
27. 12·742 kg
28. 0·084 cm
29. 0·973 g
30. 12·150 N

The greatest possible sum of measurements is found by adding the greatest possible values of these measurements.

The least possible sum of measurements is found by adding the least possible values of these measurments.

Example 2

Find (a) the greatest possible sum and
   (b) the least possible sum of
    3·8 and 6·0 cm.

(a) Greatest sum = 3·85 cm + 6·05 cm
     = 9·90 cm

(b) Least sum = 5·75 cm + 5·95 cm
    = 9·70 cm.

Exercise 6.5c

Find (a) the greatest possible sum, and
       (b) the least possible sum, of the following
           measurements.

1. 4 m + 6 m
2. 25 m + 10 m
3. 1 6m + 21 m
4. 3 cm + 15 cm
5. 20 cm + 15 cm
6. 51 m + 16 m
7. 7 l + 13 l
8. 12 mg + 36 mg
9. 3·2 cm + 3·6 cm
10. 14·5 cm + 9·8 m
11. 4·0 cm + 2·5 cm
12. 1·8 l + 0·9 l
13. 17·4 kg + 25·8 kg
14. 34·1 g + 16·8 g
15. 21·5 mg + 36·4 mg
16. 4·8 m + 127·2 m
17. 8·91 cm + 7·63 cm
18. 4·55 g + 12·86 g
19. 7·98 m + 10·42 m
20. 16·86 cm + 0·82 cm
21. 27·63 g + 14·82 g
22. 9·07 km + 6·90 km
23. 0·85 mg + 1·80 mg
24. 5·75 l + 2·00 l
25. (6 ± 0.5) g + (17 ± 0.5) g
26. $(6.3 \pm 0.05)l$ + $(12.8 \pm 0.05)l$
27. 4 g + 5 g + 8 g
28. 16 m + 17 m + 4 m
29. 3·8 m + 9·7 m + 2·4 m
30. 4·0 km + 2·5 km + 1·8 km
31. Find the greatest possible perimeter of a rectangle of length 12 cm and breadth 4 cm.
32. Find the greatest and least possible perimeter of a square of side 6·5 m.
33. Find the least possible perimeter of an equilateral triangle whose sides are all 1·50 cm long.
34. A can containing 4 litres of dye is emptied into a barrel containing 60 litres of water. Find the greatest possible total volume.
35. A steel tube 1·65 m long is welded onto the end of a tube of length 2·00 m. Find the greatest and least possible values of total length.

Exercise 6.5d

For the given measurements, find
(a) the greatest possible product, and
(b) the least possible product.
You may find a calculator helpful here.

1. 4 m × 6 m
2. 5 cm × 12 cm
3. 11 mm × 3 mm
4. 8 cm × 7 cm
5. 9 m × 20 m
6. 10 m × 2 m
7. 15 cm × 7 cm
8. 3 m × 30 m
9. 9 cm × 45 cm
10. 10 km × 5 km
11. 4·2 cm × 2·1 cm
12. 3·4 mm × 7·6 mm
13. 8·4 m × 3·2 m
14. 5·8 m × 9·1 m
15. 4·3 cm × 6·9 cm
16. 8·2 mm × 3·6mm
17. 4·0 cm × 2·5 cm
18. 6·4 m × 7·0 m
19. 10·1 mm × 8·0 mm
20. 6·9 km × 5·0 km
21. 0·05 cm × 6·02 cm
22. 1·06 cm × 0·08 cm
23. 4·20 m × 0·04 m
24. 9·55 km × 0·01 km
25. (3 ± 0.5) m × (2 ± 0.5) m
26. (12 ± 0.2) mm × (5 ± 0.2) mm
27. (7 ± 0.1) cm × (8 ± 0.1) cm
28. (1 ± 0.5) m × (14 ± 0.5) m
29. (0.8 ± 0.02) m × (0.9 ± 0.02) m
30. (1.4 ± 0.02) mm × (0.8 ± 0.02) mm
31. Find the greatest possible area of a rectangle of length 10 cm and breadth 3 cm.
32. A rectangle measures 4 m by 6 m.
    Find (a) the least possible perimeter, and
          (b) the least possible area.

To find the greatest possible product of measurements, multiply the greatest possible values of the measurements.
To find the least possible product of measurements, multiply the least possible values of the measurements.

Example 3

The length and breadth of a rectangle are 5 cm and 7 cm repectively.
Find (a) the greatest possible area, and
       (b) the least possible area of the rectangle.

(a) Greatest possible area = 5·5 × 7·5
                           = 41·25 cm²

(b) Least possible area = 4·5 × 6·5
                        = 29·25 cm²

When two measurements are subtracted,

the greatest difference =

$$\left(\begin{array}{l}\text{greatest possible value of}\\ \text{largest measurement}\end{array}\right) - \left(\begin{array}{l}\text{least possible value of}\\ \text{smallest measurement}\end{array}\right)$$

the least difference =

$$\left(\begin{array}{l}\text{least possible value of}\\ \text{largest measurement}\end{array}\right) - \left(\begin{array}{l}\text{greatest possible value of}\\ \text{smallest measurement}\end{array}\right)$$

Example 4

Two volumes are measured as 5 litres and 3 litres.
Find the greatest and least possible difference between these volumes.

Greatest difference = 5·5 − 2·5 = 3·0 litres
Least difference = 4·5 − 3·5 = 1·0 litres

Exercise 6.5e

Find the greatest and least possible differences
between the following measurements.

1. 4 m − 2 m
2. 16 cm − 11 cm
3. 13 g − 11 g
4. 47 km − 12 km
5. 74 m − 29 m
6. 27 cm − 10 cm
7. 100 cm − 17 cm
8. 36 g − 18 g
9. 29 l − 3 l
10. 270 ml − 10 ml
11. 6·8 g − 3·2 g
12. 12·8 mm − 7·4 mm
13. 19·5 l − 16·3 l
14. 27·6 g − 4·4 g
15. 12·5 km − 3·1 km
16. 97·5 N − 37·5 N
17. 4·2 l − 0·8 l
18. 13·0 kg − 5·4 kg
19. 95·4 mg − 14·0 mg
20. 0·9 m − 0·3 m

21 A lorry is loaded with 4 tonnes of sand from a
pile containing 20 tonnes. Find the greatest
possible weight of sand remaining in the pile.

22. 5·5 litres of lemonade is drained from a tank
containing 160·0 litres. Find the least possible
volume of lemonade left in the tank.

23. 2·00 m of steel cable is cut from a coil of length
100·00 m. Find the greatest and least possible
length of cable left in the coil.

24 An airline passenger is told to reduce the weight
of his luggage by 2 kg. If the luggage weighs
12 kg, what is the greatest possible weight of
luggage he can take?

25. A girl jogs round 2·7 km of a 3·8 km course.
What is the least possible distance she must jog
to complete the course?

When dividing one measurement by another,

the greatest possible value =

$$\frac{\text{greatest possible numerator}}{\text{least possible denominator}}$$

the least possible value =

$$\frac{\text{least possible numerator}}{\text{greatest possible denomiator}}$$

Example 5

500 kg of sugar is to be packed into sacks,
each containing 20 kg of sugar.
Find (a) the greatest possible number of
sacks, and
(b) the least possible number of sacks
obtained.

(a) Greatest possible number of bags = $\frac{500·5}{19·5}$

= 25·67

= 25 sacks

(b) Least possible number of bags = $\frac{499·5}{20·5}$

= 24·37

= 24 sacks

Exercise 6.5f

For the given measurements, find
(a) the greatest possible value, and
(b) the least possible value.
Give your answers correct to 2 D.P. where
appropriate. You may find a calculator helpful here.

1. 4 ÷ 2
2. 12 ÷ 4
3. 6 ÷ 3
4. 8 ÷ 2
5. 24 ÷ 6
6. 14 ÷ 7
7. 18 ÷ 9
8. 27 ÷ 3
9. 20 ÷ 5
10. 16 ÷ 4
11. 16·0 ÷ 2·0
12. 0·2 ÷ 0·2
13. 15·0 ÷ 3·0
14. 4·2 ÷ 2·1
15. 51·2 ÷ 6·4
16. 43·2 ÷ 7·2
17. 20·0 ÷ 4·0
18. 30·0 ÷ 10·0
19. 27·5 ÷ 2·5
20. 88·8 ÷ 2·2

21. Find the greatest possible number of 5·0 kg bags
of potatoes that can be filled from a van carrying
400·0 kg of potatoes.

22. Find the least possible number of round trips
a boat can make to transfer 850 kg of chemicals
from one bank of a river to the other, if the boat
carries 50 kg of chemicals at a time.

23. The area of a rectangle is 16·0 cm². If the length
is 4·0 cm, find the greatest and least possible
values of breadth.

24. A racing car travels 400 m in 50 seconds. Find the
greatest possible average speed of the car over
this distance (in metres per second).

25. A 250·00 m coil of rope is to be cut into sections
5·00 m long. What is the least possible number of
sections that can be obtained?

Area is a measure of the size of a surface; area is measured in square units.
The common metric units are:
Square millimetres (mm²)
Square centimetres (cm²)
Square metres (m²)
Hectares (ha)
Square kilometres (km²)

$1 \text{ cm}^2 = 100 \text{ mm}^2$
$1 \text{ m}^2 = 10\ 000 \text{ cm}^2$
$1 \text{ km}^2 = 1\ 000\ 000 \text{ m}^2$
$1 \text{ km}^2 = 100 \text{ ha}$
$1 \text{ ha} = 10\ 000 \text{ m}^2$

Perimeter is a measure of the distance round the outline of a shape. It is a measure of length.

Example 1

(a) Change 1·25 m² to cm²
    $1\cdot25 \text{ m}^2 = 1\cdot25 \times 10\ 000$
    $= 12\ 500 \text{ cm}^2$
(b) Change 275 mm² to cm²
    $275 \text{ mm}^2 = 275 \div 100$
    $= 2\cdot75 \text{ cm}^2$
(c) Change 43 000 m² to ha
    $43\ 000 \text{ m}^2 = 43\ 000 \div 10\ 000$
    $= 4\cdot3 \text{ ha}$
(d) Change 3·7 km² to ha
    $3\cdot7 \text{ km}^2 = 3\cdot7 \times 100$
    $= 370 \text{ ha}$

Exercise 7.1a

Change the following.

1. 3·75 m² to cm²
2. 2·75 cm² to mm²
3. 465 mm² to cm²
4. 12 500 cm² to m²
5. 150 mm² to cm²
6. 0·675 m² to cm²
7. 127 mm² to cm²
8. 4·1 m² to cm²
9. 62 cm² to m²
10. 1764 mm² to m²
11. 48 000 m² to ha
12. 21 ha to m²
13. 82·5 km² to ha
14. 4100 ha to km²
15. 6 500 000 m² to km²
16. 47 000 m² to ha
17. 0·07 ha to m²
18. 3845 ha to km²
19. 490 000 m² to ha
20. 6·329 ha to m²

Rectangle

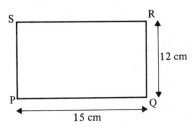

area = length × width
     = PQ × QR
     = 15 cm × 12 cm = 180 cm²
perimeter = PQ + QR + RS + SP
          = 15 + 12 + 15 + 12 = 54 cm

Square

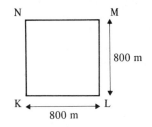

area = length × width
     = KL × LM
     = 800 m × 800 m
     = 640 000 m²
     $= \dfrac{640\ 000}{10\ 000} \text{ ha}$
     = 64 ha
perimeter = KL + LM + MN + NK
          = 800 + 800 + 800 + 800
          = 3200 m

Exercise 7.1b

1. Find (i) the area, (ii) the perimeter of the rectangle PQRS if
   (a) PQ = 16 cm, QR = 9 cm
   (b) PQ = 18 cm, QR = 11 cm
   (c) PQ = 35 mm, QR = 20 mm
   (d) PQ = 40 mm, QR = 24 mm
   (e) PQ = 25 m, QR = 12 m
   (f) PQ = 24 m, QR = 15 m

2. Find (i) the area, (ii) the perimeter of the
   square KLMN if
   (a) KL = LM = 20 cm
   (b) KL = LM = 16 cm
   (c) KL = LM = 1·4 m
   (d) KL = LM = 1·9 m
   (e) KL = LM = 18 mm
   (f) KL = LM = 25 mm

3. Find the area of rectangle PQRS, in hectares, if
   (a) PQ = 400 m, QR = 200 m
   (b) PQ = 300 m, QR = 400 m
   (c) PQ = 500 m, QR = 120 m
   (d) PQ = 60 m, QR = 1500 m
   (e) PQ = 50 m, QR = 900 m
   (f) PQ = 52 m, QR = 75 m

Example 2

(a) The length of a rectangle is 10 cm, and its
    perimeter is 36 cm. How wide is it?

   length + width $= \frac{1}{2}$ perimeter

   $\qquad\qquad\qquad = \frac{1}{2} \times$ 36 cm = 18 cm

   10 cm + width = 18 cm

   ∴ width = 18 − 10 = 8 cm

(b) The perimeter of a rectangle is 5·8 m, and
    its width is 1·2 m. What is its length?

   length + width $= \frac{1}{2}$ perimeter

   $\qquad\qquad\qquad = \frac{1}{2} \times$ 5·8 m = 2·9 m

   length + 1·2 m = 2·9 m

   ∴ length = 2·9 − 1·2 = 1·7 m

Exercise 7.1c

1. Find the width of the following rectangles.

	length	perimeter
(a)	10 cm	30 cm
(b)	15 cm	50 cm
(c)	25 mm	80 mm
(d)	6 m	18 m
(e)	12 mm	46 mm
(f)	9 m	42 m
(g)	30 cm	98 cm
(h)	17 mm	82 mm
(i)	6·5 km	18 km
(j)	4·2 m	16 m

2. Find the length of the following rectangles.

	width	perimeter
(a)	20 cm	56 cm
(b)	18 cm	54 cm
(c)	10 m	32 m
(d)	14 cm	62 cm
(e)	22 mm	70 mm
(f)	35 mm	94 mm
(g)	7 km	36 km
(h)	45 m	132 m
(i)	5·5 cm	20 cm
(j)	6·6 mm	22 mm

Example 3

The perimeter of a square is 224 mm.
How long is the square?

All four sides of a square are equal, so

4 × length = perimeter

∴ length   = perimeter ÷ 4

$\qquad\qquad$ = 224 ÷ 4

$\qquad\qquad$ = 56 mm

Exercise 7.1d

Find the length of each square, if the perimeter is
as follows:

1. 80 cm	2. 120 mm	3. 64 m
4. 76 m	5. 152 cm	6. 220 cm
7. 432 cm	8. 1024 mm	9. 6·4 km
10. 3·24 km	11. 2·72 m	12. 0·96 mm

Example 4

(a) The area of a rectangle is 156 cm². If
    it is 13 cm long, how wide is it?

   $\qquad$ area = length × width

   $\qquad$ 156 = 13 × width

   ∴ width = 156 ÷ 13 = 12 cm

(b) The area of a square is 49 cm². How
    long is its perimeter?

   $\qquad$ length of side

   $\qquad\qquad$ = 7 cm (because 7 × 7 = 49)

   ∴ perimeter = 7 + 7 + 7 + 7 = 28 cm

Exercise 7.1e

Find the width of the following rectangles.

1. area = 117 cm^2, length = 9 cm
2. area = 136 cm^2, length = 8 cm
3. area = 285 mm^2, length = 15 mm
4. area = 414 mm^2, length = 18 mm
5. area = 16·5 m^2, length = 1·1 m

Find the perimeter of the following squares.

6. area = 81 cm^2
7. area = 121 cm^2
8. area = 225 cm^2
9. area = 1·44 m^2
10. area = 2·56 m^2

Example 5

(a) A lawn measuring 10 m by 7·5 m is to be reseeded. A box of grass seed contains 500 g. If 100 g of grass seed covers 1 m^2, how many boxes will be needed?

$$area = 10 \text{ m} \times 7\cdot5 \text{ m} = 75 \text{ m}^2$$
no. of grams of seed needed
$$= 75 \times 100$$
$$= 7500 \text{ g}$$
$$no. \text{ of boxes} = 7500 \div 500$$
$$= 15$$

(b) A floor measuring 2 m by 1·5 m is to be covered with tiles measuring 25 cm square. How many tiles will be needed?

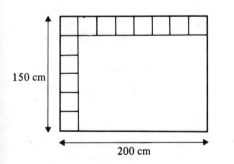

150 cm

200 cm

no. of tiles along length = 200 ÷ 25 = 8
no. of tiles along width = 150 ÷ 25 = 6
∴ tiles needed = 8 × 6 = 48

Exercise 7.1f

1. A metal plate measuring 32 cm by 12 cm is cut up into small squares of side length 4 cm. How many squares will be cut?
2. A lawn 5 m by 4 m is to be covered by pieces of square turf, each of side length 50 cm. How many pieces will be required?
3. A wall space in a bathroom measuring 2 m by 60 cm is to be covered with square tiles of side length 20 cm. How many tiles will be required? Find the cost of surfacing if the price of the tiles is £2·80 for ten.
4. A ceiling measures 6 m by 4 m. It is to be covered with square tiles of side length 40 cm. How many tiles will be required? Find the cost of the covering if the price of the tiles is £1·44 for twelve.
5. A yard measuring 10 m by 5 m is to be surfaced with concrete the dry mixture for which is supplied in 100-kg bags. If 80 kg can suitably surface 1 m^2, how many bags will be required?
6. A pathway 20 m long and 3 m wide is to be covered with shingle which is supplied in 50-kg bags. If 40 kg of shingle can suitably cover 1 m^2, how many bags will be required?
7. A small garden plot measuring 5 m by 4 m is to be covered with a special kind of soil which is supplied in 25-kg bags. If 30 kg of soil can suitably cover 1 m^2, how many bags will be required?
8. A lawn measuring 12 m by 10 m is to be covered with pieces of turf each 60 cm by 40 cm. How many pieces will be required?
9. A pavement of length 600 m and width 1 m is to be laid by using slabs of dimensions 80 cm by 50 cm. How many slabs are required?
10. A pavement of length 500 m and width 1·5 m is to be laid by using slabs of dimensions 75 cm by 40 cm. How many slabs are required?

To find the area of an irregular shape, divide it up into shapes whose area can be found.

Example 6

Find the area of the following shape.

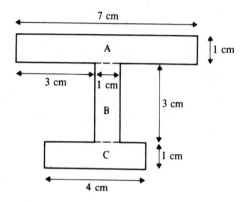

Divide the figure into three rectangles A, B, and C as shown.

area of A $= 7 \times 1 = 7$ cm²
area of B $= 3 \times 1 = 3$ cm²
area of C $= 4 \times 1 = 4$ cm²
∴ area of figure $= 7 + 3 + 4 = 14$ cm²

Exercise 7.1g

Find the area of each shape.

1.

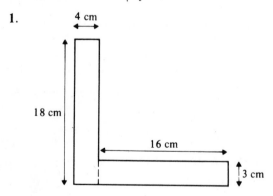

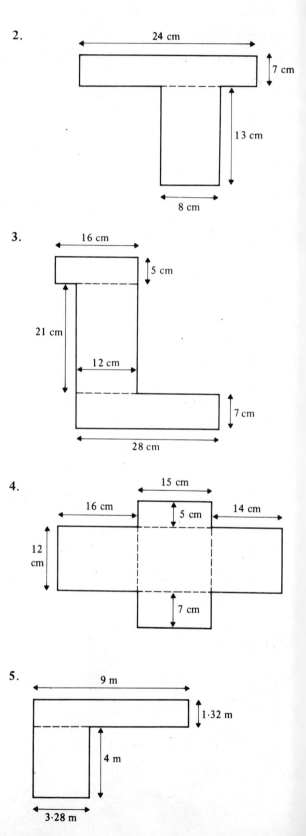

The vertical height of a triangle is sometimes called the *altitude*.

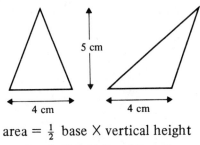

area = $\frac{1}{2}$ base X vertical height
  = $\frac{1}{2}$ X 4 X 5 = 10 cm²

Exercise 7.2a

Find the area of the following triangles:
1. base 20 cm, height 35 cm
2. base 19 cm, height 32 cm
3. base 40 mm, height 15 mm
4. base 25 mm, height 24 mm
5. base 3·6 cm, height 50 cm
6. base 2·5 cm, height 120 cm
7. base 20 m, altitude 4·5 m
8. base 1·25 cm, altitude 40 cm
9. base 12 cm, altitude 250 mm
10. base 7·6 cm, altitude 150 mm

Example 1

(a) The area of a triangle is 72 cm²
    If its vertical height is 16 cm, what is the length of its base?

  area = $\frac{1}{2}$ base X vertical height

  72 = $\frac{1}{2}$ base X 16

  72 = base X 8

  ∴ base = 72 ÷ 8 = 9 cm

(b) The area of a triangle is 6·4 mm² .
    If its base is 4 mm long, what is its altitude?

  area = $\frac{1}{2}$ base X altitude

  6·4 = $\frac{1}{2}$ X 4 X altitude

  6·4 = 2 X altitude

  ∴ altitude = 6·4 ÷ 2 = 3·2 mm

Exercise 7.2b

1. Find the length of the base of each triangle:

	area	vertical height
(a)	50 cm²	20 cm
(b)	100 cm²	10 cm
(c)	80 mm²	40 mm
(d)	90 mm²	60 mm
(e)	160 cm²	80 cm
(f)	66 m²	12 m
(g)	56 m²	14 m
(h)	108 cm²	24 cm
(i)	77 cm²	22 cm
(j)	90 mm²	36 mm

2. Find the altitude of each triangle:

	area	length of base
(a)	300 cm²	200 cm
(b)	125 cm²	50 cm
(c)	108 cm²	72 cm
(d)	75 mm²	30 mm
(e)	350 mm²	100 mm
(f)	2400 mm²	120 mm
(g)	1050 cm²	300 cm
(h)	105 cm²	84 cm
(i)	98 m²	56 m
(j)	126 cm²	35 cm

Example 2

Find the area of the following shape.

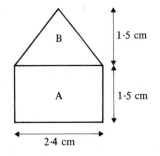

Divide the figure into the rectangle A and the triangle B.

  area of A = 2·4 X 1·5
       = 3·6 cm²

  area of B = $\frac{1}{2}$ (2·4 X 1·5)
       = $\frac{1}{2}$ X 3·6 = 1·8 cm²

area of figure = 3·6 + 1·8 = 5·4 cm²

Exercise 7.2c

Find the area of each shape.

1.

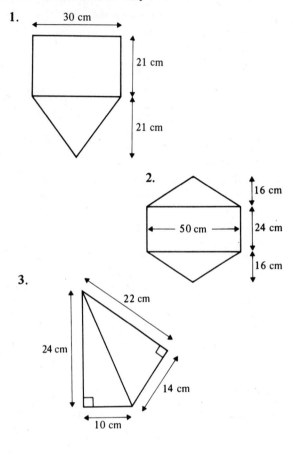

30 cm

21 cm

21 cm

2.

16 cm

50 cm

24 cm

16 cm

3.

22 cm

24 cm

14 cm

10 cm

4.

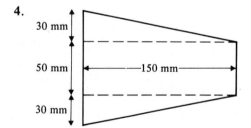

30 mm

50 mm

150 mm

30 mm

5.

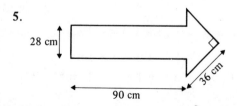

28 cm

90 cm

36 cm

6.

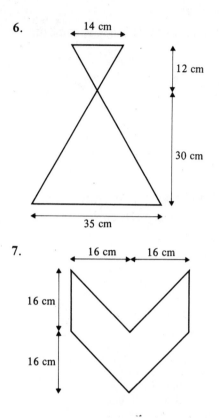

14 cm

12 cm

30 cm

35 cm

7.

16 cm   16 cm

16 cm

16 cm

8. Find the area of the shaded part.

8 cm

3 cm

6 cm

4 cm

9. Find the area of the shaded part.

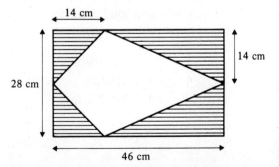

14 cm

14 cm

28 cm

46 cm

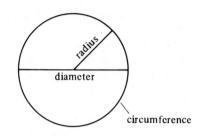

The length of the diameter D is twice the length of the radius r.

$$D = 2r$$
$$\text{circumference} = \pi D \text{ or } 2\pi r$$
$$\text{area} = \pi r^2$$
$$\text{diameter} = \text{circumference} \div \pi$$
$$\text{radius} = \sqrt{(\text{area} \div \pi)}$$

Example 1

Find the circumference of a circle when

(a) the radius is 14 cm, using $\pi = 3\frac{1}{7}$
(b) the diameter is 20 cm, using $\pi = 3 \cdot 14$

(a) circumference $= 2\pi r$

$$= \frac{2}{1} \times \frac{22}{7} \times \frac{14}{1} = 88 \text{ cm}$$

(b) circumference $= \pi D$
$$= 20 \times 3 \cdot 14 = 62 \cdot 8 \text{ cm}$$

Exercise 7.3a

Find the circumference of each circle.

In questions 1 to 10 assume $\pi = 3\frac{1}{7}$ or $\frac{22}{7}$.

1. diameter = 21 cm   2. diameter = 49 cm
3. radius = 35 cm   4. radius = 7 cm
5. radius = 2·1 m   6. radius = 2·8 m
7. diameter = 8·4 m   8. diameter = 147 mm
9. diameter = 105 mm 10. diameter = 91 mm

In questions 11 to 20 assume $\pi = 3 \cdot 14$.

11. diameter = 5 cm   12. diameter = 3 cm
13. diameter = 11 cm   14. radius = 4 cm
15. radius = 15 cm   16. radius = 200 mm
17. radius = 300 mm   18. radius = 450 mm
19. diameter = 1·5 m   20. diameter = 2·5 m

Example 2

Find (a) the diameter, (b) the radius of a circle whose circumference is 132 cm. (Take $\pi = 3\frac{1}{7}$).

(a) circumference $= \pi D$

$$132 = 3\frac{1}{7} D$$

$\therefore$ diameter $D = 132 \div 3\frac{1}{7}$

$$= \frac{\overset{6}{\cancel{132}}}{1} \times \frac{7}{\underset{1}{\cancel{22}}} = 42 \text{ cm}$$

(b) radius = (diameter) $\div$ 2
$$= 42 \div 2 = 21 \text{ cm}$$

Exercise 7.3b

Find (a) the diameter, (b) the radius of each circle, assuming $\pi = 3\frac{1}{7}$ or $\frac{22}{7}$.

1. circumference = 176 cm
2. circumference = 264 cm
3. circumference = 352 mm
4. circumference = 308 mm
5. circumference = 396 mm
6. circumference = 484 mm
7. circumference = 13·2 m
8. circumference = 4·4 m
9. circumference = 57·2 cm
10. circumference = 61·6 cm

Example 3

Find the area of a circle when

(a) the radius is 14 cm, using $\pi = 3\frac{1}{7}$
(b) the diameter is 20 cm, using $\pi = 3 \cdot 14$

(a) area $= \pi r^2 = 3\frac{1}{7} \times 14 \times 14$

$$= \frac{22}{\underset{1}{\cancel{7}}} \times \frac{\overset{2}{\cancel{14}}}{1} \times \frac{14}{1} = 616 \text{ cm}^2$$

(b) diameter = 20 cm, so radius = 10 cm
$$\text{area} = \pi r^2 = 3 \cdot 14 \times 10 \times 10$$
$$= 3 \cdot 14 \times 100$$
$$= 314 \text{ cm}^2$$

Exercise 7.3c

Find the area of each circle.
In questions 1 to 10 assume $\pi = 3\frac{1}{7}$ or $\frac{22}{7}$.

1. radius = 21 cm
2. radius = 7 cm
3. diameter = 70 mm
4. diameter = 56 mm
5. radius = 0·7 m
6. radius = 1·05 m
7. radius = $3\frac{1}{2}$ cm
8. radius = $1\frac{3}{4}$ cm
9. radius = $\frac{7}{20}$ m
10. diameter = $2\frac{4}{5}$ m

In questions 11 to 20 assume $\pi = 3·14$

11. radius = 2 cm
12. radius = 3 cm
13. diameter = 10 cm
14. diameter = 60 mm
15. diameter = 40 mm
16. diameter = 30 mm
17. diameter = 50 mm
18. radius = 1·5 m
19. radius = 0·5 m
20. radius = 2·5 m

Example 4

How many revolutions does a cycle wheel of diameter 63 cm complete in travelling 396 m? (Take $\pi = \frac{22}{7}$)

circumference of wheel = πD

$$= \frac{22}{7} \times \frac{63}{1} = 198 \text{ cm}$$

distance to travel = 396 m
$$= 396 \times 100 = 39\ 600 \text{ cm}$$
$\therefore$ no. of revolutions = $39\ 600 \div 198 = 200$

Exercise 7.3d

Assume that $\pi = 3\frac{1}{7}$ or $\frac{22}{7}$

1. The pulley on the jib of a crane has a diameter of 56 cm. Find
   (a) the circumference of the pulley,
   (b) the number of revolutions turned by the pulley when the crane raises a load through a vertical height of 44 m.
2. The pulley above a lift cage has a diameter of 14 cm. Find
   (a) the circumference of the pulley,
   (b) the number of revolutions it makes while the cage ascends 33 m.
3. At Cannock colliery the cage pulley has a diameter of 3·5 m. Find
   (a) the circumference of the pulley,
   (b) the number of revolutions it makes while the miner's cage descends to the coal seam 495 m below the ground.

4. A bowling-green roller has a diameter of 1·4 m. Find
   (a) the circumference of the roller,
   (b) the number of revolutions it makes while travelling the length of a 30·8 m green.
5. A glass cutter has a cutting wheel of diameter 9·8 mm. Find
   (a) the circumference of the wheel,
   (b) the number of revolutions it makes while cutting a pane of width 154 cm.
6. A wall tin-opener has a cutting wheel of diameter 1·68 cm. How many revolutions will it make while opening a tin of diameter 15·12 cm?
7. A photographic print is circular, with a radius of 70 mm.
   (a) Find the area of the print.
   (b) The outside edge of this print is to be trimmed with white cord which costs 12p per centimetre. Find the cost of this edging.
8. The top of a circular tank of diameter 7 m is to be galvanised. If it costs £1·50 to galvanise 1 m², find the cost of galvanizing both sides of the tank top.

Example 5

Find the area of the following shape.
(Take $\pi = 3·14$)

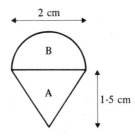

Divide the figure into the triangle A and the semi-circle B.

$$\text{area of A} = \tfrac{1}{2} \times 2 \times 1·5$$
$$= 1·5 \text{ cm}^2$$
$$\text{area of B} = \tfrac{1}{2} (3·14 \times 1 \times 1)$$
$$= 1·57 \text{ cm}^2$$
$\therefore$ area of figure $= 1·5 + 1·57 = 3·07 \text{ cm}^2$

Exercise 7.3e

Find the area of each shape.

1. (Take $\pi = 3 \cdot 14$)

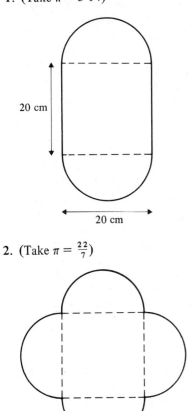

20 cm

20 cm

2. (Take $\pi = \frac{22}{7}$)

28 mm

28 mm

3. (Take $\pi = \frac{22}{7}$)   **4.** Take ($\pi = 3 \cdot 14$)

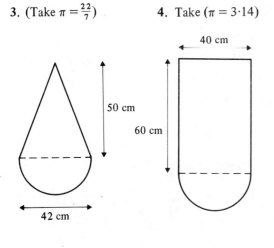

50 cm

42 cm

40 cm

60 cm

5. (Take $\pi = \frac{22}{7}$)

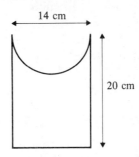

14 cm

20 cm

6. Find the area of the shaded part.
(Take $\pi = 3 \cdot 14$)

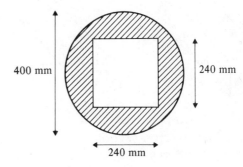

400 mm

240 mm

240 mm

7. Find the area of the shaded part.
(Take $\pi = 3 \cdot 14$)

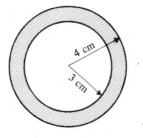

4 cm

3 cm

8. Find the area of the shaded part.
(Take $\pi = \frac{22}{7}$)

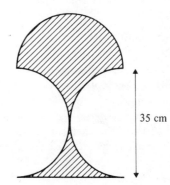

35 cm

Volume is a measure of the space occupied by a solid; volume is measured in cubic units, e.g. m^3, cm^3, mm^3.

Capacity is the volume of a liquid or a gas, measured in litres which a vessel of the same shape and size as the solid can hold, 1 litre = 1000 cm^3 or millilitres

Prism

A prism is a solid with a uniform cross section, as shown in the examples below.

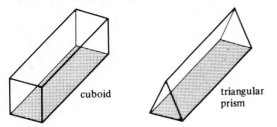

cuboid   triangular prism

volume of prism
= (area of cross section) × length
= (area of end) × length

Example 1

A cuboid measures 5 cm by 12 cm by 10 cm. What is its volume?

area of end = 5 × 12 cm^2
  volume = 5 × 12 × 10 cm^3
        = 600 cm^3

Example 2

A rectangular tank measures 2 m by 3 m by 4 m and is full of water. What is its capacity in litres?

area of end = 200 × 300 cm^2
  volume = 200 × 300 × 400 cm^3

capacity = $\dfrac{\overset{1}{\cancel{200}} \times \overset{60}{\cancel{300}} \times 400}{\underset{5}{\cancel{1000}}}$ litres

= 24 000 litres

Exercise 7.4a

Find the volume of each cuboid.

1. height = 5 cm, length = 11 cm, width = 6 cm.
2. height = 3 cm, length = 20 cm, width = 12 cm.
3. height = 5 cm, length = 16 cm, width = 15 cm.
4. height = 1·2 m, length = 5 m, width = 1·5 m.
5. height = 2·5 m, length = 3·5 m, width = 3·2 m.
6. height = 1·25 m, length = 6·4 m, width = 4·5 m.

Find the volumes of the following cubes.

7. side-length = 9 cm.   8. side-length = 12 cm.
9. side-length = 60 mm.   10. side-length = 80 mm.

Find (a) the volume of each of the following tanks and (b) the capacity of each tank in litres.

11. height = 3 m, length = 6 m, width = 4 m.
12. height = 2 m, length = 7 m, width = 6 m.
13. height = 1·6 m, length = 5 m, width = 3·5 m.
14. height = 2·4 m, length = 4·5 m, width = 2·5 m.
15. height = 1·5 m, length = 3·75 m, width = 1·6 m.
16. height = 0·15 m, length = 0·5 m, width = 0·32 m.
17. height = 0·16 m, length = 0·25 m, width = 0·2 m.
18. height = 25 cm, length = 30 cm, width = 16 cm.

Example 3

The volume of a cuboid is 175 cm^3, and the area of its cross section is 50 cm^3.
Find the length of the cuboid.

volume = (area of cross section) × length
    175 = 50 × length
∴ length = 175 ÷ 50
        = 3·5 cm

Example 4

A cuboid has width 4 mm and height 15 mm. If its volume is 144 mm^3, find its length.

area of end = 4 × 15 mm^2
    volume = (area of end) × length
      144 = 4 × 15 × length

∴ length $= \dfrac{\overset{36}{\cancel{144}}}{\underset{1}{\cancel{4}} \times 15}$

$= \frac{36}{15} = 2\cdot4$ mm

Exercise 7.4b

Find the length of each cuboid.
1. area of end $= 10$ cm^2, volume $= 80$ cm^3
2. area of end $= 5$ cm^2 volume $= 75$ cm^3
3. area of end $= 8$ mm^2, volume $= 640$ mm^3
4. area of end $= 12$ cm^2, volume $= 972$ cm^3
5. area of end $= 15$ m^2, volume $= 465$ m^3
6. area of end $= 25$ cm^2, volume $= 300$ cm^3
7. area of end $= 24$ mm^2, volume $= 312$ mm^3
8. area of end $= 16$ cm^2, volume $= 56$ cm^3
9. area of end $= 28$ m^2, volume $= 70$ m^2
10. area of end $= 35$ cm^2, volume $= 287$ cm^3

Example 5

Find the volume of the triangular prism shown.

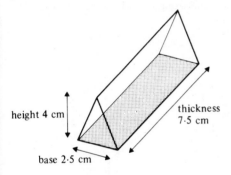

height 4 cm

thickness 7·5 cm

base 2·5 cm

area of end $= \frac{1}{2}$ $(2 \cdot 5 \times 4)$ cm^2

$$\text{volume} = \frac{2 \cdot 5 \times \cancel{4}^{2}}{\cancel{2}_{1}} \times 7 \cdot 5 \text{ cm}^3$$

$$= 5 \times 7 \cdot 5 = 37 \cdot 5 \text{ cm}^3$$

Example 6

A cylinder has a radius of $3\frac{1}{2}$ cm and is 14 cm long. Find its volume. (Take $\pi = \frac{22}{7}$).

area of end $= \pi r^2 = \frac{22}{7} \times \frac{7}{2} \times \frac{7}{2}$ cm^2

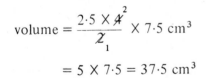

$$\text{volume} = \frac{\overset{11}{\cancel{22}}}{\cancel{7}_1} \times \frac{\overset{1}{\cancel{7}}}{\cancel{2}_1} \times \frac{7}{\cancel{2}_1} \times \frac{\overset{7}{\cancel{14}}}{1}$$

$$= 539 \text{ cm}^3$$

Exercise 7.4c

Find the volume of each triangular prism.
1. base $= 6$ cm
   height $= 5$ cm
   thickness $= 12$ cm
2. base $= 7$ cm
   height $= 4$ cm
   thickness $= 11$ cm
3. base $= 8$ cm
   height $= 9$ cm
   thickness $= 15$ cm
4. base $= 12$ cm
   height $= 10$ cm
   thickness $= 25$ cm
5. base $= 40$ mm
   height $= 50$ mm
   thickness $= 75$ mm
6. base $= 30$ mm
   height $= 48$ mm
   thickness $= 25$ mm
7. base $= 50$ mm
   height $= 32$ mm
   thickness $= 15$ mm
8. base $= 18$ mm
   height $= 25$ mm
   thickness $= 24$ mm
9. base $= 0 \cdot 8$ m
   height $= 1 \cdot 5$ m
   thickness $= 5$ m
10. base $= 4 \cdot 8$ m
    height $= 2 \cdot 5$ m
    thickness $= 1 \cdot 5$ m

Find the volume of each cylinder
(take $\pi = 3\frac{1}{7}$ or $\frac{22}{7}$).

11. radius $= 7$ cm
    length $= 12$ cm
12. radius $= 3$ cm
    length $= 21$ cm
13. radius $= 5$ cm
    length $= 14$ cm
14. radius $= 8$ cm
    length $= 35$ cm
15. radius $= 10$ mm
    length $= 28$ mm
16. radius $= 14$ mm
    length $= 15$ mm
17. radius $= 20$ mm
    length $= 56$ mm
18. radius $= 15$ mm
    length $= 70$ mm
19. radius $= 0 \cdot 7$ m
    length $= 2$ m
20. radius $= 0 \cdot 5$ m
    length $= 1 \cdot 4$ m

Example 7

The volume of a triangular prism is $55 \cdot 2$ cm^3. If the area of the cross section of this prism is 12 cm^2, find the length of the prism.

volume of prism
$$= \text{(area of cross section)} \times \text{length}$$
$$55 \cdot 2 = 12 \times \text{length}$$
$$\therefore \text{length} = 55 \cdot 2 \div 12$$
$$= 4 \cdot 6 \text{ cm}$$

Exercise 7.4d

Find the length of each prism.

	area of cross section	volume
1.	10 cm^2	50 cm^3
2.	20 cm^2	600 cm^3

3.	25 cm²	500 cm³
4.	12 mm²	240 mm³
5.	16 mm²	4800 mm³
6.	15 cm	2250 cm³
7.	11 cm²	781 cm³
8.	32 m²	2560 m³
9.	64 cm²	480 cm³
10.	18 mm²	765 mm³

Example 8

A machine for making ice freezes 5·76 litres of water into ice cubes measuring 4 cm by 3 cm by 2 cm. How many bricks will be made?

$$\text{volume of 1 ice brick} = 4 \times 3 \times 2 = 24 \text{ cm}^3$$
$$\text{volume of water} = 5 \cdot 76 \text{ litres}$$
$$= 5 \cdot 76 \times 1000 \text{ cm}^3$$
$$= 5760 \text{ cm}^3$$
$$\therefore \text{no. of bricks made} = 5760 \div 24 = 240$$

Example 9

A storage tank at a petrol station is a cuboid measuring 4 m long by 3 m wide by 2 m deep. During one day the garage sells 7200 litres of petrol from the tank. What is the fall in the level of the petrol in the tank?

$$\text{volume of petrol sold} = 7200 \times 1000 \text{ cm}^3$$
area of cross section of tank
$$= 400 \times 300 \text{ cm}^2$$

$$\therefore \text{depth fallen} = \frac{\overset{24}{\cancel{7200}} \times \overset{10}{\cancel{1000}}}{\underset{4}{\cancel{400}} \times \underset{1}{\cancel{300}}}$$

$$= \frac{240}{4} = 60 \text{ cm}$$

Exercise 7.4e

1. A rectangular block of metal has dimensions equal to 24, 12 and 8 cm. It is melted down and recast into cubes of edge length 4 cm. How many cubes will be cast?

2. A child has a number of toy bricks, all of which are cubes of edge length 6 cm. He also has a wooden box which is 30 cm long, 24 cm wide and 18 cm deep. If all of the bricks pack exactly into the box, how many bricks does he have?

3. An ice machine freezes 6 litres of water and then forms ice bricks of dimensions 5, 4 and 2 cm. How many ice bricks will be formed?

4. The platform at the base of a statue is to be built to a height of 50 cm on a square base measuring 150 cm by 150 cm. It is to be constructed from bricks of dimensions 30, 15 and 10 cm. How many bricks will be required?

5. A wall, 18 m long, 120 cm high and 45 cm thick, is to be constructed using bricks of dimensions 30, 15 and 10 cm. How many bricks will be required?

6. A coffee urn has a square base measuring 30 cm by 30 cm and it is 50 cm in height. It is used to serve coffee into cups of capacity 0·15 litre. By how many centimetres will the level of the coffee in the urn drop after 120 cups have been served?

7. The dimensions of a greenhouse are shown in the diagram. Calculate the total volume enclosed by the structure.

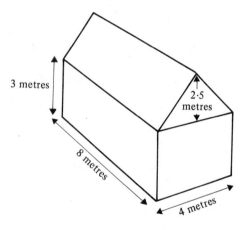

3 metres · 2·5 metres · 8 metres · 4 metres

8. A drinking trough for cattle takes the shape of an inverted triangular prism. It is 3 m long, 40 cm wide and its maximum depth is 30 cm. Find its capacity in litres. If it is filled by using a bucket of capacity 15 litres, how many buckets full of water are required?

9. A beer cask has a radius of 20 cm and is 70 cm high. Find its capacity in litres. How many glasses full of beer can it serve if the capacity of each glass is 0·55 litres? ($\pi = \frac{22}{7}$).

10. A railway tunnel 140 m in length is to be bored with a circular cross section of radius 4 metres. What volume of spoil has to be excavated? If the spoil is to be taken away in wagons of capacity 88 m³, how many wagon loads will be moved? ($\pi = \frac{22}{7}$).

Cone

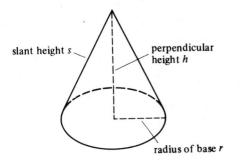

slant height *s*

perpendicular height *h*

radius of base *r*

volume of a cone

$$= \frac{(\text{area of base}) \times (\text{perpendicular height})}{3}$$

$$= \frac{\pi r^2 h}{3}$$

Example 1

Find the volume of a cone, with base radius 6 cm, and perpendicular height 5 cm (take $\pi = 3 \cdot 14$).

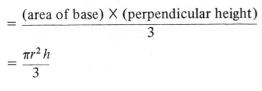

$$\text{volume} = \frac{\pi r^2 h}{3} = \frac{3 \cdot 14 \times \overset{2}{6} \times 6 \times 5}{\underset{1}{3}} \text{ cm}^3$$

$$= 3 \cdot 14 \times 60 = 188 \cdot 4 \text{ cm}^3$$

Exercise 7.5a

Find the volumes of the following cones.

1. radius = 3 cm, perpendicular height = 7 cm.
$(\pi = \frac{22}{7})$
2. radius = 9 cm, perpendicular height = 35 cm.
$(\pi = \frac{22}{7})$
3. radius = 10 cm, perpendicular height = 15 cm.
$(\pi = 3 \cdot 14)$
4. radius = 15 cm, perpendicular height = 40 cm.
$(\pi = 3 \cdot 14)$
5. radius = 60 mm, perpendicular height = 105 mm.
$(\pi = \frac{22}{7})$
6. radius = 45 mm, perpendicular height = 70 mm.
$(\pi = \frac{22}{7})$
7. radius = 20 mm, perpendicular height = 60 mm.
$(\pi = 3 \cdot 14)$
8. radius = 0·7 m, perpendicular height = 1·5 m.
$(\pi = \frac{22}{7})$

9. radius = 0·35 m, perpendicular height = 1·2 m.
$(\pi = \frac{22}{7})$
10. radius = 0·4 m, perpendicular height = 0·75 m.
$(\pi = 3 \cdot 14)$

Sphere

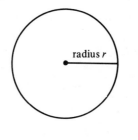

radius *r*

$$\text{volume of a sphere} = \frac{4\pi r^3}{3}$$

Example 2

Find the volume of a sphere with radius 7 mm (Take $\pi = 3\frac{1}{7}$).

$$\text{volume} = \frac{4\pi r^3}{3}$$

$$= \frac{4 \times 22 \times 7 \times 7 \times \overset{1}{7}}{3 \times \underset{1}{7}}$$

$$= \frac{88 \times 49}{3}$$

$$= \frac{4312}{3} = 1437\frac{1}{3} \text{ mm}^3$$

Exercise 7.5b

Find the volume of each sphere.

1. radius = 3 cm $(\pi = 3 \cdot 14)$
2. radius = 21 mm $(\pi = \frac{22}{7})$
3. radius = 90 mm $(\pi = 3 \cdot 14)$
4. diameter = 12 cm $(\pi = 3 \cdot 14)$
5. diameter = 3 cm $(\pi = 3 \cdot 14)$
6. radius = 1·05 cm $(\pi = \frac{22}{7})$
7. radius = $5\frac{1}{4}$ cm $(\pi = \frac{22}{7})$
8. radius = $4\frac{1}{2}$ cm $(\pi = 3 \cdot 14)$
9. radius = 2·25 cm $(\pi = 3 \cdot 14)$
10. radius = $\frac{3}{4}$ cm $(\pi = 3 \cdot 14)$

Example 3

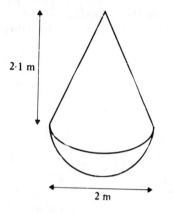

2·1 m

2 m

A navigational buoy consists of a hemispherical base of diameter 2 m with a cone
2·1 m high on top of it.
Find the volume of the buoy.
(Take $\pi = 3\cdot14$)

volume of hemispherical base

$$= \frac{2\pi r^3}{3} = \frac{2 \times 3\cdot14 \times 1 \times 1 \times 1}{3}$$

$$= 2\cdot093 \text{ m}^3$$

volume of cone $= \frac{\pi r^2 h}{3}$

$$= \frac{3\cdot14 \times 1 \times 1 \times 2\cdot1}{3}$$

$$= 2\cdot198 \text{ m}^3$$

volume of buoy $= 2\cdot093 + 2\cdot198$
$$= 4\cdot291 \text{ m}^3$$

Exercise 7.5c

1. A magician's hat is conical, with dimensions
   as shown. Find the volume of this hat.
   $(\pi = \frac{22}{7})$

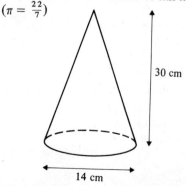

30 cm

14 cm

2. A raspberry jelly is made in the shape of a
   hemisphere of diameter 12 cm.
   Find the volume of the jelly. $(\pi = 3\cdot14)$

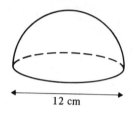

12 cm

3. The dimensions of a gate
   post are shown in the diagram.
   Find its volume.
   $(\pi = 3\cdot14)$

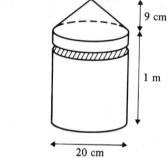

9 cm

1 m

20 cm

4. A fisherman makes a cork float from two
   identical cones with a common base.

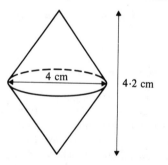

4 cm

4·2 cm

Find the volume of the float from the
dimensions shown. $(\pi = 3\cdot14)$. If 1 cm³ of
cork weighs 0·25 g, how much does this float
weigh?

5. A conical fire extinguisher has a radius of 20 cm
   and a perpendicular height of 54 cm. How many
   litres of pressurized foam does it contain when
   full? $(\pi = 3\cdot14)$.

6. A large conical flask has a radius of 10 cm and a
   perpendicular height of 30 cm. On the side it is
   marked 'Approx. 3 litres'. How much does this
   figure differ from its true capacity? (Take $\pi = 3\cdot14$).

7. A wooden cone has an outer radius of 30 cm and an inner radius of 25 cm. The outer and inner heights are 20 cm and 18 cm respectively. Find the volume of wood in the cone. ($\pi = 3 \cdot 14$).

8. Oil fills an inverted metal cone to a depth of 20 cm. If the radius of the surface of the oil is 15 cm, find the volume of oil.

   It is then poured into a rectangular can of base 25 cm by 15·7 cm. Find the depth of oil in the can. Assume that $\pi = 3 \cdot 14$.

9. A steel cuboid measuring 62·8 mm by 18 mm by 8 mm is melted down and cast into ball bearings of radius 3 mm. How many ball bearings are cast? (Take $\pi = 3 \cdot 14$).

10. A traffic bollard is made in the shape of a cone, with its apex removed.

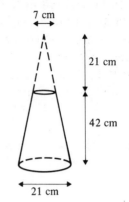

Find the volume of this bollard. ($\pi = \frac{22}{7}$)
If the bollard is made of plastic, and if 1 cm³ weighs 1·2 g, find the weight of the bollard.

7.6 SURFACE AREA

Cuboid

The surface area of a cuboid is found by adding the area of each face.

Example 1

A cuboid measures 10 cm by 5 cm by 3 cm. Find its surface area.

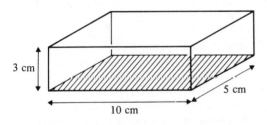

area of front face = $10 \times 3 = 30$ cm²
∴ area of rear face = 30 cm²
area of side face = $5 \times 3 = 15$ cm²
∴ area of other side face = 15 cm²
area of top = $10 \times 5 = 50$ cm²
∴ area of bottom = 50 cm²

∴ total surface area = 30 + 30 + 15 + 15
                    $+ 50 + 50$
             = 190 cm²

Example 2

The sides of a cube measure 9 mm. Find its surface area.

area of each face = $9 \times 9 = 81$ mm²
all six faces are identical, so
total surface area = $81 \times 6 = 486$ mm²

Exercise 7.6a

Find the surface area of each cuboid.
1. length = 10 cm, width = 4 cm, height = 6 cm
2. length = 4 m, width = 7 m, height = 8 m
3. length = 6 m, width = 9 m, height = 2 m
4. length = 12 cm, width = 8 cm, height = 9 cm
5. length = 20 mm, width = 1·5 mm, height = 2 mm

Find the surface area of each cube.
6. side-length = 10 cm
7. side-length = 5 mm
8. side-length = 6 m
9. side-length = 12 cm
10. side-length = 1·5 cm

Example 3

A rectangular box measures 4 m × 2·5 m × 3 m, and is open at the top. Find its surface area.

area of front face $= 4 \times 3 = 12$ m²
∴ area of rear face $= 12$ m²
area of side face $= 2·5 \times 3 = 7·5$ m²
∴ area of other side face $= 7·5$ m²
area of bottom $= 4 \times 2·5 = 10$ m²
there is no top.

∴ total surface area $= 12 + 12 + 7·5 + 7·5 + 10$
$= 49$ m²

Example 4

A cube has side-length 7 cm, and is open at the top. Find its surface area.

area of each face $= 7 \times 7 = 49$ cm²
there are 5 identical faces, so
total surface area $= 49 \times 5 = 245$ cm²

Exercise 7.6b

Each of the following cuboids is open at the top. Find the surface area of each cuboid.
 1. length $= 12$ m, width $= 2$ m, height $= 3$ m
 2 length $= 5$ cm, width $= 11$ cm, height $= 6$ cm
 3. length $= 8$ cm, width $= 13$ cm, height $= 4$ cm
 4. length $= 7$ mm, width $= 9$ mm, height $= 11$ mm
 5. length $= 10$ cm, width $= 4·5$ cm, height $= 4$ cm

Each of the following cubes is open at the top. Find the surface area of each cube.
 6. side-length $= 2$ mm
 7. side-length $= 8$ mm
 8. side-length $= 30$ cm
 9. side-length $= 1·4$ m
10. side-length $= 3·8$ cm

Cylinder

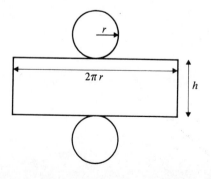

The curved surface area of a cylinder $= 2\pi rh$
The area of each end face $= \pi r^2$

Example 5

A cylinder has radius $3\frac{1}{2}$ cm and height 10 cm. Find the total surface area if the cylinder is
(a) open at both ends
(b) closed at both ends.
    (Take $\pi = \frac{22}{7}$)

(a) curved surface area $= 2\pi rh$

$$= \frac{\cancel{2}^1}{1} \times \frac{22}{\cancel{7}_1} \times \frac{\cancel{7}^1}{\cancel{2}_1} \times \frac{10}{1} \text{ cm}^2$$

$= 220$ cm²

∴ total surface area $\quad = 220$ cm²

(b) area of one end $= \pi r^2$

$$= \frac{\cancel{22}^{11}}{\cancel{7}_1} \times \frac{\cancel{7}^1}{\cancel{2}_1} \times \frac{7}{2}$$

$$= \frac{77}{2} = 38\frac{1}{2} \text{ cm}^2$$

∴ area of other end $= 38\frac{1}{2}$ cm²

∴ total surface area $= 200 + 38\frac{1}{2} + 38\frac{1}{2}$

$= 297$ cm²

Exercise 7.6c

The following cylinders are open at both ends. Find the curved surface area of each cylinder. (Take $\pi = 3·14$)
 1. base radius $= 10$ cm, height $= 15$ cm
 2. base radius $= 20$ cm, height $= 30$ cm
 3. base radius $= 5$ m, height $= 60$ m
 4. base radius $= 6$ cm, height $= 25$ cm
 5. base radius $= 1·5$ m, height $= 3$ m

The following cylinders are closed at one end only. Find the total surface area of each cylinder. (Take $\pi = \frac{22}{7}$)
 6. base radius $= 7$ cm, height $= 10$ cm
 7. base radius $= 28$ cm, height $= 2$ cm
 8. base radius $= 21$ mm, height $= 8$ mm
 9. base radius $= 0·7$ cm, height $= 5$ cm
10. base radius $= 10\frac{1}{2}$, height $= 50$ cm

The following cylinders are closed at both ends. Find the total surface area of each cylinder. (Take $\pi = 3·14$)
11. base radius $= 5$ cm, height $= 10$ cm
12. base radius $= 2$ cm, height $= 100$ cm

13. base radius = 6 m, height = 200 m
14. base radius = 100 mm, height = 28 mm
15. base radius = 50 cm, height = 0·5 cm

Cone

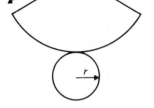

The curved surface area of a cone
    = π × (radius of base) × (slant height)
    = πrs

The area of the base of a cone = πr^2

Example 6

A cone has a slant height of 12 mm, and the radius of its base is 5 mm. If the cone is closed at the base, find its total surface area. (Take π = 3·14)

curved surface area = πrs
                    = 3·14 × 5 × 12
                    = 188·4 mm²
area of base = πr^2
             = 3·14 × 5 × 5
             = 78·5 mm²
∴ total surface area = 188·4 + 78·5
                     = 266·9 mm²

Exercise 7.6d

Find the curved surface area of each cone.
(Take $\pi = \frac{22}{7}$)

1. base radius = 7 cm, slant height = 10 cm
2. base radius = 14 cm, slant height = 40 cm
3. base radius = $3\frac{1}{2}$ m, slant height = 4 m
4. base radius = $1\frac{3}{4}$ m, slant height = 16 m
5. base radius = $10\frac{1}{2}$ cm, slant height = 12 cm

If each cone is closed at the base, find the total surface area. (Take π = 3·14)

6. base radius = 10 cm, slant height = 20 cm
7. base radius = 20 cm, slant height = 60 cm
8. base radius = 100 mm, slant height = 120 mm
9. base radius = 30 cm, slant height = 65 cm
10. base radius = 50 mm, slant height = 200 mm

Sphere

The surface area of a sphere = $4\pi r^2$

Example 7

The diameter of a sphere is $3\frac{1}{2}$ cm. Find its surface area. (Take $\pi = \frac{22}{7}$)

radius = $3\frac{1}{2} \div 2 = 1\frac{3}{4}$ cm
surface area = $4\pi r^2$

$$= \frac{4}{1} \times \frac{22}{7} \times \frac{7}{4} \times \frac{7}{4} \text{ cm}^2$$

$$= \frac{77}{2} = 38\frac{1}{2} \text{ cm}^2$$

Exercise 7.6e

Find the surface area of each sphere.

1. radius = 2 mm, take π = 3·14
2. radius = 5 cm, take π = 3·14
3. radius = $3\frac{1}{2}$ m, take $\pi = \frac{22}{7}$
4. radius = 10 cm, take π = 3·14
5. radius = $5\frac{1}{4}$ mm, take $\pi = \frac{22}{7}$
6. diameter = 14 cm, take $\pi = \frac{22}{7}$
7. diameter = 10 cm, take π = 3·14
8. diameter = 21 m, take $\pi = \frac{22}{7}$
9. diameter = $17\frac{1}{2}$ mm, take $\pi = \frac{22}{7}$
10. diameter = 40 cm, take π = 3·14

When using a formula, replace the letters by their given value.

Example 1

A cylinder of height 4 cm has a volume of 1256 cm³.

Use $r = \sqrt{\dfrac{V}{\pi h}}$ to find the radius of the cylinder. (Take $\pi = 3{\cdot}14$)

$h = 4$ cm, $V = 1256$ cm³

$$r = \sqrt{\dfrac{1256}{3{\cdot}14 \times 4}}$$
$$= \sqrt{100}$$
$$= 10$$

∴ the radius is 10 cm

Example 2

The volume of a cone is 4400 mm³.
If the radius of the base of the cone is 14 mm, find its height.

Use $h - \dfrac{3V}{\pi r^2}$, and take $\pi - \dfrac{22}{7}$

$V = 4400$ mm³, $r = 14$ mm

$$h = \dfrac{\overset{1}{\cancel{7}}}{\cancel{22}_1} \times \dfrac{3 \times \overset{\overset{50}{\overset{100}{200}}}{\cancel{4400}}}{7 \,\underset{1}{\cancel{14 \times 14}}_2}$$

$$= \dfrac{3 \times 50}{7} = \dfrac{150}{7}$$

$$= 21\tfrac{3}{7}$$

∴ the height is $21\tfrac{3}{7}$ mm

Exercise 7.7a

1. The curved surface area A of a cone is 314 cm², and the slant height s is 50 cm. Find the radius.

   Use $r = \dfrac{A}{\pi s}$, and take $\pi = 3{\cdot}14$

2. A cone has base radius $r = 10$ mm, and its surface area A is 125·6 mm².
   Find the slant height of the cone.

   Use $s = \dfrac{A}{\pi r}$, and take $\pi = 3{\cdot}14$

3. A cone has volume $V = 50{\cdot}24$ cm³, and its base radius r is 2 cm. Find the height of the cone.

   Use $h = \dfrac{3V}{\pi r^2}$, and take $\pi = 3{\cdot}14$

4. Find the height of a cylinder of radius $r = 9$ m. The surface area A is 282·6 m², and the cylinder is open at both ends.

   Use $h = \dfrac{A}{2\pi r}$, and take $\pi = 3{\cdot}14$

5. The curved surface area A of a cylinder is 880 cm². The height of the cylinder is 10 cm. Find the radius.

   Use $r = \dfrac{A}{2\pi h}$, and take $\pi = \dfrac{22}{7}$

6. The curved surface area A of a cone is 198 cm², and the slant height s is 14 cm. Find the base radius.

   Use $r = \dfrac{A}{\pi s}$, and take $\pi = \dfrac{22}{7}$

7. The volume V of a cone is 770 cm³, and its base radius r is 3·5 cm. Find its height.

   Use $h = \dfrac{3V}{\pi r^2}$, and take $\pi = \dfrac{22}{7}$

8. The height h of a cylinder is 5 m, and its curved surface area A is 314 m². Find the base radius.

   Use $r = \dfrac{A}{2\pi h}$, and take $\pi = 3{\cdot}14$

9. Find the height of an open cylinder, if its curved surface area A is 264 mm², and its base radius r is 21 mm.

   Use $h = \dfrac{A}{2\pi r}$, and take $\pi = \dfrac{22}{7}$

10. The radius r of the base of a cylinder is 20 cm, and its volume V is 6280 cm³. Find the height of the cylinder.

    Use $h = \dfrac{V}{\pi r^2}$, and take $\pi = 3{\cdot}14$

11. Find the base radius r of a cylinder of volume $V = 25{\cdot}12$ mm³, if its height h is 2 mm.

    Use $r = \sqrt{\dfrac{V}{\pi h}}$, and take $\pi = 3{\cdot}14$

12. A cylinder has height h 42 m, and volume $V = 2112$ m³. Find the radius of its base.

    Use $r = \sqrt{\dfrac{V}{\pi h}}$, and take $\pi = \dfrac{22}{7}$

13. The surface area A of a sphere is 616 cm². Find its radius.

    Use $r = \sqrt{\dfrac{A}{4\pi}}$, and take $\pi = \dfrac{22}{7}$

14. Find the radius of a sphere of surface area $A = 1256$ cm².

    Use $r = \sqrt{\dfrac{A}{4\pi}}$, and take $\pi = 3{\cdot}14$

15. A cylinder of height h 2·5 cm has volume $V = 3140$ cm³. Find its base radius.

    Use $r = \sqrt{\dfrac{V}{\pi h}}$, and take $\pi = 3{\cdot}14$

Plot a graph as follows.

(i) Draw the axes at right angles to each other.
(ii) Scale the axes according to the data given.
(iii) Label each axis clearly.
(iv) Plot the points.
(v) Join up these points either by straight lines or by a suitable curve.
(vi) Give the graph a title.

Example 1

On a holiday in France, a boy notes every hour how many litres of petrol are left in the tank of the car. The table lists his results.

Time	Number of litres
10.00	30
11.00	25
12.00	17·5
13.00	12
13.00	40
14.00	40
15.00	40
16.00	32
17.00	25

Using a scale of 2 cm to 1 hour on the horizontal axis and 1 cm to 5 litres on the vertical axis, draw a graph to show the information in the table.

From your graph, find:

(a) during which hour most petrol was used,
(b) between which times the family stopped for lunch,
(c) what was the petrol consumption in kilometres per litre between 12.00 and 13.00 if the car travelled 66 km in this hour.

The graph below is shown smaller than is required to be drawn.

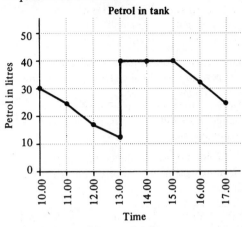

Petrol in tank

From the graph:

(a) Most petrol was used between 15.00 and 16.00, i.e. 8 litres.
(b) The family stopped for lunch between 13.00 and 15.00 because no petrol was used between 13.00 and 15.00
(c) The petrol used between 12.00 and 13.00 was $5\frac{1}{2}$ litres.
∴ petrol consumption $= 66 \div 5\frac{1}{2}$

$$= 12 \text{ km}/l.$$

Exercise 8.1a

1. The following table gives the outside temperature at hourly intervals on a certain day in Winter.

Time	Temp	Time	Temp
6 a.m.	0°C	2 p.m.	14°C
7 a.m.	0°C	3 p.m.	12°C
8 a.m.	1°C	4 p.m.	9°C
9 a.m.	1°C	5 p.m.	5°C
10 a.m.	2°C	6 p.m.	4°C
11 a.m.	4°C	7 p.m.	2°C
12 noon	9°C	8 p.m.	1°C
1 p.m.	10°C		

Draw a graph to show this information.
Use a scale of 2 cm to 1 hour on the horizontal axis, and a scale of 1 cm to 1°C on the vertical axis.

(a) From your graph, estimate the temperature
at the following times:

(i) 10.30 a.m.   (ii) 2.30 p.m.
(iii) 4.30 p.m.   (iv) 6.30 p.m.
(v) 1.15 p.m.   (vi) 4.15 p.m.

(b) During which one-hour interval does the
temperature rise most quickly, and by how
many degrees?
(c) During which one-hour interval does the
temperature fall most quickly, and by how
many degrees?

2. The following table gives the voltage of the
electrical mains supply at hourly intervals for a
certain day.

Time	Voltage	Time	Voltage
5 a.m.	250 V	3 p.m.	253 V
6 a.m.	247 V	4 p.m.	255 V
7 a.m.	243 V	5 p.m.	245 V
8 a.m.	235 V	6 p.m.	232 V
9 a.m.	239 V	7 p.m.	235 V
10 a.m.	247 V	8 p.m.	237 V
11 a.m.	250 V	9 p.m.	239 V
12 noon	249 V	10 p.m.	240 V
1 p.m.	245 V	11 p.m.	245 V
2 p.m.	250 V	12 midnight	251 V

Draw a graph to show this information. Use a
scale of 1 cm to 1 h on the horizontal axis, and
1 cm to 2 V on the vertical axis starting at 230 V.

(a) From your graph, estimate the voltage at the
following times:

(i) 8.30 p.m.   (ii) 12.30 p.m.
(iii) 4.30 p.m.   (iv) 6.15 a.m.
(v) 9.15 a.m.   (vi) 7.45 a.m.

(b) During which one-hour interval does the
voltage drop most quickly and by how much?

3. Haweswater in Cumbria is used as a reservoir. The
depth of the water in the reservoir on the first day
of each month in a certain year is given in the
table.

Date	Depth	Date	Depth
1 Jan.	56 m	1 July	42 m
1 Feb.	54 m	1 Aug.	35 m
1 Mar.	58 m	1 Sept.	25 m
1 Apr.	56 m	1 Oct.	33 m
1 May	50 m	1 Nov.	44 m
1 June	45 m	1 Dec.	50 m

Draw a graph to show this information. Use a
scale of 2 cm to 1 month on the horizontal axis,
and 1 cm to 2 m on the vertical axis starting at
24 m.
From your graph, estimate the depth of water on
the following dates:

(i) 16 Jan.   (ii) 15 Apr.
(iii) 16 Aug.   (iv) 15 Nov.
(v) 7 Feb.   (vi) 23 Sept.
(vii) 8 Mar.   (viii) 23 Nov.

4. The atmospheric pressure of 12 midday and
12 midnight for a certain week is given in the
table in mm of mercury.

Day	Pressure	Day	Pressure
Mon.	750	Fri.	756
	752		752
Tues.	753	Sat.	755
	754		759
Wed.	762	Sun.	758
	765		764
Thurs.	766		
	762		

Draw a graph to show this information using
a scale of 2 cm to 12 h on the horizontal axis,
and 1 cm to 1 mm on the vertical axis starting
at 750 mm.

(a) From your graph, estimate the atmospheric
pressure at the following times:

(i) 6 p.m. Mon.   (ii) 6 a.m. Fri.
(iii) 6 p.m. Sat.   (iv) 6 p.m. Sun.
(v) 3 a.m. Wed.   (vi) 3 p.m. Fri.

(b) During which 12-hour interval does the
pressure rise most quickly, and by how
many mm?
(c) During which 12-hour interval does the
pressure fall most quickly, and by how
many mm?

5. A man walks from Romsley to Hagley over the summit of both Walton Hill and Adam's Hill. The table gives his distance from Romsley and his altitude at various points.

	Distance	Altitude
Romsley	0 km	230 m
Whitehall Farm	1·5 km	230 m
Walton Hill Summit	2·0 km	315 m
St. Kenelm's Pass	3·0 km	250 m
Adam's Hill Summit	3·5 km	305 m
Hagley Hall	5·0 km	155 m
Hagley Village	5·5 km	140 m

Draw a graph to show this information using a scale of 4 cm to 1 km on the horizontal axis, and 1 cm to 10 m on the vertical axis starting at 130 m.

(a) From your graph, estimate his altitude at the following points:

  (i) 2·5 km from Romsley,
  (ii) 4 km from Romsley,
  (iii) 1 km from Hagley Village.

(b) Is his first climb or his second climb the steeper?
(c) Find how many metres of altitude he loses for every kilometre he walks down from Adam's Hill to Hagley Hall.

6. Draw a smooth curve on a graph to show the length of the perimeter in cm of a square of a given area in cm^2 from the table below. Use a scale of 2 cm to 5 cm on the horizontal axis, and 2 cm to 10 cm^2 on the vertical axis.

Area cm^2	Perimeter cm	Area cm^2	Perimeter cm
1	4	36	24
4	8	49	28
9	12	64	32
16	16	81	36
25	20	100	40

From your graph find:

(a) the area of a square having a perimeter of length: (i) 15 cm (ii) 22 cm (iii) 29 cm (iv) 35 cm
(b) the length of perimeter of a square having an area: (i) 20 cm^2 (ii) 45 cm^2 (iii) 60 cm^2 (iv) 90 cm^2

7. Draw a smooth curve from the information in the table. This shows the distance at which an image is formed by a convex lens of an object at various distances from the lens.

Object distance in cm	Image distance in cm
12	60·0
13	42·0
14	35·0
18	22·5
20	20·0
30	15·0
50	12·5

Show the object distance on the horizontal axis, and the image distance on the vertical axis. Use a scale of 2 cm to 10 cm for each axis.

From your graph find:
(a) the image distance for an object distance of (i) 15 cm (ii) 23 cm (iii) 25 cm (iv) 41 cm.
(b) the object distance for which the image distance is:
  (i) twice as long (ii) five times as long (iii) half as long.

8. The table shows the temperature of water in a kettle at various times after the electric current has been switched on.

Time	Temp.	Time	Temp.
start	10°C	90s	83°C
15s	36°C	120s	91°C
30s	50°C	150s	96°C
45s	61°C	180s	99°C
60s	70°C	210s	100°C

Draw a graph to show this information. Use a scale of 2 cm to 30 seconds on the horizontal axis, and 2 cm to 10°C on the vertical axis.

(a) Find the temperature of the water after (i) 75s (ii) 105s (iii) 2 min 15s
(b) Find the time to heat the water:

  (i) from 10°C to 55°C
  (ii) from 55°C to 100°C

and suggest a reason for your answers.

Straight line graphs are the result of plotting quantities which are directly proportional to one another.

Further information can be obtained from points which lie in between those already plotted.

Example 1

For his holiday in France, a boy produced the following table to help him change £'s into francs or francs into £'s.

Number of £'s	1	2	3	4	5	10
Number of francs	9	18	27	36	45	90

Using a scale of 1 cm to £1 on the horizontal axis and 1 cm to 10 francs on the vertical axis, draw the graph.

From your graph determine:

(a)   how many francs he would get for £7·50,

(b)   the value in £'s of 32 francs.

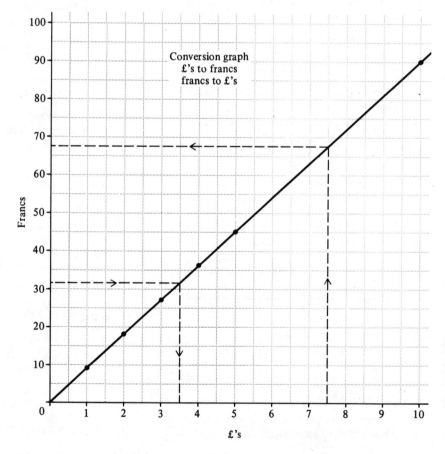

From the above graph:

(a)   he would get approximately 67 francs for £7·50,

(b)   32 francs are worth approximately £3·50.

Exercise 8.2a

1. The table shows the conversion of speed in metres per second to speed in kilometres per hour.

Speed m/s	0	20	40	60
Speed km/h	0	72	144	216

Draw a graph to show this information. Use a scale of 2 cm to 10 m/s on the horizontal axis and 2 cm to 10 km/h on the vertical axis.

(a) From your graph, convert these speeds to km/h:
   (i) 15 m/s; (ii) 35 m/s; (iii) 55 m/s.
(b) From your graph, convert these speeds to m/s:
   (i) 36 km/h; (ii) 90 km/h; (iii) 162 km/h.

2. The speed of a car at various times is shown in the table.

Time s	0	3	6	9	12
Speed km/h	0	27	54	81	108

Show this information on a graph using a scale of 1 cm to 1 second on the horizontal axis and a scale of 2 cm to 10 km/h on the vertical axis.

(a) Find the speed of the car after:
   (i) 5s; (ii) 8s; (iii) 11s.
(b) Find the time when the speed of the car is:
   (i) 18 km/h; (ii) 63 km/h; (iii) 90 km/h.

3. The current in amps in a wire for a given potential difference in volts is shown in the table.

p.d. V	0	3	6	9	12
Current A	0	1·2	2·4	3·6	4·8

Draw a graph of this information, using a scale of 1 cm to 1 V on the horizontal axis and 2 cm to 1 A on the vertical axis.

(a) Find the current for a p.d. of:
   (i) 1 V; (ii) 4 V; (iii) 7 V; (iv) 11 V.
(b) Find the p.d. for a current of:
   (i) 0·8 A; (ii) 2 A; (iii) 3·2 A; (iv) 4·2 A

4. The average train fare for a journey of a given distance is shown in the table.

Distance km	50	100	150
Fare £'s	2	4	6

Show this information on a graph using a scale of 1 cm to 10 km on the horizontal axis and 4 cm to £1 on the vertical axis.

(a) Find the expected fare for a journey of:
   (i) 40 km; (ii) 90 km; (iii) 120 km.
(b) Find the expected journey distance for a fare of:
   (i) £1·20; (ii) £2·80; (iii) £4·40.

5. The real depth of water in a beaker is compared with its apparent depth when viewed from above in the table below.

Real depth cm	3	6	9	12
Apparent depth cm	2·25	4·5	6·75	9

Draw a graph for this, showing the real depth to a scale of 1 cm to 1 cm on the horizontal axis and the apparent depth on the vertical axis to a scale of 2 cm to 1 cm.

(a) Find the apparent depth for a real depth of:
   (i) 1 cm; (ii) 8 cm; (iii) 10 cm.
(b) Find the real depth for an apparent depth of:
   (i) 1·5 cm; (ii) 3·75 cm; (iii) 8·25 cm.

6. A department store is selling everything at a discount dependent on the normal purchase price, as shown in the table.

Price £'s	100	400	600	1000
Discount £'s	8	32	48	80

Draw a graph of this information showing the normal purchase price on the horizontal axis to a scale of 2 cm to £100, and the discount on the vertical axis to a scale of 2 cm to £10.

(a) Find the discount offered on
   (i) a cooker costing £150,
   (ii) a colour television costing £350,
   (iii) a three-piece suite at £800.
(b) Find the normal purchase price of
   (i) a washing machine at a discount of £8,
   (ii) a set of dining chairs at a discount of £36,
   (iii) a bedroom suite at a discount of £60.

Example 2

The table shows the cost of buying gas for cooking and heating.

No. of therms used	5	8	10	12	16	18
Cost (£'s)	£1·75	£2·20	£2·50	£2·80	£3·40	£3·70

The cost consists of a fixed amount (the Standing Charge) plus an amount which is proportional to the number of therms used.
Draw a straight line graph to illustrate the above data. Use a scale of 1 cm to represent 2 therms on the horizontal axis and 1 cm to represent 50p on the vertical axis.

From your graph, estimate:
(a) the cost when 14 therms are used,
(b) the number of therms used when the cost is £3·55,
(c) the Standing Charge,
(d) the average cost per therm excluding the Standing Charge.

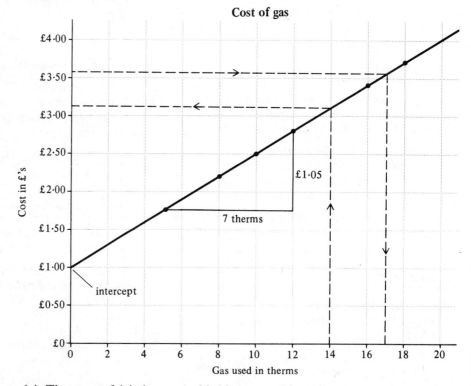

Cost of gas

(a) The cost of 14 therms is £3·10.
(b) The number of therms used is 17.
(c) The Standing Charge is obtained from the *intercept* (the point where the line cuts the vertical axis). This is the cost when no gas is used. The Standing Charge is £1.
(d) Without the Standing Charge, the cost per therm is the *gradient* of the line.

$$\text{Gradient} = \frac{2·80 - 1·75}{12 - 5} = \frac{1·05}{7} = 0·15 \qquad \therefore \text{Cost per therm} = 15\text{p.}$$

Exercise 8.2b

1. The temperature of the water in a hot water tank is recorded at 15 minute intervals after the heater is switched on.

Time min.	15	30	45	60	75	90
Temp. °C	20	30	40	50	60	70

Draw a graph of this information. Show the time from 0 min on the horizontal axis to a scale of 2 cm for 10 min. Show the temperature from 0°C on the vertical axis to a scale of 2 cm for 10°C.

(a) Find the temperature after 27 minutes.
(b) Find the time taken for the temperature to reach 64°C.
(c) Find the temperature when the heater was first switched on.
(d) Find the gradient of the line in °C per min.

2. A spring is stretched by hanging various weights from it, as recorded in the table below.

Weight g	125	250	375	500
Length of spring cm	40	50	60	70

Show this information on a graph using a scale of 4 cm to 100 g from 0 g to 500 g on the horizontal axis, and a scale of 2 cm to 10 cm for the length of spring from 0 cm to 70 cm on the vertical axis.

From your graph, find:
(a) the extended length of the spring when stretched by a 100 g weight.
(b) the weight that will stretch the spring to a length of 58 cm,
(c) the natural length of the spring (from the intercept),
(d) the gradient of the line in cm per g.

3. A gas in an enclosed vessel is heated, and the pressure and temperature are recorded.

Temp. °C	20	40	60	80	100
Pressure mm Hg	800	850	900	950	1000

Draw a graph showing the temperature from 0°C on the horizontal axis to a scale of 2 cm for 10°C and the pressure in millimetres of mercury from 600 mm on the vertical axis to a scale of 4 cm for 100 mm Hg.

From your graph, find:
(a) the pressure at 32°C,
(b) the temperature at 920 mm Hg,
(c) the pressure at 0°C,
(d) the increase in pressure for every °C increase in temperature.

4. The altitude (height above sea level) of a railway incline is given in the table.

Distance from start km	3	6	9	12	15
Altitude m	150	190	230	270	310

Draw a graph of this information. Show the altitude from sea level on the vertical axis to a scale of 5 cm for every 100 m. Show the distance from 0 km to a scale of 1 cm for every 1 km on the horizontal axis.

From your graph, find:
(a) the altitude 4·5 km from the start,
(b) the distance from the start at an altitude of 290 m.
(c) the altitude of the track at the start,
(d) the gradient of the incline.

5. The length of a steel rod is accurately measured at various temperatures. The results are shown in the table.

Temp. °C	Length mm
20	1002·2
40	1002·4
60	1002·6
80	1002·8
100	1003·0

Show this information on a graph. Use a scale from 0°C of 2 cm to 10°C on the horizontal axis, and a scale from 1000 mm of 5 cm to 1 mm on the vertical axis.

From your graph, find:
(a) the length of the rod at 70°C,
(b) the temperature when the length of the rod is 1002·3 mm,
(c) the length of the rod at 0°C,
(d) the increase in length of the rod for every °C rise in temperature.

Travel graphs are usually straight lines.
The horizontal axis is always used for the
time taken and the vertical axis for the
distance travelled.

Note:

(a) Units must correspond.
    If the speed is given in km/h, then the
    distance is measured in km and the
    time in h.
(b) distance travelled = speed × time.

Example 1

At 13.00 h, a boy starts to cycle to visit
a friend who lives 25 km away. If he can
cycle at an average speed of 10 km/h, at
what time will he arrive?

1. Draw suitable axes, with the time on the
   horizontal axis scaled from 13.00 h and
   the distance on the vertical axis scaled to
   25 km.
2. The cyclist travels at 10 km/h. So after
   1 hour (i.e. at 14.00 h), he will be 10 km
   away from his starting point. After 2 hours
   (i.e. at 15.00 h), he will be 20 km away
   from his starting point.
   Using the information, plot the straight
   line to show his journey.
3. Read off his time of arrival from the
   graph, i.e. 15.30 h.

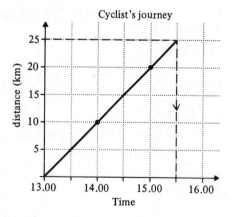

Exercise 8.3a

1. A man travels from Auchtermuchty to Carstairs
   on his motor cycle at an average speed of 40
   km/h.
   If he starts at 5.00 p.m., find from a travel graph
   the time when he reaches:
   (a) Gairney Bank, 20 km from Auchtermuchty
   (b) Old Philpstoun, 50 km from Auchtermuchty
   (c) Mid Calder, 60 km from Auchtermuchty
   (d) Forth, 80 km from Auchtermuchty
   (e) Carstairs, 90 km from Auchtermuchty.

2. A train to Leeds leaves London at 8.00 p.m. Its
   average speed is 100 km/h. Draw a travel graph
   to show this journey and find when the train
   reaches:

   (a) Hatfield, 25 km from London
   (b) Hitchin, 50 km from London
   (c) Peterborough, 125 km from London
   (d) Retford, 225 km from London
   (e) Doncaster, 250 km from London
   (f) Leeds, 300 km from London

3. A train to Plymouth leaves London at 4.00 p.m.
   Its average speed is 120 km/h. Draw a travel
   graph to show this journey and find when the
   train reaches:

   (a) Reading, 60 km from London
   (b) Newbury, 90 km from London
   (c) Westbury, 150 km from London
   (d) Castle Cary, 180 km from London
   (e) Exeter, 270 km from London
   (f) Plymouth, 360 km from London

4. A delivery van travels from Peterhead to Mundurno. The van leaves at 2.00 p.m. and travels at
   an average speed of 28 km/h. Draw a suitable
   graph to find when it reaches:
   (a) Mintlaw, 14 km from Peterhead
   (b) Kinknockie, 21 km from Peterhead
   (c) Newburgh, 42 km from Peterhead
   (d) Foveran, 56 km from Peterhead
   (e) Balmedia, 63 km from Peterhead
   (f) Mundurno, 70 km from Peterhead.

5. A man leaves Rannoch Station at 6.00 a.m. and
   cycles to Bankfoot at an average speed of 16
   km/h.
   Draw a suitable graph to find when he reaches:
   (a) Dalchalloch, 32 km from Rannoch Station
   (b) Struan Station, 40 km from Rannoch Station
   (c) Blair, 44 km from Rannoch Station
   (d) Pass of Killiecrankie, 52 km from Rannoch
       Station

(e) Ballinluig, 64 km from Rannoch Station
(f) Little Dunkeld, 76 km from Rannoch Station
(g) Bankfoot, 84 km from Rannoch Station

6. An athlete runs at an average speed of 12 km/h from Kidderminster to Birmingham. If he leaves at 3.00 p.m., find from a suitable graph the time he arrives at:

(a) Blakedown, 6 km from Kidderminster
(b) Halesowen, 15 km from Kidderminster
(c) Quinton, 18 km from Kidderminster
(d) Bearwood, 21 km from Kidderminster
(e) Birmingham, 27 km from Kidderminster.

7. A party of walkers leave Newton Stewart at 11.00 a.m.
If their average speed is 4 km/h, find from a suitable graph their time of arrival at:
(a) Palnure burn, 7 km from Newton Stewart
(b) Blairs, 9 km from Newton Stewart
(c) Kirkbride, 16 km from Newton Stewart
(d) Carsluith Castle, 18 km from Newton Stewart
(e) Kirkdale, 20 km from Newton Stewart.

8. A man leaves Birmingham by car at 2.00 p.m. and travels along the M6 motorway towards Preston at an average speed of 96 km/h. Find from a travel graph the time at which he passes the service areas:

(a) 24 km from Birmingham
(b) 72 km from Birmingham
(c) 144 km from Birmingham
(d) 168 km from Birmingham

The graph shows a motorist's journey from Cambridge to Bury St. Edmunds and back.

From this graph, find:

(a) the time at which the motorist reached Newmarket,
(b) the distance from Newmarket to Bury St. Edmunds,
(c) how long the motorist stayed in Bury St. Edmunds,
(d) his average speed for the journey from Bury St. Edmunds back to Cambridge.

From the graph:

(a) he reaches Newmarket at 10.30 h

Example 2

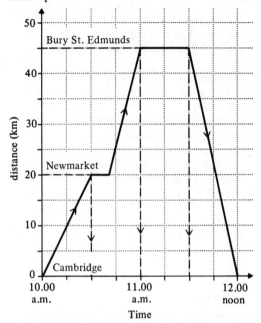

(b) distance from Cambridge to Newmarket = 20 km
distance from Cambridge to Bury St. Edmunds = 45 km.
so distance from Newmarket to Bury St. Edmunds = 45 − 20 = 25 km

(c) He reached Bury St. Edmunds at 11.00h
He left Bury St. Edmunds at 11.30 h
So he stayed 30 minutes.

(d) Time taken from Bury St. Edmunds to Cambridge = 30 minutes
Distance from Bury St. Edmunds to Cambridge = 45 km

$$\text{So speed} = \frac{\text{distance in km}}{\text{time taken in hours}}$$

$$= \frac{45}{30/60} = \frac{45}{\frac{1}{2}}$$

$$= 90 \text{ km/h}$$

Exercise 8.3b

1. The graph opposite shows a return journey from Hawick to Selkirk that a man made by bus. He stopped in Ashkirk on the way and caught the next bus.
   (a) At what time did he reach Selkirk?
   (b) Find the distance from Ashkirk to Selkirk.
   (c) How long did he stop in Ashkirk?
   (d) How long did he stop in Selkirk?
   (e) Find the average speed of his journey
        (i)   from Hawick to Ashkirk,
        (ii) from Ashkirk to Selkirk,
        (iii) from Selkirk back to Hawick.

2. The graph opposite shows a return journey from Sevenoaks to London that a man made by train. He changed trains at Orpington on the way in order to meet a friend. He returned from London to Sevenoaks direct.

   (a) At what time did he reach Orpington?
   (b) At what time did he reach London?
   (c) How far is it from Sevenoaks to Orpington?
   (d) Find the distance from Orpington to London.
   (e) Find how long he waited at Orpington.
   (f) How long did he stay in London?
   (g) Find the average speed of his train
        (i)   from Sevenoaks to Orpington,
        (ii) from Orpington to London,
        (iii) from London back to Sevenoaks.

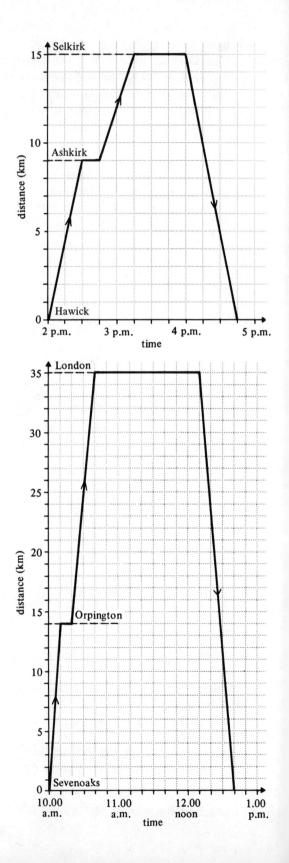

3. The graph opposite shows the return journey of a delivery van from Dalry to Knockentiber. The journey to Knockentiber is direct, but the van stops in Irvine on the way back.
   (a) When did the van reach Knockentiber?
   (b) When did the van leave Irvine?
   (c) Find the distance from Knockentiber to Irvine.
   (d) How long did the van stop in Irvine?
   (e) Find the average speed of the van:
       (i)   from Dalry to Knockentiber,
       (ii) from Knockentiber to Irvine,
       (iii) from Irvine to Dalry.

4. The graph opposite shows the return journey a man made by car from Dundee to Aberdeen. On the way out, he stopped in Brechin, but returned from Aberdeen direct.
   (a) When did he reach Aberdeen?
   (b) How far is Aberdeen from Brechin?
   (c) How long did he stay in Aberdeen?
   (d) Find the average speed:
       (i)   from Aberdeen back to Dundee,
       (ii) from Dundee to Brechin,
       (iii) from Brechin to Aberdeen.

5. The graph opposite shows a return train journey a man made from Manchester to Carlisle. He caught a through train on the outward journey, but on the way back he had to change at Preston.

   (a) Estimate the time he left Preston.
   (b) How long did he stay in Carlisle?
   (c) Estimate the distance from Carlisle to Preston.
   (d) How long do you think he waited in Preston?
   (e) Estimate the average speed of his train:
       (i)   from Manchester to Carlisle,
       (ii) from Carlisle to Preston,
       (iii) from Preston back to Manchester.

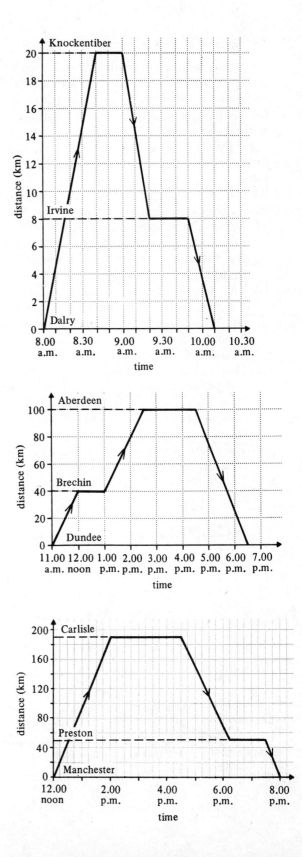

Example 3

At noon a man starts cycling from Saltcoats
to Wemyss Bay 30 km away, at a steady speed
of 10 km/h. He stops for lunch at 1.00 p.m.
for 30 minutes, and then resumes his journey
at the same speed.
A car leaves Wemyss Bay at 12.30 p.m. and
drives to Saltcoats at a steady speed of 50
km/h.
When does the cyclist meet the car, and how
far are they from Saltcoats?

1. Draw suitable axes with the time on the
   horizontal axis scaled from 12 noon and
   the distance on the vertical axis scaled
   to 30 km.
2. The cyclist travels at 10 km/h. After
   1 hour (i.e. at 1 p.m.) he will be 10 km
   from Saltcoats. He then stops for lunch.
   From this information, plot the two
   straight lines to show his journey.
3. The car leaves Wemyss Bay at 12.30 p.m.
   and travels 25 km in $\frac{1}{2}$ hour, i.e. it is 5 km
   from Saltcoats at 1.00 p.m. Plot the
   straight line for the car from this information.
4. The point of intersection of the two lines
   gives the required information.
   They meet at 12.55 p.m. 9 km from
   Saltcoats.

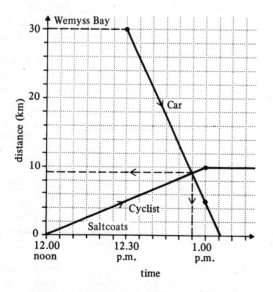

Exercise 8.3c

1. At 2.00 p.m. Jane leaves for her house and walks
   along a main road at 6 km/h. At 2.30 p.m. her
   sister Anne leaves the house and cycles along the
   same main road at 15 km/h. Find from a graph:
   (a) the time at which they meet,
   (b) the distance that they both are from their
   house at that time.
2. At 10.00 a.m. Peter leaves his house and cycles
   along a main road at 20 km/h. At 10.15 a.m. his
   father leaves the house and drives his car along
   the same main road at 45 km/h. Find from a
   graph:
   (a) the time at which they meet and,
   (b) the distance that they both are from their
   house at that time.
3. At 9.00 a.m. a lorry leaves a warehouse in
   Glasgow, and sets off for a shop in Edinburgh
   at an average speed of 50 km/h. At 9.10 a.m.
   the haulage firm's manager starts the same
   journey in his car and his average speed is 75 km/h.
   Find from a graph:
   (a) the time at which the manager catches up
       with the lorry driver,
   (b) the distance from Glasgow that they have
       both covered.
4. At 8.00 a.m. a local train leaves London
   (St. Pancras) towards Kettering at an average
   speed of 75 km/h. At 8.10 a.m. an express
   departs from St. Pancras and travels at an average
   speed of 100 km/h on a parallel track to the one
   used by the local train. Find from a graph:
   (a) the time at which the express overtakes the
       local train,
   (b) the distance from London that both trains
       have then reached.
5. At 9.00 a.m. a boy starts to cycle from Ringford
   to Little Duchrae at an average speed of 18 km/h.
   At exactly the same time another boy starts to
   walk from Little Duchrae to Ringford at an
   average speed of 6 km/h. If the distance between
   the two places is 12 km, find from a graph
   (a) the time when they meet.
   (b) how far they both are from Ringford at
       that time.
6. At 3.00 p.m. a man begins to drive his car from
   Oxford to Banbury at an average speed of
   50 km/h. At exactly the same time another man
   begins to cycle from Banbury to Oxford at an
   average speed of 10 km/h. If the distance between
   the two places is 36 km, find from a graph:
   (a) the time when they meet
   (b) how far they both are from Oxford at that
       time.

A *tally chart* is a convenient way of recording the frequency of an event, as shown below.

choice	tally	frequency
mashed potatoes	ⅢⅡ ⅢⅡ ⅢⅡ ⅢⅡ Ⅱ	22
roast potatoes	ⅢⅡ ⅢⅡ ⅢⅡ ⅢⅡ ⅢⅡ ⅢⅡ ⅢⅡ ⅢⅡ ⅢⅡ ⅢⅡ	29
chipped potatoes	ⅢⅡ ⅢⅡ ⅢⅡ ⅢⅡ ⅢⅡ ⅢⅡ ⅢⅡ ⅢⅡ ⅢⅡ ⅢⅡ	49
	total	100

The tally chart shows the choice of potatoes by 100 pupils in a school canteen.

Example 1

A dice was rolled 50 times and the scores were recorded as follows.

```
2  5  1  3  1  3  4  1  6  4
4  4  4  2  5  5  5  3  5  4
3  5  2  3  4  2  3  2  6  6
1  6  6  1  6  6  2  3  2  5
2  6  5  4  4  1  4  5  6  1
```

Record this information on a suitable tally chart.

score	tally	frequency
6	ⅢⅡ ⅢⅡ	9
5	ⅢⅡ ⅢⅡ	9
4	ⅢⅡ ⅢⅡ	10
3	ⅢⅡ Ⅱ	7
2	ⅢⅡ ⅢⅡ	8
1	ⅢⅡ Ⅱ	7
	total	50

Exercise 9.1a

1. The details below show the first 30 names on a voting list for a certain district, together with how they voted.

name	vote	name	vote	name	vote
Adams P.	Conservative	Andrews M.	Conservative	Beasley O.	Conservative
Adams T.	Conservative	Andrews S.	Conservative	Benson D.	Liberal
Aitken R.	did not vote	Ashton B.	did not vote	Benson J.	Liberal
Alder C.	Labour	Ashton N.	did not vote	Benson P.	Labour
Alder W.	Labour	Atkinson H.	Labour	Brown P.	Conservative
Allerton V.	Labour	Atkinson M.	Labour	Brown V.	Conservative
Alsop L.	Conservative	Bailey W.	did not vote	Burns C.	Labour
Anderson G.	Liberal	Barker W.	Labour	Burns M.	did not vote
Anderson K.	Liberal	Bassett A.	Conservative	Butler J.	Conservative
Anderson R.	Labour	Bates L.	Labour	Butler L.	Conservative

Record the numbers of each kind of vote (including those who did not vote) on a tally chart.

2. A card was withdrawn from a pack forty times and replaced. The results were
   recorded as Picture, Ace, or Number. The details are given below.

Numbers	Number	Ace	Picture	Ace	Ace
Ace	Picture	Number	Number	Number	Number
Picture	Number	Ace	Number	Picture	Picture
Picture	Number	Picture	Number	Number	Number
Number	Picture	Ace	Number	Ace	Number
Number	Ace	Number	Picture	Picture	
Number	Number	Number	Number	Number	

   Record this information on a suitable tally chart.

3. A football team played fifty matches in one season. The number of goals they
   scored in each match is shown below.

1	0	2	3	4	0	3	2	2	0
3	2	5	1	3	1	0	0	0	1
0	1	1	0	1	0	1	4	1	2
0	1	3	2	1	2	2	2	6	0
1	4	2	0	5	1	4	3	3	4

   Show on a tally chart the frequency of goals scored.

4. The destination and time of departure of all inter-city trains leaving King's Cross
   station in London on a weekday is shown below.

4.05	Leeds	11.35	Newcastle	16.04	Bradford
5.50	Aberdeen	11.55	Aberdeen	16.08	York
7.40	Newcastle	12.04	Hull	16.42	Hull
7.45	Bradford	12.25	York	17.00	Edinburgh
8.00	Edinburgh	12.45	Bradford	17.04	Leeds
8.04	Hull	13.00	Edinburgh	17.35	Newcastle
8.25	Cleethorpes	13.04	Cleethorpes	18.00	Edinburgh
9.00	Edinburgh	13.25	Newcastle	18.04	Bradford
9.04	Leeds	14.00	Edinburgh	18.09	York
9.30	Newcastle	14.04	York	18.20	Cleethorpes
10.00	Edinburgh	14.45	Leeds	19.00	Newcastle
10.04	York	15.00	Aberdeen	19.04	Newcastle
10.45	Leeds	15.45	Leeds	19.40	Leeds
10.55	Edinburgh	16.00	Edinburgh		

   Show on a tally chart the number of trains for each destination.

5. The year letter on the registration number plate of fifty cars that passed along a
   main road on a day in August was recorded as follows.

J	P	N	M	S	S	R	G	N
R	M	B	R	G	N	P	R	H
N	S	T	J	R	F	M	J	
L	G	R	N	P	R	R	S	
R	N	S	S	T	H	T	S	
S	S	N	J	P	N	S	M	

   Draw up a tally chart to show how many cars of each year letter were recorded.

Class intervals are often used to reduce the size of a tally chart.

score	tally	frequency
0–4	//	2
5–9	ЖК	5
10–14	ЖК ///	8
15–19	ЖК /	6
20–24	///	3
25–29	/	1
total		25

The tally chart shows the scores of contestants in a shooting competition.
The *class limits* are the largest and smallest possible results in each class interval.
So, from the *second* class in the tally chart

> 5 is the lower class limit, and
> 9 is the upper class limit.

Example 1

The points scored by 50 entrants in a disco-dancing championship were:

43	38	23	4	33	44	44	58	34	44
3	6	39	23	14	31	14	27	42	46
38	22	12	41	21	5	24	2	18	57
10	30	41	28	30	30	37	34	29	45
40	52	19	30	14	33	16	34	2	45

Read this information on a suitable tally chart.

score	tally	frequency
0–9	ЖК /	6
10–19	ЖК ///	8
20–29	ЖК ///	8
30–39	ЖК ЖК ////	14
40–49	ЖК ЖК /	11
50–59	///	3
total		50

Exercise 9.1b

1. 50 people each stated the number of hours they spent watching TV, one week.

0	9	13	20	21	10	27	2	28	12
6	24	5	2	0	12	20	7	8	23
13	13	8	8	25	22	26	13	9	14
2	20	23	5	26	11	3	29	28	14
23	24	4	21	7	4	26	6	22	10

Record this information on a tally chart, beginning with class intervals 0–4, 5–9, 10–14.

2. The points scored by 45 contestants in a general knowledge quiz were:

20	58	28	46	29	21	23	30	9
32	24	3	31	0	52	31	27	22
23	43	20	47	26	28	9	34	28
32	5	41	33	41	33	28	21	50
25	44	25	44	21	29	49	34	23

Record this information on a tally chart, beginning with class intervals 0-9, 10-19, 20-29.

3. The times taken by 48 rally cars to cross a moor were measured in seconds, correct to the nearest second.

220	215	213	219	205	224	211	217
221	201	215	221	215	223	207	220
220	226	210	222	219	224	216	223
224	214	222	216	210	217	229	209
219	228	218	223	221	220	217	215
226	222	223	219	224	227	212	216

Record these times on a tally chart, beginning with class intervals 200-204, 205-209.

4. The number of goals scored by each of 60 amateur football teams in a season is noted.

9	30	14	32	29	7	23	37	25	14	7	16
44	22	42	0	56	41	50	11	35	41	34	24
19	3	17	21	30	15	37	20	2	20	15	34
38	24	32	46	10	48	2	39	24	28	13	21
10	46	14	23	38	17	27	46	22	4	33	13

Record these results on a tally chart, beginning with class intervals 0-9, 10-19.

5. The number of pearls recovered by each diver in a team is noted, during one week.

30	35	41	25	33	27	37	35	25	34	40	31
21	26	38	29	43	27	44	40	25	28	46	39
42	38	42	40	35	23	43	32	44	36	33	48
30	36	26	28	36	41	26	36	43	32	26	49

Record this information on a tally chart, beginning with a class interval of 20-24.

6. The number of cars in the showroom of several garages is noted, one Saturday.

12	15	19	17	13	19	17	15
16	10	15	19	16	11	25	19
14	19	12	16	17	16	13	17
16	18	20	14	12	14	17	24
18	20	14	20	17	17	20	15

Record this information on a tally chart, beginning with a class interval of 10-11.

Diagrams are often used to illustrate statistical information, to make it easier to understand.

Bar chart

In this type of diagram bars are drawn to represent the results.

Example 1

A school 'snack bar' sells on one day the following varieties of crisps:

Variety	Plain	Smoky Bacon	Cheese and Onion	Salt 'n' Vinegar	Roast Beef	Roast Chicken
Number sold	42	28	20	16	4	10

Show this information on a bar chart.

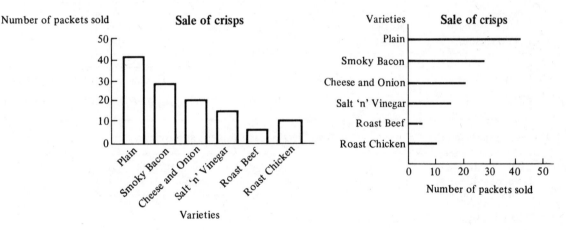

Exercise 9.2a

Draw suitable bar charts to display the following data:

1.
Day of the week	Sunday	Monday	Tuesday	Wednesday	Thursday	Friday	Saturday
Hours of sunshine	5	10	12	8	2	1	7

2.
Favourite lesson	Maths	English	History	Geography	French	Science	Art
Number of pupils	21	19	18	6	2	24	15

3.
Goals scored	0	1	2	3	4	5	6
Number of teams	1	17	12	8	4	3	2

4.
Mark	0	1	2	3	4	5	6	7	8	9	10
Number of pupils	1	4	6	5	8	14	7	3	5	2	1

5.

Favourite game	Billiards	Tennis	Golf	Judo	Karate	Football
Number of pupils	4	17	9	3	11	14

6.

Hobby	Stamps	Athletics	Painting	Photography	Reading	Music
Number of girls	10	12	4	25	19	50

7.

Species of fish	Salmon	Trout	Perch	Pike	Eel	Bream
Number caught	14	36	17	6	48	20

8.

Type of car	Saloon	Estate	Sports	Convertible	Budget	Luxury
Number sold	180	74	13	9	95	56

9.

Tape length (min)	10	12	15	45	60	90	120
Number sold	14	12	2	28	16	25	10

10.

Day of the week	Sunday	Monday	Tuesday	Wednesday	Thursday	Friday	Saturday
Number of campers	70	25	25	40	54	75	150

11.

Type of record	Rock	Soul	Reggae	Punk	Folk	Disco
Number sold	140	85	92	138	25	200

12.

Day of the week	Sunday	Monday	Tuesday	Wednesday	Thursday	Friday	Saturday
No. of ice creams sold	60	50	80	144	359	500	482

13. The table shows the number of subjects studied by each student at a college.

No. of subjects	1	2	3	4	5	6	7	8
No. of students	180	275	190	482	370	425	210	76

14. The table shows the number of books in a school library.

Subject	Mathematics	History	Geography	Science	Art	Languages	Fiction
Number	240	400	640	320	960	800	1440

15. The table gives details of the instruments played in a group of 240 pupils

Instrument	Clarinet	Guitar	Violin	Trumpet	Drums
Number	36	40	24	58	82

16. The table shows the number of children per family from a sample of 100 families.

Number of children	0	1	2	3	4	5	6	7	8	9
Number of families	7	15	21	12	18	11	12	2	0	2

Histogram

A frequency distribution is usually shown as a histogram.

In this book the height of each rectangle is a measure of the frequency of each event. There are no spaces between the rectangles in a histogram.

Example 2

Four coins are each tossed 64 times. The number of times a 'head' appeared on all four coins is shown on the tally chart. Show this information on a histogram.

'heads'	tally	frequency
4	///	3
3	₩₩ ₩₩ ₩₩ /	16
2	₩₩ ₩₩ ₩₩ ₩₩ ///	23
1	₩₩ ₩₩ ₩₩ //	17
0	₩₩	5
	total	64

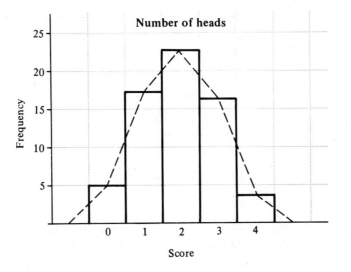

A *frequency polygon* is formed by joining the mid-points of each rectangle on the histogram by straight lines. This is shown by the broken lines on the graph above. The polygon is completed by continuing the lines to the horizontal axis as shown.

Exercise 9.2b

In questions **1** to **6**, illustrate the details on a histogram.

1. A farmer records on a tally chart for each of his chickens the number of eggs per week laid by the chicken. The chart for one of his chickens over a period of a year is shown below.

no. of eggs per week	tally	frequency (no. of weeks)
7	//	2
6	////	4
5	##L ##L	10
4	##L ##L ##L /	16
3	##L ##L /	11
2	////	4
1	///	3
0	//	2
	total	52

2. The tally chart shows the number of wet days in each week over a period of one year.

no. of wet days	tally	frequency (no. of weeks)
7	//	2
6	##L	5
5	##L //	7
4	##L ##L //	12
3	##L ##L	10
2	##L ////	9
1	##L /	6
0	/	1
	total	52

3. The finishing position of a sprinter in 50 different races is recorded in the tally chart. In each race there were eight runners.

position	tally	frequency
8th	////	4
7th	##L /	6
6th	##L ///	8
5th	##L ##L //	12
4th	##L ##L	10
3rd	##L //	7
2nd	//	2
1st	/	1
	total	50

4. A football club plays 50 matches in a season. The number of goals they score for each match in one season is shown below.

No. of goals	0	1	2	3	4	5	6
No. of matches	6	10	11	12	6	3	2

5. The results in marks out of 10 of an English test for sixty children is shown below.

Marks scored	0	1	2	3	4	5	6	7	8	9	10
No. of pupils	1	2	3	6	10	11	9	8	5	3	2

6. The size of shoe worn by each of 30 children is shown in the table.

Size of shoe	2	3	4	5	6
No. of children	3	6	11	8	2

In questions 7 to 11, illustrate the details on a frequency polygon.

7. A cricketer bowls in twenty matches in one season. The number of wickets he took in each match is shown in the table.

No. of wickets	0	1	2	3	4	5
No. of matches	3	5	6	3	2	1

8. A lorry driver can make a total of 6 deliveries between a depot and a shop in one working day. The actual number of deliveries he makes each day over a period of 30 days is shown in the table.

No. of deliveries	0	1	2	3	4	5	6
No. of days	1	3	4	7	8	5	2

9. The career of a batsman covered 25 seasons. The table records the number of centuries he scored over these seasons.

No. of centuries	0	1	2	3	4	5
No. of seasons	2	5	7	6	4	1

10. A football club entered a five-round knockout competition for 20 seasons. The club never won the trophy. The table shows the round in which the club was eliminated for each of the seasons.

Round when club was eliminated	1	2	3	4 (semi-final)	5 (final)
No. of seasons	4	5	6	3	2

11. A taxi has seating capacity for six passengers. The actual number of passengers carried for each of 40 journeys is shown.

No. of passengers	1	2	3	4	5	6
No. of journeys	7	8	10	9	4	2

A histogram may be drawn from a tally chart which uses class intervals. The horizontal axis shows the class limits instead of single results.

Example 3

The number of members attending a car club each week is noted. Show this information on a histogram.

No. of members	tally	frequency
20–29	<< /	6
30–39	<< //	7
40–49	<< << ////	14
50–59	<< << //	12
60–69	<< /	6
70–79	///	3
	total	48

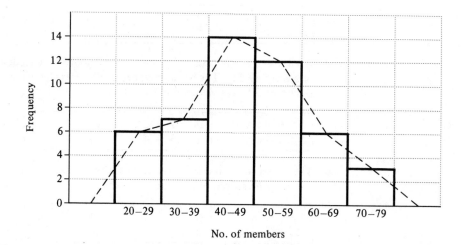

A frequency polygon may be drawn as before (Example 2) by joining the mid-points of each rectangle on the histogram by straight lines. The polygon is completed by continuing the lines to the horizontal axis as shown.

Exercise 9.2c

In questions **1** to **5**, illustrate the details on a histogram.

1. The weight of grain sold each week by a merchant is shown below.

Weight (tonnes)	0-4	5-9	10-14	15-19	20-24	25-29
No. of weeks	5	5	15	20	10	5

2. A milkman notes the number of customers who order eggs delivered each morning.

No. of customers	0-9	10-19	20-29	30-39	40-49	50-59
No. of days	4	6	7	10	8	3

3. The manager of an exhaust repair centre records the number of complete exhaust systems fitted each day.

No. of exhausts	0-2	3-5	6-8	9-11	12-14	15-17	18-20
No. of days	7	9	7	15	20	14	8

4. A cinema owner notes the number of tickets sold each afternoon.

No. of tickets	120-129	130-139	140-149	150-159	160-169	170-179
No. of afternoons	12	0	14	15	8	11

5. A garage owner counts the number of customers buying petrol before 9 a.m. every morning.

No. of customers	25-29	30-34	35-39	40-44	45-49
No. of days	16	28	8	0	4

In questions **6** to **10**, illustrate the details on a frequency polygon.

6. The time a number of villagers spend travelling to work each day is shown below.

Time (minutes)	10-12	13-15	16-18	19-21	22-24	25-27	28-30
No. of villagers	5	15	21	25	10	10	6

7. A group of students recorded the number of hours they each spent studying during one week.

No. of hours	2-5	6-9	10-13	14-17	18-21	22-25
No. of students	7	10	14	15	12	2

8. A microcumputer repair specialist notes the number of computers he repairs each week.

No. of computers	28-30	31-33	34-36	37-39	40-42	43-45
No. of weeks	11	10	14	9	4	2

9. The Manager of a mail order company records the number of orders received, each
   day for 4 months.

No. of orders	0–49	50–99	100–149	150–199	200–249	250–299
No. of days	15	19	24	12	0	6

10. A laboratory supervisor notes the weight of each sample of metal analysed during
    one week.

Weight (grams)	1·0–1·9	2·0–2·9	3·0–3·9	4·0–4·9	5·0–5·9
No. of samples	16	35	50	24	10

Pie chart

A pie chart can be used to compare the types of items in a set of results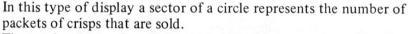

Example 4

The following numbers of bags of crisps were sold in a school snack bar
one day:

Variety	Plain	Smoky Bacon	Cheese and Onion	Salt 'n'Vinegar	Roast Beef	Roast Chicken
Number sold	42	28	20	16	4	10

Show this information on a pie chart.

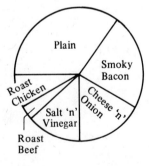

In this type of display a sector of a circle represents the number of
packets of crisps that are sold.
The total number of packets sold is 120; this is represented by the
angle at the centre of the circle, 360°.
So 1 packet is represented by $360 \div 120 = 3°$ and the sector angle
for *Plain Crisps* = $42 \times 3 = 126°$.
The other angles are:

Smoky bacon = $28 \times 3 = 84°$
Cheese 'n' onion = $20 \times 3 = 60°$
Salt 'n' vinegar = $16 \times 3 = 48°$
Roast beef = $4 \times 3 = 12°$
Roast chicken = $10 \times 3 = 30°$

Exercise 9.2d

Draw pie charts to represent the following information.

1.
Sport chosen	Football	Rugby	Hockey	Cross-country	Volley ball
Number	40	32	16	24	8

2.
Transport used	Train	Aeroplane	Bus	Car	Cycle
Number	20	16	8	12	4

3.
Team supported	Spurs	Arsenal	Chelsea	West Ham	Fulham
Number	30	24	18	12	6

4.
Department	Hi fi	Food	Clothes	Furniture	Carpets	Shoes	Stationery
No. of customers	15	90	60	45	40	25	85

5.
Shoe size	1	2	3	4	5	6	7	8	9	10
Number in year	6	9	30	51	60	45	39	18	9	3

6.
Time of bus	8.00 a.m.	10.00 a.m.	12.00 noon	2.00 p.m.	4.00 p.m.	6.00 p.m.	8.00 p.m.
No. of passengers	36	27	18	9	27	45	18

7.
Party	Conservative	Labour	Liberal	Independent	Did not vote
No. of voters	300	240	60	20	100

8.
Day of week	Monday	Tuesday	Wednesday	Thursday	Friday	Saturday	Sunday
No. of cars using ferry	75	60	45	30	90	135	105

9.
No. of items bought	1	2	3	4	5 or more
No. of customers	180	120	96	24	60

10.
Career chosen	Army	Banking	Police	Industry	Teaching
No. of pupils	36	48	72	42	18

Example 5

The pie chart shows the sales of five flavours
of ice cream in a cafe.
What fraction of the total sales is due to each
flavour?

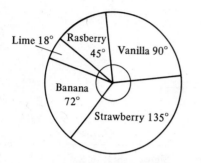

The total angle at the centre is always 360°.

The Vanilla sector represents $\dfrac{\overset{1}{90}}{\underset{4}{360}} = \dfrac{1}{4}$ of the

total, so Vanilla ice creams account for $\dfrac{1}{4}$ of

the total sales.

$$\text{Strawberry} = \dfrac{\overset{\overset{3}{27}}{135}}{\underset{\underset{8}{72}}{360}} = \dfrac{3}{8} \qquad \text{Banana} = \dfrac{\overset{1}{72}}{\underset{5}{360}} = \dfrac{1}{5}$$

$$\text{Lime} = \dfrac{\overset{1}{18}}{\underset{20}{360}} = \dfrac{1}{20} \qquad \text{Raspberry} = \dfrac{\overset{1}{45}}{\underset{8}{360}} = \dfrac{1}{8}$$

Exercise 9.2e

In each question, find the fraction of the total
represented by each sector.

1. A gents' outfitter notes the types of garments he
   sells.

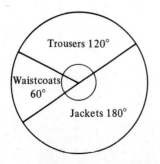

2. The books in a library are classified in three
   categories.

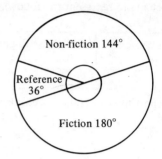

3. A guide book lists the places of entertainment in
   an area.

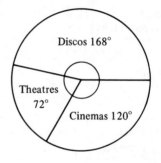

4. The manager of a holiday camp lists the sources
   of his profits.

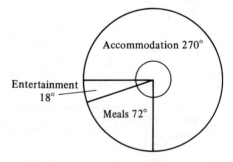

5. An accountant investigates payments made to a
   salesman.

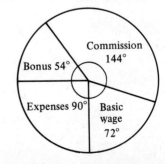

Example 6

The pie chart shows the types of dog food sold in a supermarket in one week.

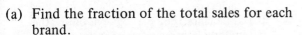

(a) Find the fraction of the total sales for each brand.

(b) If 4800 tins were sold altogether, find the number of tins of each brand sold.

(a) 360° represents the total sales,

so 1° represents $\dfrac{1}{360}$ of the total.

$$\text{Krunch} = 90° = \dfrac{\overset{1}{\cancel{90}}}{1} \times \dfrac{1}{\underset{4}{\cancel{360}}} = \dfrac{1}{4} \text{ of the total}$$

$$\text{Oh Boy!} = 60° = \dfrac{\overset{1}{\cancel{60}}}{1} \times \dfrac{1}{\underset{6}{\cancel{360}}} = \dfrac{1}{6} \text{ of the total}$$

$$\text{Doggie Meal} = 210° = \dfrac{\overset{7}{\cancel{210}}}{1} \times \dfrac{1}{\underset{12}{\cancel{360}}}$$

$$= \dfrac{7}{12} \text{ of the total}$$

(b) Total sales = 4800 tins

$$\text{Krunch} = \dfrac{1}{\underset{1}{\cancel{4}}} \times \dfrac{\overset{1200}{\cancel{4800}}}{1} = 1200 \text{ tins}$$

$$\text{Oh Boy!} = \dfrac{1}{\underset{1}{\cancel{6}}} \times \dfrac{\overset{800}{\cancel{4800}}}{1} = 800 \text{ tins}$$

$$\text{Doggie Meal} = \dfrac{7}{\underset{1}{\cancel{12}}} \times \dfrac{\overset{400}{\cancel{4800}}}{1} = 2800 \text{ tins}$$

Exercise 9.2f

1. The pie chart shows the hairdressers used by women in a town.

(a) Find the fraction of the total number of women who use each hairdresser.

(b) Altogether 300 women visit these hairdressers. How many visit each one?

2. A grocer's shop sells four brands of tea bags.

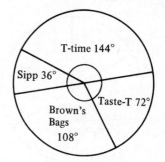

(a) Find the fraction of the total sales due to each brand.

(b) If the grocer sells 500 boxes of tea bags altogether, how many of each brand does he sell?

3. Wooden bowls sold in a craft shop are of 3 types.

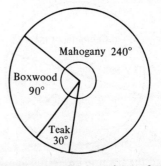

(a) What fraction of the total are of each type?

(b) If the shop sells 240 bowls altogether, how many are of each type?

4. The pie chart shows the types of illustrations used in a magazine.

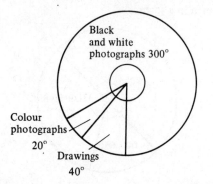

(a) What fraction of the total are of each type?
(b) If the magazine uses 54 illustrations, how many will be of each type?

5. The pie chart shows the kinds of pets owned by people in a street.

(a) Find the fraction of the total for each kind of pet.
(b) If there are 60 pets altogether, how many are there of each kind?

6. The main interests of members of an electronics club are as shown.

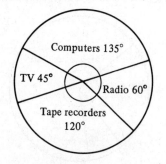

(a) Find the fraction interested in each.
(b) If there are 120 members, find how many are interested in each.

7. A professional photographer takes photographs in one month of various types as shown.

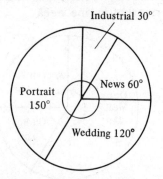

(a) Find the fraction of the total which are of each type.
(b) If he takes 1860 photographs during the month, how many are of each type?

8. A toothpaste manufacturer analyses the cost of his advertising as shown.

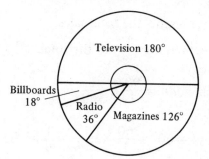

(a) Find the fraction of the total cost for each method.
(b) If the total cost was £40 000, find the cost of each method.

9. A manufacturer records the cost of making wrought iron gates.

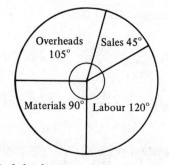

(a) Find the fraction of the total cost represented by each category.
(b) If it costs £60 to make a gate, find how much is due to each category.

Range measures the 'spread' of a set of results.
Range = (largest result) − (smallest result)

Example 1

The number of Valentine cards received by each of seven girls is noted.
Find the range of the results.

2   2   4   6   9   9   11

largest number of cards = 11
smallest number of cards = 2
∴ range = 11 − 2 = 9 cards

Example 2

The number of pupils in a class is noted, on various days. Find the range.

No. of pupils	22	23	24	25	26	27	28	29	30
No. of days	1	0	3	7	9	5	2	1	2

largest number of pupils = 30
smallest number of pupils = 22
∴ range = 30 − 22 = 8 pupils

Exercise 9.3

Find the range of each set of results.

1. The number of telephone calls received in a shop is noted, each day.

   2   4   4   8   10   10   12   15

2. The number of letters delivered each day by a postwoman is recorded.

   27   35   84   85   94   103   105

3. The number of watches repaired each week in a jeweller's shop is:

   3   3   7   8   8   12   15   20   21

4. A dentist records the number of patients he treats, each day.

   8   12   13   15   17   17   17   20   22   24

5. A photographer writes down the number of photographs he takes each week.

   240   280   320   355   405   428   471

6. Every morning, a newsagent notes the number of newspapers he sells.

   172   160   160   166   165   166   160   164

7. An air traffic controller notes the number of private aircraft using an airport each day.

No. of aircraft	0	1	2	3	4	5	6
Frequency (No. of days)	4	6	8	6	0	4	2

8. A parking warden records the number of tickets he writes each week.

No. of tickets	50	51	52	53	54	55
No. of weeks	6	4	6	15	11	10

9. The number of matches in each box on a shelf is counted.

No. of matches	47	48	49	50	51	52	53
No. of boxes	24	29	29	30	28	25	22

10. The number of tracks on various records is noted.

No. of tracks	6	7	8	9	10	11	12
No. of records	4	4	5	6	8	6	5

9.4 MODE

The mode is 'the most popular' item, i.e. the one result that occurs most frequently in a set of results.

Example 1

Find the mode of the following:

4, 3, 2, 1, 4, 3, 3, 2, 1, 3, 4, 2, 3, 4, 3, 3, 2, 2, 2, 3, 4, 2, 3, 3, 1.

Number	Tally	Frequency
1	///	3
2	₶ //	7
3	₶ ₶	10
4	₶	5

∴ 3 is the mode.

Exercise 9.4a

1. The temperature in °C on 20 winter days was:
5, 3, 3, 2, 0, 0, 1, 2, 5, 4, 3, 1, 5, 4, 2, 2, 4, 3, 4, 3.
What was the modal temperature?

2. In a local football league, the goals scored one Saturday were 4, 0, 2, 6, 3, 0, 5, 2, 5, 7, 0, 1, 4, 2, 1, 7, 6, 6, 5, 3, 2, 3, 7, 6, 2, 0, 1, 4.
What is the modal score?

3. The times (in minutes) taken by a man going to work were 54, 57, 55, 57, 56, 58, 54, 53, 54. 55, 56, 57, 53, 58, 54, 58, 54, 57, 55, 54.
What is the modal time?

4. The shoe sizes of a class are 5, 7, 6, 7, 5, 5, 6, 6, 7, 9, 9, 6, 6, 6, 8, 9, 8, 9, 5, 7, 7, 8, 5, 8.
What is the modal size?

5. In various shops, 1 kg of tomatoes was priced in pence as follows.
39, 39, 40, 43, 43, 44, 37, 44, 37, 39, 40, 42, 43, 43, 44, 43, 44, 37, 39, 38, 39, 40, 43.
What is the modal price?

6. The attendance of a class is as follows.
30, 32, 28, 28, 29, 30, 31, 28, 27, 27, 31, 28, 29, 32, 32, 28, 29, 30, 31, 30, 32, 31, 29, 29, 30, 30, 28, 27, 28, 32.
What is the modal attendance?

7. Two dice are thrown together 20 times and the results are shown below.

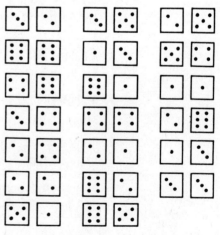

Find the modal score.

8. Three coins A, B and C are tossed simultaneously 20 times over. Each either lands Heads (H) or Tails (T) as shown in the table below.

A	B	C	A	B	C
H	H	T	T	H	T
H	T	T	T	T	H
T	T	H	H	T	H
H	T	T	H	H	H
T	H	T	T	H	T
T	T	T	T	T	T
H	H	T	H	T	H
H	T	H	T	H	T
H	T	T	T	H	H
H	H	H	T	T	T

Which is the modal set of appearances:
   3 Heads
   3 Tails
   2 Heads and 1 Tail
or 1 Head and 2 Tails?

9. A hockey club takes part annually in a league tournament with five other clubs. The table below shows the record of this club over fifteen seasons. Find the modal number of league points.

	Games won (2 points)	Games drawn (1 point)	Games lost (0 point)
1964	3	4	3
1965	4	0	6
1966	8	0	2
1967	6	2	2
1968	4	2	4
1969	6	0	4
1970	2	4	4
1971	6	2	2
1972	6	1	3
1973	5	4	1
1974	5	1	4
1975	7	2	1
1976	5	0	5
1977	3	2	5
1978	7	0	3

10. The scoring record of a rugby club over fifteen matches is given in the table. Find the modal number of match points.

Tries (4 points)	Goals (3 points)	Conversions (2 points)
1	1	1
4	5	1
3	1	0
3	3	2
1	2	1
3	3	3
4	2	1
2	5	2
3	5	3
4	3	1
2	0	1
2	4	2
4	4	2
2	5	1
0	3	0

In a frequency table containing class intervals, the *modal class* is the class which occurs with the greatest frequency.

Example 2

The ages of workers in a factory are as shown.
Find the modal class.

Age (years)	Frequency
20–24	8
25–29	7
30–34	10
35–39	15
40–44	12
45–49	8

$\therefore$ 35–39 is the modal class.

Exercise 9.4b

Find the modal class in each of the following distributions.

1. The widths of steel sections in a stockist's yard are:

Width (mm)	3–5	6–8	9–11	12–14	15–17
No. of lengths	20	25	15	10	10

2. The number of people in a cafeteria is noted, at various times.

No. of people	0–4	5–9	10–14	15–19	20–24	25–29
Frequency	6	12	8	7	3	1

3. The number of grasshoppers in 1 m² is recorded, in areas of a country park.

No. of grasshoppers	0–2	3–5	6–8	9–11	12–14	15–17
No. of areas	4	8	11	12	9	6

4. The gas consumption of a furnace is noted, every hour.

Consumption (litres/h)	1000–1999	2000–2999	3000–3999	4000–4999	5000–5999	6000–6999
No. of hours	4	10	5	1	2	2

5. The following are results from a weightlifting contest:

Weight (kg)	70–74	75–79	80–84	85–89	90–94	95–99
No. of competitors	1	4	8	5	3	1

6. A girl records the rainfall each day.

Rainfall (mm)	0–0·9	1·0–1·9	2·0–2·9	3·0–3·9	4·0–4·9	5·0–5·9
No. of days	8	10	7	4	0	2

7. The total number of tickets issued by a bus conductor, on each run, is noted.

No. of tickets	500–599	600–699	700–799	800–899	900–999
No. of runs	6	12	8	10	14

8. The number of football spectators passing through a turnstile is recorded on a counter, at each match.

No. of spectators	0–999	1000–1999	2000–2999	3000–3999	4000–4999	5000–5999
No. of matches	2	5	6	4	1	2

9. An exam board publishes the number of candidates sitting each of its exams.

No. of candidates	0–149	150–299	300–449	450–599	600–749
No. of exams	7	4	6	2	1

10. The secretary of a photography club writes down the number of members at each meeting, during a year.

No. of members	0–9	10–19	20–29	30–39	40–49	50–59
No. of meetings	10	15	12	8	4	3

9.5 MEDIAN

The *median* is the middle value when the results are placed in order of size.

Example 1

Find the median of:

(a) 2, 4, 6, 8, 10, 12, 14
(b) 3, 2, 5, 7, 4, 6, 9, 2

(a) In order of size, the numbers are:

2, 4, 6, [8,] 10, 12, 14

Therefore the median is 8.

(b) In order of size, the numbers are:

2, 2, 3, [4, 5,] 6, 7, 9

Therefore the median is $\frac{4+5}{2} = 4\frac{1}{2}$

Exercise 9.5a

1. The attendance of a class during one week was 32, 30, 29, 33, 34. What is the median attendance?

2. The number of potatoes needed to fill five 5-kg bags were 69, 84, 76, 71, 73. Find the median.

3. In one week the midday temperatures were 26°C, 25°C, 20°C, 19°C, 23°C, 22°C, 17°C. What is the median temperature?

4. In one week the daily tips received by a waitress were 75p, 120p, 90p, 95p, 60p, 70p, 80p. Find the median.

5. A 200-m sprinter ran in six races and his times were 23·7s, 23·0s, 23·5s, 24·1s, 22·9s, and 23·3s. What is his median time?

6. The times in minutes taken by a girl walking to work were 39, 37, 42, 40, 35, 36. What is the median time?

7. The price per $\frac{1}{2}$ kg of apples on various market stalls was 27p, 27p, 30p, 28p, 32p, 33p. What is the median price?

8. The weights in kg of eight boys were 40, 36, 33, 36, 45, 38, 41, 43. What is the median weight?

9. The waist measurements of eight girls measured in cm were 69, 68, 59, 70, 72, 57, 62, 64. What is the median of these measurements?

The median is the middle result in a frequency table.

Example 2

The number of goals scored by 15 football teams one Saturday was as follows. What is the median score?

Goals scored	Frequency
0	4
1	2
2	4
3	3
4	1
5	1
	total = 15

The middle result is the 8th team's score.
i.e. there are 7 teams below the 8th team, and 7 teams above the 8th team, so the 8th team is the middle team.
The first 4 teams scored 0 goals;
the first 6 teams (i.e. 4 + 2) scored 0 goals or 1 goal;
the first 10 teams (i.e. 4 + 2 + 4) scored 0, 1 or 2 goals.
So, teams 7, 8, 9 and 10 scored 2 goals each.
$\therefore$ team 8 scored 2 goals.

Example 3

The number of cars on hire from a depot is noted, each day, Find the median.

No. of cars	No. of days
10	3
11	7
12	5
13	2
14	2
15	1
	total = 20

The middle result lies between the 10th and 11th results.
i.e. there are 10 results below the middle result, and 10 results above the middle result.
On the first 3 days, 10 cars were on hire;
on the first 10 days (i.e. 3 + 7), 10 or 11 cars were on hire;
on the first 15 days (i.e. 3 + 7 + 5), 10, 11 or 12 cars were on hire.
So, on the 10th day 11 cars were on hire,
and on the 11th day 12 cars were on hire.
$\therefore$ the median is $\frac{10 + 11}{2} = \frac{21}{2} = 10.5$ cars.

Exercise 9.5b

Find the median of each set of results.

1. The number of trees in each garden in a street was:

No. of trees	0	1	2	3	4
No. of gardens	3	6	4	1	1

2. The number of lens tissues in each sachet in a camera shop was:

No. of tissues	47	48	49	50	51	52
No. of sachets	2	4	3	4	0	2

3. The number of passengers on a long-distance coach is recorded, each journey.

No. of passengers	35	36	37	38	39	40
No. of trips	4	2	1	2	5	3

4. The number of faulty tyres belonging to each car visiting a testing station is noted.

No. of faulty tyres	0	1	2	3	4	5
No. of cars	3	8	5	3	4	2

5. The number of species of birds seen in a garden every morning is:

No. of species	2	3	4	5	6
No. of mornings	12	14	4	3	2

6. The number of people in each group visiting a small museum is noted, one day.

No. of people	0	1	2	3	4	5	6
No. of groups	1	2	4	6	6	4	2

7. A photographer writes down the aperture she uses for each photograph.

Aperture (f number)	1·8	2·8	4	5·6	8	11	16
No. of photographs	6	3	3	8	2	1	4

8. An estate agent notes the number of enquiries about each house during its first 7 days on sale.

No. of enquiries	12	13	14	15	16	17
No. of houses	4	14	16	6	4	1

9. A designer of electronic circuits notes the diameter of wire required in each design.

Diameter (mm)	0·13	0·18	0·25	0·35	0·5	0·7
No. of designs	1	2	4	1	1	1

The median lies in the class containing the middle result.

Example 4

The number of long distance calls made on a telephone was recorded, every day for a month. In which class does the median lie?

No. of calls	Frequency
0–4	10
5–9	4
10–14	3
15–19	8
20–24	3
25–29	2

total = 30

The median lies between the 15th and 16th results.
The 1st class contains 10 calls;
the 1st and 2nd classes contain 10+ 4 = 14 calls;
the 1st, 2nd and 3rd classes contain 10 + 4 + 3 = 17 calls;
so, the 15th and 16th results must both lie in the 3rd class.
∴ the median lies in class 10–14.

Exercise 9.5c

State the class in which the median lies, in each of the following distributions.

1. The height of each man in a Social Club is recorded.

Height (cm)	155–159	160–164	165–169	170–174	175–179
No. of men	3	6	7	4	5

2. After a typing exam, the number of mistakes on each page is counted.

No. of mistakes	0–4	5–9	10–14	15–19	20–24	25–29
No. of pages	3	7	2	4	2	1

3. The time taken by each car in one stage of a rally is recorded.

Time (min)	10–11	12–13	14–15	16–17	18–19	20–21
No. of cars	1	2	4	6	1	3

4. A video game in an airport records the number of times it is used, each day, for a month.

No. of games	0–9	10–19	20–29	30–39	40–49	50 or more
No. of days	4	3	10	7	6	1

5. A manageress analyses the sales in a record shop.

No. of records	0–9	10–19	20–29	30–39	40–49
No. sold	8	6	3	2	1

6. A farmer counts the number of eggs collected from a hen-house, daily.

No. of eggs	0–4	5–9	10–14	15–19	20–24	25–29
No. of days	5	3	2	8	7	5

7. The distribution of ages of women employed in a shirt factory is shown.

Age (years)	16–25	26–35	36–45	46–55	56–65
No. of women	12	10	8	15	5

9.6 MEAN

The *mean* of a set of results is the sum of all the results divided by the number of results.

Example 1

The heights of six women in cm are 162, 156, 165, 153, 150, 156. What is the mean height?

$$\text{mean} = \frac{162 + 156 + 165 + 153 + 150 + 156}{6}$$

$$= \frac{942}{6} = 157 \text{ cm}$$

Exercise 9.6a

1. The number of wet days during four months were:
   August, 11
   September, 8
   October, 13
   November, 24
   Find the mean number of wet days.
2. A salesman earns the following amounts over a four-week period: £53·32, £60·21, £55·27, £61·08. Find the mean wage.
3. The times in minutes of five railway journeys from Edinburgh to Glasgow were 43, 47, 49, 41, 50. Find the mean journey time.

4. The times of five coach journeys from London to Birmingham were 2 h 45 min, 2 h 33 min, 3 h 5 min, 2 h 32 min, 3 h 10 min. Find the mean journey time.
5. The number of spectators at a football club for five rounds of a cup competition was: 5200, 8130, 13 205, 18 055, 24 030. Find the mean attendance.
6. A small business spends the following on postal charges in one week: 98p, £1·10p, $86\frac{1}{2}$p, 79p, $95\frac{1}{2}$p, 35p. Find the mean.
7. A 400-m sprinter ran in six races and his times were 47·4s, 46·8s, 46·6s, 45·9s, 47·5s, 47·8s. Find the mean of these times.
8. Two dice were thrown together six times and the results are shown below.

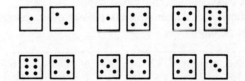

   Find the mean of these scores.
9. The heights in cm of eight women are 163, 160, 166, 159, 167, 170, 162, 165. Find the mean of these heights.
10. The heights in cm of eight men are 185, 179, 178, 180, 186, 183, 180, 177. Find the mean of these heights.

Example 2

The frequency table shows the number of Heads when four coins are tossed 64 times. What is the mean?

Heads	frequency
4	3
3	16
2	23
1	17
0	5

$$\text{mean} = \frac{(4 \times 3) + (3 \times 16) + (2 \times 23) + (1 \times 17) + (0 \times 5)}{64}$$

$$= \frac{12 + 48 + 46 + 17 + 0}{64} = \frac{123}{64}$$

$$= 1 \cdot 92 \text{ (to 3 s.f.)}$$

Exercise 9.6b

1. The table shows the different prices a woman paid for six eggs and the frequency of her purchases.

Price in p	27	26	25	24	23
No. of times paid	3	5	5	3	4

Find the mean price that she paid.

2. A girls' sports club has fifty members. The table shows the frequency of girls of each age.

Age in years	11	12	13	14	15	16
No. of girls	4	8	5	12	9	12

Find the mean age of the girls.

3. A class contains thirty-two pupils. The table shows the frequency of attendance over a period of 30 days.

No. present	32	31	30	29	28	27	26
No. of days	5	6	9	7	1	1	1

Find the mean attendance figure.

4. The table shows the number of passengers who left on the 15.30 bus from Stourbridge to Bridgnorth over a period of 60 days.

No. of passengers	8	7	6	5	4	3
No. of days	6	8	9	10	11	16

Find the mean number of passengers on this bus over the period.

5. A motorist buys petrol for his car daily. The table shows how many gallons he bought each day over a period of 42 days.

No. of gallons	6	5	4	3	2
No. of days	8	5	14	9	6

Find the mean number of gallons he bought each day.

6. The temperature at midday is recorded daily over a period of 30 days as shown in the table.

Temperature (°C)	24	23	22	21	20	19	18
No. of days	2	2	3	5	10	3	5

Find the mean temperature for this period.

7. The table shows the number of goals a football team scored for each of 50 matches.

No. of goals	6	5	4	3	2	1	0
No. of matches	1	2	4	8	10	12	13

Find the mean number of goals per match.

8. When four dice were thrown together a total of 200 times, the number of sixes scored per throw is shown in the table.

No. of sixes	4	3	2	1	0
No. of throws	1	2	7	40	150

Find the mean number of sixes scored per throw.

9. Four playing cards are removed from a pack at random. This is repeated 200 times. The number of picture cards drawn each time is shown in the table.

No. of picture cards	4	3	2	1	0
No. of draws	2	5	15	59	119

Find the mean number of picture cards removed per draw.

10. The table shows the atmospheric pressure in mm of mercury at midday over a period of 20 days.

Pressure	750	752	754	756	758	760
No. of days	1	0	3	2	1	3

Pressure	762	764	766	768	770
No. of days	3	2	0	3	2

Find the mean pressure over the period.

To find the mean when class intervals are given, use mid-values to represent each class.

For each class,

$$\text{mid-value} = \frac{\text{sum of class limits}}{2}$$

Example 3

The number of boxes in each of 35 crates delivered to a supermarket is shown below. Find the mean.

No. of boxes	0–4	5–9	10–14	15–19	20–24
No. of crates	10	6	8	7	4

Mid-values;

1st class:	$(0 + 4) \div 2 = 4 \div 2 = 2$
2nd class:	$(5 + 9) \div 2 = 14 \div 2 = 7$
3rd class:	$(10 + 14) \div 2 = 24 \div 2 = 12$
4th class:	$(15 + 19) \div 2 = 34 \div 2 = 17$
5th class:	$(20 + 24) \div 2 = 44 \div 2 = 22$

No. of boxes	Mid-value	Frequency
0–4	2	10
5–9	7	6
10–14	12	8
15–19	17	7
20–24	22	4

$$\text{mean} = \frac{(2 \times 10) + (7 \times 6) + (12 \times 8) + (17 \times 7) + (22 \times 4)}{35}$$

$$= \frac{20 + 42 + 96 + 119 + 88}{35} = \frac{365}{35}$$

$$= 10 \cdot 4 \text{ (to 3 s.f.)}$$

Exercise 9.6c

Find the mean of each of the following distributions.

1. The number of calls a salesman makes each day during one month is shown.

No. of calls	0–4	5–9	10–14	15–19	20–24	25–29
No. of days	6	8	2	1	2	1

2. A researcher notes the number of people in a Post Office queue, every hour, on the hour.

No. of people	0–4	5–9	10–14	15–19	20–24	25–29
Frequency	4	6	10	8	6	6

3. The height of plants in a green house is recorded.

Height (cm)	4-6	7-9	10-12	13-15	16-18
No. of plants	3	4	8	10	5

4. A training officer lists the size of his classes.

No. of pupils	1-5	6-10	11-15	16-20	21-25
No. of classes	2	5	5	7	1

5. The weight of each package in a van is recorded.

Weight (kg)	0-2	3-5	6-8	9-11	12-14	15-17
No. of packages	6	4	5	3	1	1

6. The manager of a luxury hotel notes the number of occupied rooms, each night.

No. of rooms	20-24	25-29	30-34	35-39	40-44	45-49
Frequency	10	20	9	8	7	6

7. The thickness of each piece of rosewood in a kiln is measured.

Thickness (cm)	2-4	5-7	8-10	11-13	14-16	17-19
No. of pieces	18	16	6	0	8	2

8. The number of sheets sent to a laundry by various hotels one Tuesday is as shown.

No. of sheets	1-5	6-10	11-15	16-20	21-25
No. of hotels	2	3	8	10	2

9. A manager in a secretarial agency lists the number of applicants received for similar jobs, at various times.

No. of applications	0-3	4-7	8-11	12-15	16-19
Frequency	10	18	11	9	2

10. The time taken by each competitor in a cycle race is noted.

Time (min)	30-31	32-33	34-35	36-37	38-39	40-41
No. of competitors	1	3	6	12	4	2

The probability p of an event happening is:

$$p = \frac{\text{number of 'successful' results}}{\text{total number of all possible results}}$$

This is usually written as a fraction, except that a certainty is 1 and an impossibility is 0.

Example 1

Seven counters numbered 1, 2, 3, 4, 5, 6, 7, are placed in a box. If one counter is drawn out at random, what is the probability that it is a counter with a number divisible by 3?

number of 'successes' = 2: these are counters 3 and 6.

total number of possibilities = 7

$$\therefore p = \frac{2}{7}$$

Exercise 9.7a

1. Six counters numbered 1, 3, 4, 5, 8, 9, are placed in a box. If one counter is drawn out at random, what is the probability that it is a counter

   (a) with an odd number,
   (b) with an even number?

2. If a dice is thrown, what is the probability that the score is

   (a) a prime number,
   (b) a square number?

3. Twelve counters labelled A, B, C, D, E, F, G, H, I, J, K, L, are placed in a box. If one counter is drawn out at random, what is the probability that it is a counter

   (a) with a consonant letter,
   (b) with a vowel letter?

4. If a card is withdrawn at random from a pack of 52 playing cards, what is the probability that it is:

   (a) an ace,
   (b) a picture card,
   (c) any number from 2 to 10?

5. A class contains 15 boys and 9 girls. If the class votes for one pupil to be the class captain and all of them stand an equal chance of election, what is the probability that the elected captain

will be   (a) a boy,
          (b) a girl?

6. A farmer has 40 white sheep and 32 black sheep. If they are rounded up in any order for shearing, what is the probability that the first to be sheared is   (a) a white sheep,
          (b) a black sheep?

7. A class of 30 boys contains 18 with dark hair, 8 with blonde hair and 4 with red hair. If the class proceeds to the assembly hall in random order, what is the probability that the first to enter the hall has

   (a) dark hair,
   (b) blonde hair,
   (c) red hair?

8. A television set has four channel-selectors, one for BBC 1, one for BBC 2, one for ITV and one which is not tuned to any channel. If I press any one of the four selectors at random, what is the probability

   (a) that I will receive a programme,
   (b) that I will receive a programme broadcast by the BBC?

9. A note-paper manufacturer makes pads in four colours: blue, white, pink, and green. In a pack of 36, there are always 12 blue pads, 10 white pads, 8 pink pads and 6 green pads. If I open a new pack, what is the probability that the pad is:

   (a) blue,
   (b) white,
   (c) pink,
   (d) green?

10. If two coins are tossed simultaneously, what is the probability of:

   (a) two heads,
   (b) two tails,
   (c) one head and one tail?

Combined probabilities

Addition rule If events A and B are *mutually exclusive* (i.e. they cannot happen at the same time) then

$$p(\text{A } or \text{ B}) = p(\text{A}) + p(\text{B})$$

Example 2

A box contains 6 blue balls, 4 red balls, and 5 white balls. What is the probability of picking out from the box either a red ball or a blue ball?

$$p \text{ (red)} = \frac{4}{15}; \ p \text{ (blue)} = \frac{6}{15}$$

$$\therefore p \text{ (red } or \text{ blue)} = \frac{4}{15} + \frac{6}{15} = \frac{10}{15} = \frac{2}{3}$$

Note:

The probability of picking out either a red ball or a blue ball or a white ball

$$= \frac{4}{15} + \frac{6}{15} + \frac{5}{15} = 1$$

Exercise 9.7b

1. A bag contains 8 red discs, 4 blue discs, and 1 white disc. What is the probability of picking out:
   (a) either a red disc or a blue disc,
   (b) either a red disc or a white disc,
   (c) either a blue disc or a white disc,
   (d) either a blue disc or a white disc or a red disc?

2. A small box of chocolates contains 4 hard centres, 6 soft centres and 2 foil-wrapped chocolates. What is the probability of picking out:
   (a) either a hard centre or a soft centre,
   (b) either a hard centre or a foil-wrapped chocolate,
   (c) either a soft centre or a foil-wrapped chocolate,
   (d) not a soft centre?

3. On a shelf in a supermarket, there are 8 packets of Sudso, 10 packets of Foamo and 12 packets of Washo soap powder. What is the probability that a housewife takes off the shelf:
   (a) either a packet of Foamo or a packet of Washo,
   (b) either a packet of Sudso or a packet of Foamo,
   (c) not a packet of Foamo?

4. A £1 cash bag contains four 10p coins, eight 5p coins and ten 2p coins. If one coin is picked out, what is the probability that the coin is:
   (a) a 10p coin,
   (b) not a 5p coin,
   (c) not a copper coin,
   (d) either a 5p coin or a 2p coin?

5. In a class of 30 boys, 15 choose Rugby, 12 choose Soccer and 3 choose Hockey to play during the winter term. What is the probability that the youngest:
   (a) does not play Rugby,
   (b) does not play Soccer,
   (c) does not play Hockey?

6. An aviary contains 12 white birds, 8 black birds and 5 yellow birds. If one of the birds escapes, what is the probability that it will be:
   (a) either a white bird or a black bird,
   (b) either a white bird or a yellow bird,
   (c) not a white bird?

7. A newspaper rack contains 7 Daily Tales, 3 Local News and 5 National News. If one of these papers is sold, what is the probability that it is:
   (a) either a Daily Tales or a National News,
   (b) either a National News or a Local News,
   (c) not a National News?

8. In a gymnasium there are 16 heavyweight wrestlers, 6 middleweights and 2 lightweights What is the probability that the only left-handed wrestler is:
   (a) either a heavyweight or a middleweight,
   (b) either a heavyweight or a lightweight,
   (c) not a lightweight?

9. A car park contains 40 cars. 20 are blue, 14 are red and 6 are white. What is the probability that the first car driven out of the exit is:
   (a) either blue or white,
   (b) either red or blue,
   (c) not red?

10. A tin holds 35 pencils. 16 are HB, 8 are H and the remainder are 2H. What is the probability that a pencil taken from the tin is:
    (a) either H or 2H,
    (b) either H or HB,
    (c) either HB or H or 2H,
    (d) 4H,
    (e) not 2H?

Multiplication rule If A and B are *independent* events (i.e. they do not affect each other) then

$$p \text{ (A } and \text{ B)} = p \text{ (A)} \times p \text{ (B)}$$

Example 3

A bag contains 6 blue balls and 4 red balls. A ball is taken out at random and replaced; then a second ball is taken out at random. What is the probability that:
(a) both balls are red,
(b) both balls are a different colour,
(c) both are red if the first ball is not replaced?

(a) p (first ball is red) $= \dfrac{4}{10}$

p (second ball is red) $= \dfrac{4}{10}$

$\therefore p$ (both are red) $= \dfrac{4}{10} \times \dfrac{4}{10}$

$= \dfrac{16}{100} = \dfrac{4}{25}$

(b) Method 1.

p (both are blue) $= \dfrac{6}{10} \times \dfrac{6}{10}$

$= \dfrac{36}{100} = \dfrac{9}{25}$

p (both are red) $= \dfrac{4}{10} \times \dfrac{4}{10}$

$= \dfrac{16}{100} = \dfrac{4}{25}$

$\therefore p$ (both are different) $= 1 - \left(\dfrac{9}{25} + \dfrac{4}{25} \right)$

$= \dfrac{12}{25}$

Method 2.

p (first ball is red) $= \dfrac{4}{10}$

p (second ball is blue) $= \dfrac{6}{10}$

$\therefore p$ (first red, second blue) $= \dfrac{4}{10} \times \dfrac{6}{10}$

$= \dfrac{24}{100} = \dfrac{6}{25}$

Similarly p (first blue,

second red) $= \dfrac{6}{25}$

$\therefore p$ (both are different) $= \dfrac{6}{25} + \dfrac{6}{25}$

$= \dfrac{12}{25}$

(c) p (first ball is red) $= \dfrac{4}{10}$

p (second ball is red) $= \dfrac{3}{9}$

$\therefore p$ (both are red) $= \dfrac{4}{10} \times \dfrac{3}{9}$

$= \dfrac{12}{90} = \dfrac{2}{15}$

Exercise 9.7c

1. A box contains 6 red socks and 4 white socks. A sock is taken out at random and put back in the box; a second sock is then taken out. What is the probability that:
   (a) both socks are red,
   (b) both socks are white,
   (c) both socks are a different colour,
   (d) both are red if the first is not replaced,
   (e) both are white if the first is not replaced,
   (f) both are different colours if the first is not replaced?

2. Eight counters numbered 1, 2, 3, 4, 5, 6, 7, 8, are placed in a box. One is taken out at random and replaced; a second one is then taken out. What is the probability that:
   (a) both are number 5,
   (b) both are even numbers,
   (c) both are prime numbers,
   (d) both are divisible by 3,
   (e) both are divisible by 3 if the first is not replaced,
   (f) the sum of both counters is 15,
   (g) the sum of both counters is 15 if the first is not replaced?

3. Two dice are thrown. What is the probability that:
   (a) one is a three and one is a four,
   (b) the total score is 12,
   (c) only one is a six,
   (d) both are even numbers,
   (e) the total score is 7,
   (f) neither is a 1?

4. In a geography quiz, cards are made with the names LONDON, LIVERPOOL, LEEDS, MANCHESTER, GLASGOW, EDINBURGH printed on them. If a card is picked at random and replaced, then a second is picked, what is

the probability that:
(a) both cards have names beginning with L,
(b) both cards have names of cities in England,
(c) both cards have names of cities in Scotland,
(d) one has an English city and the other a Scottish city,
(e) both cards have the names of English cities if the first is not replaced,
(f) one has an English city and one has a Scottish city if the first is not replaced?

5. From a pack of 52 playing cards, a card is chosen and its suit is noted; it is then put back and a second card is chosen. What is the probability that:
(a) both cards are Hearts,
(b) both cards are Spades,
(c) both cards are black suits,
(d) neither card is a Diamond,
(e) the first card is a red suit and the second is a black suit,
(f) one card is a Heart and the second is a Spade?

6. A domino is chosen at random from the following group; it is then replaced and a second is chosen.

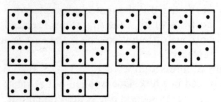

What is the probability that:
(a) both dominoes have six dots,
(b) both dominoes have seven dots,
(c) both dominoes have five dots,
(d) the first domino has six dots and the second five,
(e) both dominoes have six dots if the first is not replaced?

7. The names of 6 women and 4 men are each written on separate slips of paper and are placed in a hat. A slip is chosen and the name noted; it is then replaced and a second slip is chosen. What is the propability that:
(a) the first name is a woman's and the second is a man's,
(b) both are women's names,
(c) the first name is a man's and the second a woman's,
(d) both are men's names,
(e) both are men's names if the first is not replaced?

8. A coin is tossed and a die is rolled. What is the probability of obtaining:
(a) a head and a 6,
(b) a tail and a 1,
(c) a head and an even number,
(d) a head and a number less than 5,
(e) a tail and a number greater than 4?

9. Five cards, lettered A, B, C, D and E lie face down on a table. One of the cards is chosen, and then a die is tossed. What is the probability of obtaining:
(a) A and 6,
(b) B and an odd number,
(c) C and a multiple of 3,
(d) D and any number from 1 to 6,
(e) E and 7?

10. There are 25 tins of soup in a box; 15 are Tomato flavour and the others are Oxtail. A shopper chooses two tins from the box, at random. The first tin is not replaced before the second is chosen. What is the probability that:
(a) two tins of Oxtail are chosen,
(b) two tins of Tomato are chosen,
(c) the first tin contains Oxtail and the second Tomato,
(d) neither tin contains Tomato,
(e) the first tin does not contain Oxtail, and the second does not contain Tomato?

The values of a measure which changes with time can be compared by means of *index numbers*.

One year is chosen as the *base* year. The value in the base year is given an index number of 100. The values in all other years are then expressed as a percentage of the base year value: these percentages are index numbers.

Example 1

The cost of a table lamp was £5 in 1974, and £8 in 1978. Find an index number representing the 1978 cost compared to the 1974 cost.

Base year 1974 = 100

$$1978 \text{ index number} = \frac{\text{cost in 1978}}{\text{cost in 1974}} \times \frac{100}{1}$$

$$= \frac{8}{5} \times \frac{\overset{20}{\cancel{100}}}{1}$$

$$= 160$$

Example 2

Fertilizer costs:

Year	1972	1973	1974	1975
Cost (£/tonne)	50	60	70	75

(a) Find index numbers representing the cost of fertilizer each year, compared with the cost in 1972.

(b) By what percentage had costs risen between 1972 and 1975?

(a) Base year 1972 = 100

Year	Cost (£/tonne)	Index Number
1972	50	100
1973	60	$\frac{60}{\cancel{50}} \times \frac{\overset{2}{\cancel{100}}}{1} = 120$
1974	70	$\frac{70}{\cancel{50}} \times \frac{\overset{2}{\cancel{100}}}{1} = 140$
1975	75	$\frac{75}{\cancel{50}} \times \frac{\overset{2}{\cancel{100}}}{1} = 150$

(b) Percentage rise, between 1972 and 1975

$$= 150\% - 100\% = 50\%$$

Exercise 9.8a

1.

Year	1973	1974	1975	1976
Cost (£)	20	22	24	28

Find index numbers representing the cost in each year, compared with the cost in 1973.

2.

Year	1974	1975	1976	1977
Price (p)	40	60	66	80

By taking 1974 as base year, find index numbers representing the given prices.

3.

Year	1960	1961	1962	1963
Value (£)	800	1000	1200	1600

Calculate index numbers showing the value in each year, taking 1961 as base year.

4. The following table shows the number of members in a golf club.

Year	1970	1971	1972	1973
Members	75	120	180	360

Construct an index from these figures, taking 1970 as base year.

5.

Year	1950	1951	1952	1953
Cost (£)	125	175	150	180

Find index numbers representing the cost in each year. (Base year 1950 = 100)

6. A builder notes the quantity of cement he uses each year.

Year	1972	1973	1974	1975
Quantity (tonnes)	60	80	76	92

Taking 1973 as base year, find index numbers for the years 1972-5.

7. A tailor records the cost of a typical alteration, over a period of time.

Year	1975	1976	1977	1978
Cost (£)	10	12	18	24

Construct an index of cost, using 1975 as base year.

8. The number of trucks sold by a manufacturer is shown.

Year	1976	1977	1978	1979
Trucks (thousands)	24	120	96	180

(a) Find an index number for each year. (Base year 1977 = 100)
(b) Find the percentage increase in sales between 1977 and 1979.

9. The concentration of a chemical in a river is noted, in parts per million.

Year	1968	1969	1970	1971
Concentration	42	45	60	66

(a) Using 1970 as base year, find index numbers for each year.
(b) State the percentage rise between 1970 and 1971.

10. A sales manager records the number of drums of industrial cleaner sold by his company.

Year	1976	1977	1978	1979
Drums	300	225	250	175

(a) Find index numbers representing the number of drums sold each year. (Base year 1978 = 100)
(b) State the percentage decrease between 1976 and 1978.

11. The cost of components in a radio is:

Year	1977	1978	1979	1980
Cost (£)	6·30	7·50	16·50	17·10

(a) Using 1978 as base year, construct an index of cost.
(b) State the percentage rise in cost between 1978 and 1979.

12. The number of sheep on a small farm is noted.

Year	1960	1961	1962	1963
Sheep	160	176	208	248

(a) Construct an index using this information. (Base year 1960 = 100)
(b) Calculate the percentage increase between 1960 and 1963.

13. The profit from a grocer's shop is calculated each year.

Year	1975	1976	1977	1978
Profit	£19 800	£22 500	£23 400	£20 700

(a) Taking 1976 as base year, calculate index numbers to represent these profits.
(b) Find the percentage decrease in profits from 1976 to 1978.

14. The number of employees in a Regional Finance Department was:

Year	1969	1970	1971	1972
Employees	140	154	168	196

(a) Construct an index from these figures. (Base year 1969 = 100)
(b) Calculate the percentage increase between 1969 and each of the other years shown.

15. The number of pairs of a rare species of bird found in a sanctuary was:

Year	1958	1959	1960	1961
No. of pairs	90	96	120	150

(a) Taking 1961 as base year, find index numbers representing the number of pairs of birds seen each year.
(b) How much smaller was the number of pairs in 1958, compared to the number in 1961 (as a percentage)?

Example 3

In 1968, the cost of a train journey was £1·80 when the Index of Rail Prices was 120.
(a) If the index was 150 in 1970, how much did the journey cost then?

The index has increased in the ratio 150:120, so the cost must be increased in the same ratio.

$$\text{Cost} = \frac{\overset{75}{\cancel{150}}}{\underset{2}{\cancel{120}}_1} \times \frac{\overset{3}{\cancel{180}}}{1} \text{ pence}$$

$$= 225 \text{ pence} = £2·25$$

(b) If the index was 80 in 1964, how much did the journey cost then?

The index has decreased in the ratio 80:120, so the cost must be decreased in the same ratio.

$$\text{Cost} = \frac{\overset{40}{\cancel{80}}}{\underset{2\cdot 1}{\cancel{120}}} \times \frac{\overset{3}{\cancel{180}}}{1} \text{ pence}$$

$$= 120 \text{ pence} = £1\cdot20$$

Exercise 9.8b

1. In 1970, a furniture price index was 100, and the price of a chair was £25. Find the price of the chair when the index was
   (a) 120   (b) 160   (c) 72   (d) 200
2. A cinema charged 60p admission when the Admission Price Index was 80. Find the cost of admission when the index was
   (a) 100   (b) 60   (c) 52   (d) 120
3. A shipping company prepares an index each week from the average weights of containers. If the average weight was 15 tonnes when the index was 120, find the average weight if the index is
   (a) 160   (b) 232   (c) 88   (d) 136
4. A tyre company prepares an Index of Grip by measuring the force applied to a tyre on a test rig before it skids. When the force is 2·5 tonnes, the index is 100. Find the force when the index is
   (a) 200   (b) 120   (c) 60   (d) 250
5. In 1970 a Paint Price Index was 150 when the price of a quantity of paint was £1·20. Find the cost of the same quantity of paint if the index is
   (a) 100   (b) 120   (c) 75   (d) 125
6. The price of a ferry journey was £1·45 in 1970. Find the cost of the same journey in 1971, 1972 and 1973.

Year	1970	1971	1972	1973
Price Index	100	120	140	220

7. The index reflects the cost of cloth used by a dressmaker. If the cost of a bolt of cloth was £25 in 1972, find the cost in 1971, 1973 and 1974.

Year	1971	1972	1973	1974
Index Number	120	125	135	160

8. An index is devised from the power output of a motorcycle.

Year	1976	1977	1978	1979
Index Number	170	180	210	240

If the power output was 68 h.p. in 1976, find the output in the other years shown.

9. The following index represents the quantity of artificial flavouring in a tinned food.

Year	1966	1967	1968	1969
Index Number	40	55	60	100

If the food contained 5 mg of artificial flavouring in 1969, find the quantity in 1966, 1967 and 1968.

10.

Year	1973	1974	1975	1976
Productivity Index	120	115	125	135

In 1973 the average number of workpieces completed in a day was 24. Find the average number completed daily in 1974, 1975 and 1976 by using the Productivity Index.

11. An index represents the volume of oil used in a factory.

Year	1958	1959	1960	1961
Index Number	80	108	124	116

If the volume of oil used in 1958 was 12 000 litres, find the volume used in 1959, 1960 and 1961. 1961.

12. The following index represents the sales of a machine tool.

Year	1963	1964	1965	1966
Sales Index	93	108	132	141

If 720 of these machine tools were sold in 1964, find the number sold in 1963, 1965 and 1966.

13.

Year	1975	1976	1977	1978
Wage Index	110	117	135	162

In 1975 a man's salary was £1980 per year. If his salary increases at the same rate as the Wage Index shown, what should he be paid in 1976, 1977 and 1978?

14. In 1978, a decorator charged £434 to repaint an office. If the decorator's prices change according to the following index, find the charge for painting the same office in 1976, 1977 and 1979.

Year	1976	1977	1978	1979
Price Index	170	206	248	317

15.

Year	1960	1961	1962	1963
Weight Index	378	426	500	376

The Weight Index shows the changes in the weight of rock taken from a quarry. If 8460 tonnes was taken in 1963, how much was taken in 1960, 1961 and 1962?

Example 4

Year	1975	1976	1977	1978
Index of Production	100	125	80	150

(Base year 1975 = 100)
Change the above index, making 1976 the new base year.

(New base year 1976 = 100)

Year	*Old Index Number*	*New Index Number*
1976	125	100
1975	100	$\frac{100}{1} \times \frac{100}{125} = 80$
1977	80	$\frac{100}{1} \times \frac{80}{125} = 64$
1978	150	$\frac{100}{1} \times \frac{150}{125} = 120$

Exercise 9.8c

1.

Year	1975	1976	1977	1978
Index Number	50	100	80	120

(Base year 1976 = 100)
Change the base year of the above index to 1975 = 100.

2.

Year	1965	1966	1967	1968
Index Number	100	80	92	104

(Base year 1965 = 100)
Using base year 1966 = 100, find new Price Index numbers.

3.

Year	1972	1973	1974	1975
Index Number	100	125	150	160

(Base year 1972 = 100)
Find new index numbers taking base year 1973 = 100.

4. A marketing executive constructs a Sales Index.

Year	1965	1966	1967	1968
Sales Index	180	150	100	200

(Base year 1967 = 100)
Find new index numbers taking base year 1968 = 100.

5. A transport clerk prepares an index from the number of shipments sent each year.

Year	1973	1974	1975	1976
Index Number	100	76	40	58

(Base year 1973 = 100)
Using base year 1975 = 100, find new index numbers.

6. The staff of an employment office prepare an index from the number of people placed in jobs.

Year	1974	1975	1976	1977
Index Number	98	112	126	140

Find new index numbers using base year 1977 = 100.

7. An international airline prepares an index of costs per passenger.

Year	1969	1970	1971	1972
Index Number	96	108	120	144

Prepare a new index taking 1971 as base year.

8. An engineer devises an index showing the change in maintainance costs for a machine.

Year	1973	1974	1975	1976
Index Number	75	87	111	126

Write down a new set of index numbers taking base year 1973 = 100.

Test Paper 1

This paper should be completed in 1 hour, or less.
Do NOT use mathematical tables, slide rules or calculators.

1. Evaluate
   (a) $12 \cdot 6 + 5 \cdot 4 - 7 \cdot 63$ (b) $5 \cdot 4 \times 7 \cdot 3$ (c) $0 \cdot 082 \times 70$
   (d) $249 \cdot 1 \div 47$ (e) $0 \cdot 04 \times 0 \cdot 5$ (f) $23 - 4(4 - 2)$

2. Express as a fraction in its simplest form:
   (a) $\frac{2}{5} + \frac{1}{3} - \frac{1}{2}$ (b) $1\frac{3}{8} - \frac{4}{5}$ (c) $1\frac{15}{16} - 1\frac{7}{8}$ (d) $2\frac{1}{2}\%$

3. Express in the form $a \times 10^n$, where $1 \leqslant a < 10$ and n is an integer:
   (a) $42\ 700$ (b) $16 \cdot 3$ (c) $0 \cdot 0044$

4. (a) Express $4 \cdot 95$ correct to 1 decimal place.
   (b) Express $17 \cdot 851$ correct to 3 significant figures.

5. (a) Express $\frac{3}{8}$ as a percentage.
   (b) 11 batteries cost £8·03. Find the cost of 9 batteries.
   (c) Find $\sqrt{360\ 000}$ by factorisation.
   (d) Calculate $12\frac{1}{2}\%$ of £5·52.
   (e) Express 72_{eight} as a base ten number.

6. In each part of this question, write the letter A, B, C, D, or E corresponding to the correct answer.
   (i) Which of the following is nearest in value to $287 \cdot 43 \times 34 \cdot 7$?
       A 9   B 90   C 900   D 9 000   E 90 000
   (ii) Which of the following is nearest in value to $\dfrac{0 \cdot 096 \times 378 \cdot 6}{0 \cdot 990}$?
       A 0·04   B 0·4   C 4   D 40   E 400
   (iii) Calculate which of the following is nearest in value to $(4 \cdot 8 \times 10^3) \div (6 \times 10^{-2})$
       A $2 \cdot 88 \times 10^2$ B $2 \cdot 88 \times 10^6$ C 8×10^0 D 8×10^4 E 8×10^{-6}
   (iv) Which of the following is nearest in value to $\sqrt{2400}$?
       A 1200   B 120   C 4800   D 500   E 50
   (v) The total surface area of a cube of side 7 cm, expressed in square centimetres, is
       A 49   B 28   C 84   D 294   E 343

Test Paper 2

This paper should be completed in 1 hour, or less.
Do NOT use mathematical tables, slide rules or calculators.

1. Evaluate
   (a) $3{\cdot}07 + 6 - 4{\cdot}2$ (b) 47×13 (c) $0{\cdot}365 \times 40$
   (d) $17{\cdot}7 \div 300$ (e) $16{\cdot}328 \div 5{\cdot}2$ (f) $12 - 3(8 - 4)$

2. Express as a single fraction, in its simplest form:
   (a) $\frac{2}{3} + \frac{1}{4} - \frac{3}{5}$ (b) $\frac{1}{5} - 1\frac{1}{8} + 1\frac{3}{4}$ (c) $\frac{7}{8} \div 1\frac{5}{16}$

3. (a) Express $4{\cdot}7023$ correct to 2 significant figures;
   (b) Express $0{\cdot}0498$ correct to 2 decimal places.
   (c) Express $694{\cdot}71$ correct to the nearest whole number.

4. Express the following in standard form; i.e. in the form $a \times 10^n$, where $1 \leqslant a < 10$ and n is an integer.
   (a) 4080 (b) $170\,000$ (c) $0{\cdot}00042$

5. (a) Express 28% as a fraction in its simplest form.
   (b) Express $\frac{5}{8}$ as a percentage.
   (c) Find 7% of £35.
   (d) Find $\sqrt{90\,000}$, by factorisation.
   (e) 9 fish suppers cost £4·95. Find the cost of 13 fish suppers.

6. In each part of this question, write the letter A, B, C, D or E corresponding to the correct answer.
   (i) The length of a rectangle is 14 cm, and its breadth is 3 cm. If both of these measurements are correct to the nearest 1 cm, what is the LEAST possible area of the rectangle, in square centimetres?
   A $14{\cdot}5 \times 3{\cdot}5$ B 14×3 C $13{\cdot}5 \times 2{\cdot}5$ D $13{\cdot}5 \times 3{\cdot}5$ E $14{\cdot}5 \times 2{\cdot}5$
   (ii) Which of the following is nearest in value to $\sqrt{6460}$?
   A 3230 B $32{\cdot}30$ C 80 D 800 E $12\,920$
   (iii) Which of the following is nearest in value to $899 \times 0{\cdot}049$?
   A 4 B 45 C 450 D 4500 E $0{\cdot}45$
   (iv) Which of the following is nearest in value to $\dfrac{49{\cdot}6 \times 76}{4208}$?
   A 1 B 10 C 100 D $0{\cdot}1$ E $0{\cdot}01$
   (v) 15% of $0{\cdot}5$ is
   A 75 B $7{\cdot}5$ C $0{\cdot}75$ D $0{\cdot}075$ E $0{\cdot}0075$

Test Paper 3

This paper should be completed in 1 hour, or less.
Do NOT use mathematical tables, slide rules or calculators.

1. Evaluate
   (a) $1\frac{3}{4} - 2 + 1\frac{1}{3}$ (b) $1\frac{7}{8} \times \frac{4}{5}$ (c) $2\frac{1}{2} \div 1\frac{3}{8}$
   (d) $5\frac{3}{5} - 2\frac{2}{5} \times \frac{1}{6}$ (e) $3\frac{1}{3} \div 2(\frac{3}{4} - \frac{1}{3})$ (f) $1\frac{2}{3} \times 2\frac{5}{8} \div 1\frac{1}{6}$

2. (a) Find $16\frac{2}{3}\%$ of £4·80.
   (b) Express 145% as a mixed number.
   (c) Express 0·65 as a percentage.
   (d) A bicycle is bought for £32 and sold for £36.
       Express the profit as a percentage of the cost price.

3. Express in the form $a \times 10^n$, where $1 \leqslant a < 10$ and n is an integer.
   (a) 47 800   (b) 0·0039   (c) 5·29

4. (a) Express 21.30 hours using a.m. and p.m. notation.
   (b) How many hours are there from 4.30 a.m. on Tuesday until 5.30 p.m. the
       following Thursday?
   (c) Will 2040 A.D. be a leap year?
   (d) If the first day of March is a Tuesday, what day of the week is the last day of
       March?
   (e) Write, in figures, thirty thousand and fourteen.

5. (a) 7 maps cost £6·79. How much will 15 maps cost?
   (b) At 60 km/h, a journey takes 25 minutes. How long will the same journey
       take, at 75 km/h?

6. In each part of this question, write the letter A, B, C, D or E
   corresponding to the correct answer.
   (i)   The maximum sum of two measurements, 4 cm and 12 cm, is
         A 4 cm + 12 cm   B 3 cm + 11 cm   C 4·1 cm + 12·1 cm
         D 4·5 cm + 12·5 cm   E 5 cm + 13 cm
   (ii) $(8 \times 10^3) \times (4 \times 10^{-2}) =$
         A 2×10^1 B $3\cdot2 \times 10^2$ C $3\cdot2 \times 10^1$ D 2×10^{-1} E $3\cdot2 \times 10^5$
   (iii) Which of the following is nearest in value to $\dfrac{4\cdot9 \times 1\cdot89}{18 \times 0\cdot95}$?

         A 0·05   B 0·5   C 5   D 50   E 500
   (iv) Which of the following is nearest in value to $\sqrt{0\cdot0228}$?
         A 0·0015   B 0·015   C 0·15   D 1·5   E 15
   (v)   Which of the following is nearest in value to $5\cdot2 \times 31\cdot4 \div 1\cdot901$?
         A 7·5   B 75   C 750   D 300   E 3000

Test Paper 4

This paper should be completed in 1 hour, or less.
Do NOT use mathematical tables, slide rules or calculators.

1. Evaluate
   (a) $3 \cdot 2 \times 200$ (b) $4 \cdot 7 \times 0 \cdot 4$ (c) $1 \cdot 3 - 2 \cdot 4 + 1 \cdot 9$
   (d) $1 \cdot 8 + 2 \times 0 \cdot 35$ (e) $3 \cdot 5 \div 1 \cdot 4$ (f) $126 \div 2 \cdot 8$

2. (a) Find the product of $\frac{1}{2}$ and $0 \cdot 8$.
   (b) Express $0 \cdot 76$ as a common fraction in its simplest form.
   (c) Express $2\frac{3}{8}$ as a percentage.
   (d) Find 85% of £42·40.

3. Express 164·52 km correct to
   (a) the nearest 10 km,   (b) 3 significant figures,   (c) 1 decimal place.

4. (a) VAT at 15% is added to a bill of £6·20. Find the total amount to be paid.
   (b) A boy scores 14 out of 40 in a test. Express his mark as a percentage.
   (c) I travel 48 kilometres in 40 minutes.
        Find my speed, in kilometres per hour.
   (d) A bus journey starts at 08.15 hours and ends at 13.22 hours.
        How long does the journey take?

5. Find, by factorisation:
   (a) $\sqrt{6400}$ (b) $\sqrt{6\frac{1}{4}}$ (c) $\sqrt{0 \cdot 0225}$

6. In each part of this question, write the letter A, B, C, D, or E
   corresponding to the correct answer.
   (i)   By how much does the product of 8 and 4 exceed the sum of 2 and 3?
         A 6   B 7   C 26   D 27   E 38
   (ii) Which of the following is nearest in value to $\dfrac{37 \cdot 1 \times 42 \cdot 8}{19 \cdot 2 \times 77 \cdot 7}$?
         A 0·01   B 0·1   C 1   D 10   E 100
   (iii) $\dfrac{4 \cdot 8 \times 10^4}{(1 \cdot 2 \times 10^3) \times (1 \cdot 6 \times 10^{-2})} =$
         A 0·25   B 2·5   C 25   D 250   E 2500
   (iv) Which of the following is nearest in value to $\dfrac{3 \cdot 91 \times 11 \cdot 4 \times 15 \cdot 5}{4 \cdot 74 \times 0 \cdot 09 \times 0 \cdot 791}$?
         A 2   B 20   C 200   D 2 000   E 20 000
   (v)   A rectangle measures 10 cm by 4 cm.
         Its maximum area, in square centimetres, is given by
         A 10×4 B $10 \cdot 5 \times 4$ C $10 \times 4 \cdot 5$ D $10 \cdot 5 \times 4 \cdot 5$ E $9 \cdot 5 \times 4 \cdot 5$

Test Paper 5

This paper should be completed in 1 hour, or less.
Do NOT use mathematical tables, slide rules or calculators.

1. Evaluate
   (a) $6 \cdot 42 + 3 \cdot 5 - 1 \cdot 64$ (b) $32 \cdot 3 \times 400$ (c) $3 \cdot 864 \div 0 \cdot 7$
   (d) $3 \cdot 264 \div 1 \cdot 28$ (e) $42 \cdot 7 \times 3 \cdot 1$ (f) $12 - 4(6 - 4)$

2. Evaluate
   (a) $\frac{6}{7} + \frac{1}{2} - \frac{2}{3}$ (b) $4\frac{2}{3} \times 2\frac{1}{2}$ (c) $2\frac{5}{6} - 1\frac{3}{4}$

3. (a) Express $7 \cdot 4251$ correct to 2 decimal places.
   (b) Express $0 \cdot 063$ correct to 1 significant figure.

4. Express in the form $a \times 10^{n}$, where $1 \leqslant a < 10$ and n is an integer.
   (a) 4270 (b) $0 \cdot 098$ (c) $(4 \cdot 2 \times 10^{3}) \times (2 \times 10^{3})$
   (d) $(1 \cdot 23 \times 10^{4}) \div (4 \cdot 1 \times 10^{2})$

5. (a) Calculate 12% of £27·50.
   (b) Express $\frac{9}{14}$ as a percentage, correct to 1 decimal place.
   (c) Find, by factorisation, or otherwise;
       (i) $\sqrt{22500}$ (ii) $\sqrt[3]{0 \cdot 027}$
   (d) 9 copies of a magazine cost £3·33.
       Find the cost of 23 copies.

6. In each part of this question, write the letter A, B, C, D or E
   corresponding to the correct answer.
   (i) Which of the following is nearest in value to $\sqrt{6360}$?
       A 320   B 32   C 800   D 80   E 8
   (ii) Which of the following is nearest in value to 479×6237 ?
       A 300   B 30 000   C 3 000 000   D 200 000   E 2 000 000
   (iii) Which of the following is nearest in value to $\dfrac{4 \cdot 27 \times 0 \cdot 0098}{0 \cdot 0242}$?
       A 20   B 2   C 0·2   D 0·02   E 0·002
   (iv) Which of the following is nearest in value to $\sqrt[3]{0 \cdot 00784}$?
       A 20   B 2   C 0·2   D 0·02   E 0·002
   (v) Which of the following is nearest in value to 11% ?
       A $\frac{1}{11}$ B $\frac{1}{9}$ C $\frac{1}{5}$ D $\frac{1}{4}$ E 11

Test Paper 6

This paper should be completed in 1 hour, or less.
Do NOT use mathematical tables, slide rules or calculators.

1. Evaluate
   (a) $16 - 20 + 9$ (b) $3 \cdot 6 + 0 \cdot 04 - 1 \cdot 95$ (c) 270×13
   (d) $4 \cdot 72 \times 0 \cdot 31$ (e) $99 \cdot 12 \div 21$ (f) $4 + 12(6 - 2)$

2. Evaluate
   (a) $1\frac{2}{3} \times \frac{4}{15}$ (b) $\frac{1}{2} + \frac{2}{3} - \frac{3}{4}$ (c) $2\frac{1}{4} \div 1\frac{4}{5}$

3. Express
   (a) $0 \cdot 092$ as a common fraction in its simplest form.
   (b) 47800 in the form $a \times 10^n$, where $1 \leqslant a < 10$ and n is an integer.
   (c) $9 \cdot 8246$ correct to 2 decimal places.
   (d) $0 \cdot 04070$ correct to 2 significant figures.
   (e) $7 \cdot 23 \times 10^{-4}$ as a single number.

4. (a) Find $12\frac{1}{2}\%$ of £64·80.
   (b) Calculate $\sqrt{25\,600}$ by factorisation.
   (c) Write $\frac{3}{8}$ as a decimal fraction.

5. (a) A car travels for 40 minutes at 50 km/h.
       How long would the same journey take, at 80 km/h ?
   (b) What time elapses between 22.11 hours on Thursday and 04.09 hours the following Saturday?
   (c) Express $\frac{9}{15}$ as a percentage.

6. In each part of this question, write the letter A, B, C, D, or E corresponding to the correct answer.
   (i) Which of the following is nearest in value to $\dfrac{11 \cdot 4 \times 18 \cdot 63}{4 \cdot 64 \times 0 \cdot 76}$?
       A 0·5   B 5   C 50   D 500   E 5000
   (ii) Which of the following is nearest in value to $\dfrac{0 \cdot 059 \times 8147}{32 \cdot 4 \times 9 \cdot 65}$?
       A 0·02   B 0·2   C 2   D 20   E 200
   (iii) Which of the following is nearest in value to $\sqrt{0 \cdot 00636}$?
       A 0·00318   B 0·318   C 0·008   D 0·08   E 0·8
   (iv) The sides of a rectangle are given as $(4 \cdot 8 \pm 0 \cdot 05)$ metres and $(2 \cdot 7 \pm 0 \cdot 05)$ metres. Which of the following represents the *minimum* possible value of area for this rectangle, in square metres?
       A $4 \cdot 8 \times 2 \cdot 7$ B $4 \cdot 85 \times 2 \cdot 75$ C $4 \cdot 75 \times 2 \cdot 65$ D $4 \cdot 8 \times 2 \cdot 75$
       E $4 \cdot 85 \times 2 \cdot 7$
   (v) $(6 \cdot 9 \times 10^{-2}) \div (2 \cdot 3 \times 10^{-3}) =$
       A 0·03   B 0·3   C 3   D 30   E 300

Test Paper 7

This paper should be completed in 1 hour, or less.
Do NOT use mathematical tables, slide rules or calculators.

1. Evaluate
   (a) $6\frac{1}{2} - 8 + 3\frac{2}{3}$ (b) $4(2\cdot3 - 0\cdot7)$ (c) $0\cdot091 \times 300$
   (d) $16 \div (2 + 1\frac{1}{3})$ (e) $36\cdot4 \div 0\cdot13$ (f) $\frac{4}{5}$ of $7\frac{1}{2}$

2. (a) If £1 = 9·68 francs, find the value of £24·25, in francs.
   (b) How much must be paid in rates, for property with rateable value £165, if the rate is 109p per £ ?
   (c) During a sale, a discount of 35% is given. What is the sale price of a coat which normally costs £63·60 ?
   (d) A bag of cement cost £18 when the price index was 120. Find the cost when the index was 85.

3. Express in the form $a \times 10^n$, where $1 \leqslant a < 10$ and n is an integer.
   (a) 47 000   (b) 0·00604   (c) $(3\cdot57 \times 10^2) \div (2\cdot1 \times 10^4)$

4. (a) An athlete runs 10 metres in 2 seconds. Find his speed in kilometres per hour.
   (b) A square has area 144 cm^2. Find its length.
   (c) What is the time taken to travel 1·8 km at a speed of 90 metres per second ?

5. State
   (a) 4·6502 correct to 1 decimal place.
   (b) 17 596 correct to 2 significant figures.
   (c) 0·0604 correct to 2 significant figures.
   (d) 47 206 km correct to the nearest 10 km.

6. In each part of this question, write the letter A, B, C, D or E corresponding to the correct answer.
   (i) A map has scale 1:200 000. What distance on the map represents 4 km on the ground?
       A 2 cm   B 20 cm   C 200 cm   D 8 cm   E 80 cm
   (ii) Between which two consecutive whole numbers does $\sqrt{218}$ lie?
       A 217 and 219   B 14 and 15   C 108 and 109   D 435 and 437   E 6 and 7
   (iii) Find the mean of 5, 6, 6, 8, 11, 12
       A 6   B 7   C 8   D 48   E 288
   (iv) In base ten, the answer to $1011_{two} + 11_{two}$ is
       A 1022   B 1110   C 14   D 10   E 8
   (v) Which of the following is nearest in value to $4\cdot85 \times 37\cdot6 \div 0\cdot19$?
       A 0·1   B 1   C 10   D 100   E 1000

Test Paper 8

This paper should be completed in 1 hour, or less.
Do NOT use mathematical tables, slide rules or calculators.

1. Evaluate
    (a) $\frac{5}{6} + 2\frac{3}{4} - 1\frac{2}{3}$ (b) $\frac{5}{8} \times \frac{4}{15}$ (c) $\frac{7}{16} \div \frac{5}{8}$

    (d) $\frac{3}{4} \times \frac{4}{5} + \frac{1}{2}$ (e) $(\frac{7}{8} + \frac{1}{3}) \div \frac{5}{12}$ (f) $5 - \frac{3}{4}(\frac{5}{6} + \frac{3}{5})$

2. (a) Evaluate
    $3 \cdot 95 \times 0 \cdot 02$ (b) $8 \cdot 64 \div 1 \cdot 6$ (c) $3 - 2 \cdot 7 \times 0 \cdot 8$

3. (a) Express 2476 correct to 2 significant figures.
   (b) Write $6 \cdot 3 \times 10^{-2}$ as a single number.
   (c) Write $473 \cdot 6$ in the form $a \times 10^{n}$, where $1 \leqslant a < 10$ and n is an integer.

4. (a) If £2·97 is spent from a £10 note, what amount remains ?
   (b) What is the next leap year after 1994 ?
   (c) How many days are there from 4th May to 21st July, inclusive of both days?
   (d) What percentage of 750 is 360?

5. (a) Travelling at 46 km/h, how long will a journey of 437 km take?
   (b) How many minutes between 8.15 a.m. and 2.05 p.m.?
   (c) A discount of $6\frac{3}{4}\%$ is given on a bill of £20. How much must be paid?
   (d) Find the arithmetic mean of 4, 6, 6, 8, 10, 14

6. In each part of this question, write the letter A, B, C, D or E
   corresponding to the correct answer.
   (i)   How many square millimetres in one square metre?
         A 100   B 1000   C 10 000   D 100 000   D 1 000 000
   (ii) 1 kg of meat costs £1·80 when the price index is 160. If the price rises to
         £2·25 for 1 kg, the index will be
         A 100   B 125   C 128   D 200   E 320
   (iii) Which of the following is nearest in value to $\dfrac{4768}{246 \times 0 \cdot 181}$?
         A 1   B 10   C 100   D 1000   E 10 000
   (iv) $\sqrt{0 \cdot 0144}$ is equal to
         A 0·0072   B 0·0288   C 0·012   D 0·12   E 1·2
   (v)   The sides of a rectangle are measured as 4·50 cm and 6·35 cm, both measure-
         ments being correct to 2 decimal places.
         Which of the following gives the maximum possible area of this rectangle, in
         square centimetres?
         A 4·50 × 6·35   B 4·6 × 6·4   C 4·5 × 6·4   D 4·505 × 6·355
         E 4·51 × 6·36

Test Paper 9

This paper should be completed in 1 hour, or less.
Do NOT use mathematical tables, slide rules or calculators.

1. Evaluate
   (a) $4\frac{1}{2} \times 3\frac{1}{3}$ (b) $12 - 2(4\frac{3}{4} - 1\frac{1}{2})$ (c) $16\frac{1}{3} \div 1\frac{1}{6}$

   (d) $2\frac{2}{3} - 1\frac{3}{4}$ (e) $\frac{5}{6} - \frac{7}{8} + 1\frac{3}{4}$ (f) $2\frac{1}{2} + 3\frac{1}{3} \times \frac{1}{2}$

2. Express
   (a) $37\frac{1}{2}\%$ as a fraction in its simplest form.
   (b) $\frac{3}{16}$ as a decimal, correct to 2 significant figures.
   (c) $0 \cdot 85$ as a percentage.
   (d) $3 \cdot 45 \times 10^{-1}$ as a single number.

3. Express $0 \cdot 0465$
   (a) correct to 3 decimal places
   (b) correct to 2 significant figures
   (c) in the form $a \times 10^n$, where $1 \leqslant a < 10$ and n is an integer.

4. (a) If 9 men can fell a wood in 25 days, how long would 5 men take?
   (b) Find, in hours and minutes, the time taken to travel 144 km at 120 km/h.
   (c) A rug is bought for £24 and sold for £27.
       Express the profit as a percentage of the cost price.
   (d) A holiday which normally costs £186 is subject to a surcharge of 15%.
       Find the total cost of the holiday.

5. Express 72_{ten} as a number in (a) base two, (b) base three, (c) base five.

6. In each part of this question, write the letter A, B, C, D or E
   corresponding to the correct answer.
   (i) Which of the following is nearest in value to $\sqrt{6314}$?
        A $31 \cdot 57$ B 3157 C 8 D 80 E 800

   (ii) Which of the following is nearest in value to $\dfrac{76}{19 \cdot 3 \times 0 \cdot 054}$?
        A $0 \cdot 08$ B $0 \cdot 8$ C 8 D 80 E 800

   (iii) $\dfrac{1 \cdot 2 \times 10^3}{(2 \cdot 4 \times 10^2) \times (2 \times 10^{-2})}$ is equal to

        A 4×10^2 B 4×10^3 C $2 \cdot 5 \times 10^3$ D $2 \cdot 5$ E $2 \cdot 5 \times 10^2$

   (iv) Which of the following is nearest in value to $\sqrt{0 \cdot 092}$?
        A $0 \cdot 046$ B $0 \cdot 184$ C $0 \cdot 3$ D $0 \cdot 03$ E $0 \cdot 003$

   (v) Which of the following is nearest in value to $0 \cdot 049 \times 16 \cdot 46 \div 3 \cdot 5$?
        A $0 \cdot 002$ B $0 \cdot 02$ C $0 \cdot 2$ D 2 E 20

Test Paper 10

This paper should be completed in 1 hour, or less.
Do NOT use mathematical tables, slide rules or calculators.

1. Evaluate
   (a) $6 - 0.95 + 1.2$ (b) 3.8×400 (e) 0.76×0.2
   (d) $3.8 \times 0.4 \div 0.19$ (e) $5.72 - 4(0.2 + 0.93)$ (f) $6(1 - 0.81) + 2.3$

2. Express as fractions in their simplest form:
   (a) 0.875 (b) $7\frac{1}{2}\%$ (c) two-thirds of three-quarters.

3. Express in the form $a \times 10^n$, where $1 \leqslant a < 10$ and n is an integer.
   (a) $107\,000$ (b) 0.0191 (c) $(4 \times 10^2) \div (1.6 \times 10^{-1})$

4. Express 146.50 km correct to
   (a) 1 decimal place,   (b) 1 significant figure,   (c) the nearest 10 km.

5. (a) Express 215_{eight} as a number in base three.
   (b) A car insurance premium of £86·50 is subject to a surcharge of 20%. Find the total cost.
   (c) Property valued at £28 500 is insured at the rate of 37p per £100. Find the premium.
   (d) A car travels at 65 km/h for 5 hours 12 minutes. How far does it travel?
   (e) 120 firelighters cost £3·80. Find the cost of 150 firelighters.

6. In each part of this question, write the letter A, B, C, D or E corresponding to the correct answer.
   (i) Which of the following is nearest in value to $\sqrt{8000}$?
       A 16 000   B 4000   C 100   D 90   E 9
   (ii) The scale of a map is 1:50 000. What distance on the ground is represented by a distance of 2·5 cm on the map?
       A 1·25 km   B 12·5 km   C 125 km   D 1 250 km   E 12 500 km
   (iii) Which of the following is nearest in value to $\dfrac{486 \times 0.8}{3714}$?
       A 0·001   B 0·01   C 0·1   D 1   E 10
   (iv) The median of the numbers 5, 4, 7, 3, 5, 9, is
       A 7   B 5·5   C 5   D 6   E 3
   (v) A square has sides of length (8 ± 0.5) cm.
       Which of the following expressions represents the *minimum* possible perimeter of this square, in centimetres ?
       A 8×4 B 8.5×4 C 7.5×4 D 8.5×4.5 E 7.5×3.5

Test Paper 1

This paper should be completed in 1½ hours, or less.
Mathematical tables, slide rules or calculators may be used.
Show the working for every question. Attempt ALL the questions.

1. A shopkeeper offers to sell a radio at a cash price of £46·40, or on hire purchase. The hire purchase terms are: a deposit of £9·85 and 15 monthly payments of £3·21.
   (a) Find the extra cost of buying by hire purchase.
   (b) Express the extra cost as a percentage of the cash price.

2. The rateable value of a man's house is £865.
   (a) If the rate is 42p in the £, find the amount he must pay in rates.
   (b) If the rate rises to 46p in the £, find how much more he must now pay in rates.

3. (a) Find the simple interest earned by £640 invested at $6\frac{1}{4}\%$ per year, for 9 months.
   (b) A bank lends a woman £1000 and charges interest at 8% of the total amount owed at the beginning of each year.
       At the end of the first year the woman pays back £280.
       Find the total amount she owes at the end of the second year.

4. (a) A car travels for 4 hours 15 minutes at 60 km/h.
       Find the total distance travelled.
   (b) A train begins a journey at 19.50 hours. It travels 399 km at 76 km/h. At what time during the next day does its journey end?

5. A short length of brass bar has a circular cross section, as shown. Running through the centre of the bar is a square hole of side 15 cm.

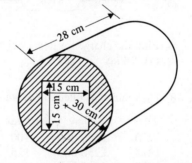

   (a) If the radius of the bar is 30 cm, calculate the area of the shaded cross section of the bar.
   (b) If the bar is 28 cm long, find the volume of the brass.
       [Area of a circle $= \pi r^2$. Take $\pi = 3\cdot14$]

6. If you are using a calculator, do both parts of this question.
   If you are not using a calculator, do part (a) only.
   (a) Calculate the volume of a cone of height 15·0 cm in which the radius of the base is 5 cm.
       Give your answer correct to 2 significant figures.
       [Volume of a cone $= \frac{1}{3}\pi r^2 h$. Take $\pi = 3\cdot14$]
   (b) Evaluate $\dfrac{4\cdot63 \times 1\cdot7 - 0\cdot04}{63\cdot63 + 17\cdot93}$ giving the answer correct to 2 significant figures.

7. The flowchart shows how to calculate the charge levied for express delivery of extremely fragile goods.

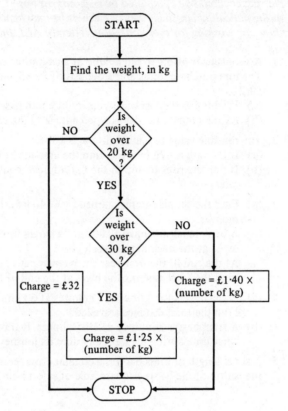

Find the charge for items weighing
(a) 16 kg   (b) 24 kg   (c) 30 kg

8. The value of a collection of watches, and the corresponding index, is shown below.

Year	Value	Index
1970	£200 000	100
1971	£250 000	125
1972	£292 000	
1973		160

(a) Calculate the value of the collection in 1973.
(b) Find the index number for 1972.
(c) If the base year is changed to 1971, calculate a new index number for 1973.

9. The readings from an electricity meter were:
        March 31st   23462 units
        June 30th   24967 units
Units of electricity cost 1·4p each for the first 250 units, and 0·8p each for any remaining units.
There is also a standing charge of £4·80 per quarter.
Calculate the cost of the electricity used between March 31st and June 30th.

10. The bar chart shows the number of cars carried on a ferry each day, during one week.

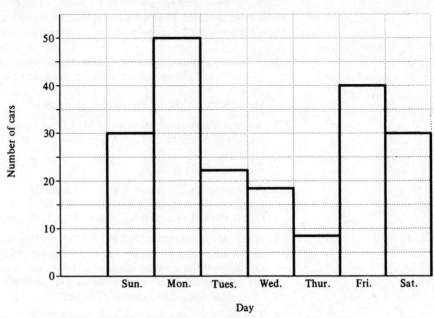

(a) Altogether, how many cars did the ferry carry during the week?

(b) Find the mean number of cars carried per day, correct to 1 decimal place.

(c) On which day did the ferry carry the smallest number of cars?

(d) On how many days did the ferry carry more than 25 cars?

(e) If one of the cars carried was chosen at random, find the probability that it was carried on Friday.

Test Paper 2

This paper should be completed in 1½ hours, or less.
Mathematical tables, slide rules or calculators may be used.
Show the working for every question. Attempt ALL the questions.

1. A woman changed £95 into dollars, at the rate of $2·45 to the pound. She spent $185·39, and changed the remainder back into pounds at the rate of $2·56 to the pound.
   How much did she finally receive, in pounds?

2. A man insures property worth £24 000, and contents worth £6 500. The premium is charged at the rate of 12·5p per £100 for property, and 24p per £100 for contents.
   Find the total premium.

3. A train travels 273 km in 4 hours 12 minutes.
   (a) Find the average speed of the train, in kilometres per hour.
   (b) If the train starts this journey at 13.57 hours, when does the journey end?

4. A wooden crate measures 2 m by 1·5 m by 0·8 m.
   Boxes measuring 20 cm by 5 cm by 7 cm are placed inside this crate.
   (a) What is the greatest number of boxes the crate can hold?
   (b) What is the volume of the unused space inside the crate, when it contains the greatest possible number of boxes?

5. The graph shows a van's journey from Carlisle to Stirling, and a car's journey from Stirling to Carlisle.

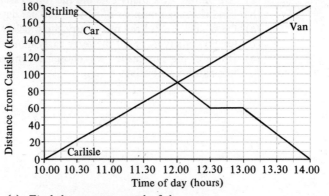

   (a) Find the average speed of the van.
   (b) When did the car leave Stirling?
   (c) Find the average speed of the car, during the first two hours of its journey.
   (d) How far was the car from Carlisle, when it stopped?
   (e) For how long did the car stop?
   (f) When did the van and the car pass each other?
   (g) How far were the van and the car from Stirling, when they passed each other?

6. If you are using a calculator, do both parts of this question.
   If you are not using a calculator, do part (a) only.
   (a) Find the radius of a sphere of surface area 154 cm^2.

   $$[\text{Radius} = \sqrt{\frac{A}{4\pi}}. \text{ Take } \pi = \frac{22}{7}]$$

   (b) Calculate $\dfrac{70·6 \times 0·12}{4·6 + 0·95 \times 2·4}$ giving your answer correct to 1 decimal place.

7. A blacksmith works from 07.30 hours until 17.15 hours, Monday to Thursday.
   On a Friday he stops work at 16.30 hours.
   His basic rate of pay is £1·84 per hour.
   (a) If he has a 1 hour lunch break each day, how many hours does he work in a
       week?
   (b) In a normal week, what will he earn?
   (c) On a Saturday he is paid time-and-a-half.
       If he works from 07.45 hours until 12.15 hours, how much will he earn on
       Saturday?
   (d) The blacksmith spends £9·20 each week on petrol, travelling to work. At the
       basic rate, how many hours must he work to pay for his petrol?

8. A lifeguard notes the number of people who go boating on a lake, each day,
   during September.

0	4	3	4	3	5	3
2	5	1	3	7	1	7
9	2	6	5	4	7	3
1	3	4	7	9	8	7
4	8	4	8	5	3	6

   (a) Construct a frequency table showing this information.
   (b) How many people went boating altogether?
   (c) Calculate the mean number of people who went boating each day.
   (d) State the mode.

9. 200 welders and 300 sheet metalworkers attend a conference.
   50 of the welders are women.
   If a person is chosen at random from those attending the conference, find the
   probability that this person is
   (a) a welder,
   (b) a female welder.

10. The following flowchart shows how to calculate the charge for hiring scaffolding.

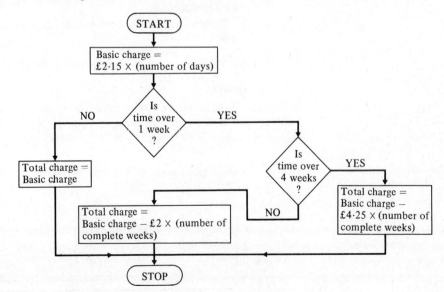

   Find the total charge if the scaffolding is hired for
   (a) 3 days,   (b) 12 days,   (c) 2 weeks,   (d) 30 days

Test Paper 3

This paper should be completed in 1½ hours, or less.
Mathematical tables, slide rules or calculators may be used.
Show the working for every question. Attempt ALL the questions.

1. The cash price of an outboard motor is £242. If the motor is bought on hire purchase, there is a deposit of £65 and 18 monthly payments of £11·80. Find the extra cost of buying by hire purchase.

2. Find the simple interest due on £760 invested at 15% per annum, for 3 years.

3. (a) Express 635_{ten} as a number in base eight.
   (b) Express 323_{four} as a number in base ten.

4. The members of a golf team buy a meal:
   > 8 soups @ 75p each
   > 3 fish @ £1·82 each
   > 5 fruit cocktails @ £1·28 each
   > 6 chicken salads @ £2·87 each
   > 2 curries @ £3·10 each
   > 8 coffees @ 60p each

   (a) Find the total cost of the meal.
   (b) If a service charge of $12\frac{1}{2}\%$ is added, what is the final bill?

5. A sports car is 2·1 m long and 1·4 m high. A model of this car is 4 cm high.
   (a) Find the scale of the model.   (b) Find the length of the model.

6. If you are using a calculator, do both parts of this question.
   If you are not using a calculator, do part (a) only.
   (a) A solid is made from a cylinder and a hemisphere.
      If the height of the cylinder is 18·6 cm and the radius of the cylinder and the radius of the hemisphere are both 3 cm, find the volume of this solid, correct to 1 decimal place.
      [Volume of a cylinder $= \pi r^2 h$; volume of a hemisphere $= \frac{2}{3}\pi r^3$; take $\pi = 3\cdot14$]
   (b) Calculate $\dfrac{6\cdot39 \times 0\cdot045}{(9\cdot763 - 0\cdot989)}$ expressing the answer correct to 3 significant figures.

7. The flowchart shows how to calculate the cost of a weekly rail ticket, according to the number of return journeys made during the week.

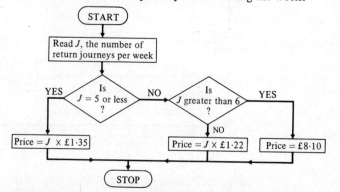

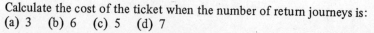

Calculate the cost of the ticket when the number of return journeys is:
(a) 3   (b) 6   (c) 5   (d) 7

8. (a) A motorist's basic insurance premium is £96. However, there is a $33\frac{1}{3}\%$
no-claims discount, because the motorist has had 2 years' accident-free driving.
There is a 10% surcharge (calculated on the cost of the premium after
deduction of the no-claims discount) because the car is not kept in a garage
overnight.
Calculate the final premium.

   (b) During 5 consecutive years, the motorist travels 11 000, 14 000, 13,000,
18 000 and 23 000 km.
Calculate the mean distance travelled per year, over this period.

9. If two dice are thrown, find the probability that the result is:
   (a) two even numbers   (b) two numbers whose sum is 7.

10. The histogram shows the number of candidates sitting an examination during the
years 1976-1979.

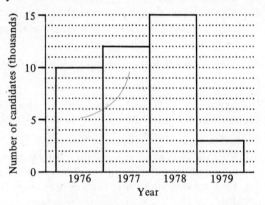

   (a) How many candidates sat the exam during the period 1976-1979 (inclusive
of both years) ?
   (b) During which year did the greatest number of candidates sit the exam?
   (c) What percentage of the total number of candidates shown sat the exam in
1977 ?
   (d) Taking the years 1978 and 1979 together, what percentage of the total
number of candidates shown sat the exam in 1978 or 1979 ?
   (e) What is the probability that a candidate, chosen at random, sat the exam in
1977 ?

Test Paper 4

This paper should be completed in 1½ hours, or less.
Mathematical tables, slide rules or calculators may be used.
Show the working for every question. Attempt ALL the questions.

1. An aircraft takes off from a landing strip and flies for 3 hours 20 minutes at an
    average speed of 420 km/h.
    (a) How far does it fly?
    (b) Its destination is 1904 km from the landing strip.
         If it flies at the same speed, how much longer will the flight take?

2. A modeller making a model aircraft estimates he will need 18 sheets of balsa at
    66p each.
    (a) Find the total cost of the balsa.
    (b) The modeller finds he can order the balsa by post, at 48p per sheet. If he does
         this, he must pay 85p for postage and packing. How much would he save, if
         he ordered the balsa by post?

3. The table shows the cost of accommodation at two hotels, per week.

	Seaview	Royale
single room	£63·80	£94·90
double room	£109·20	£182·60
suite	£268·70	£471·70

    Supplementary charge during July and
    August: £13·75 per person per week.

    (a) Find the cost of accommodation for a party of 16 people, staying in single
         rooms in the *Royale* during February for 1 week.
    (b) Find the cost of 3 suites for 7 people at the *Seaview* for a fortnight during
         July.

4. The scale of a map is 1:50 000.
    (a) Find the actual distance corresponding to 3·75 cm on the map.
    (b) What distance on the map will represent a true distance of 4 km ?
    (c) The area of a field, on the map, is 5 cm².
         What is the actual area of this field, in square metres?

5.

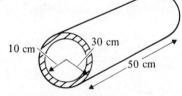

    (a) Calculate the area of a circle of radius 10 cm.
    (b) Find the area of the shaded cross section of the pipe shown above.
    (c) Find the volume of metal used to make a 50 cm length of this pipe.
         [Area of a circle $= \pi r^2$. Take $\pi = 3\cdot14$]

6. If you are using a calculator, do both parts of this question.
    If you are not using a calculator, do part (a) only.
    (a) Find the radius of the base of a cone of height 37·5 mm and volume 3925 mm³.

         $$\left[\text{Radius of base} = \sqrt{\frac{3V}{\pi h}}\,.\ \text{Take } \pi = 3\cdot14\right]$$

    (b) Evaluate $\dfrac{46 \times 0\cdot75}{3 - 9\cdot76 \times 0\cdot042}$ writing the answer correct to 2 significant figures.

7. A man borrows £1 580 from a bank, to pay for a central heating installation. The interest charged on this loan is 15% of the amount owed at the beginning of each year.
   At the end of the first year, the man pays back £617, to cover the interest and repay part of the loan.
   (a) How much does he owe after making the payment of £617 ?
   (b) If he pays back £850 at the end of the second year, how much will he still owe?

8. Changes in the price of 'Nylonothene' plastic are recorded in the following index.

Year	1977	1978	1979	1980
Index number	100		144	160
Price per tonne	£4000	£4800		

(Base year 1977 = 100)

Calculate
(a) the price per tonne in 1979.
(b) the price per tonne in 1980.
(c) the index number for 1978.
(d) It is proposed to change the base year of the index to 1980 = 100. Calculate new index numbers for 1977, 1978 and 1979, correct to the nearest whole number.

9. The flowchart shows how to calculate the charge for running a program on a small computer.

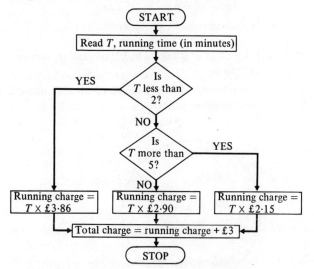

Calculate the charge for programs with running times of
(a) 1·5 minutes   (b) 7·4 minutes   (c) 5·6 minutes   (d) 2 minutes.

10. The following table shows the number of faults found during pre-delivery inspection of a number of cars.

Number of faults x	0	1	2	3	4
Number of cars f	15	8	6	6	5

(a) Altogether, how many cars were inspected?
(b) State the mode.
(c) Calculate the mean number of faults per car.

Test Paper 5

This paper should be completed in 1½ hours, or less.
Mathematical tables, slide rules or calculators may be used.
Show the working for every question. Attempt ALL the questions.

1. (a) A motorcyclist travels a distance of 132 km between 10.55 a.m. and 1.07 p.m.
What is his average speed for the journey?
   (b) He starts the next stage of the journey at 2.15 p.m., and travels 171 km at an
average speed of 54 km/h.
When will he complete this stage of his journey?

2. (a) An automatic potato peeling machine peels 450 potatoes in $1\frac{1}{2}$ hours. At the
same rate, how many potatoes will the machine peel in $6\frac{1}{2}$ hours?
   (b) The supplies on a weathership are sufficient to feed 18 men for 25 days. If 12
extra men join the ship at the start of the voyage, how long will the supplies
last (without reducing the quantity of food each person eats)?

3. An electricity generating station charges:
3·6p per unit for electricity consumed between 6 a.m. and 6 p.m.
2·2p per unit for electricity consumed between 6 p.m. and 6 a.m.
In addition, there is a standing charge of £6·45, every quarter.
A recording company consumes 3645 units in one quarter, between 9 a.m. and
4 p.m.
   (a) Calculate the electricity bill for this quarter.
   (b) How much would have been saved, if all the electricity had been consumed
between 6 p.m. and 6 a.m.?

4. A caravan owner pays insurance at the rate of £0·30 per cent for his caravan, and
£0·48 per cent for the contents.
If his caravan is valued at £3600, and its contents at £725, find his total insurance
premium.

5. The pie chart shows the costs involved in manufacturing and selling tapes containing computer games.

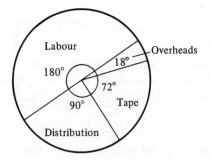

   (a) If the total cost of producing each tape is £12·60, find how much of this is
due to
(i) labour,   (ii) distribution,   (iii) tape,   (iv) overheads.
   (b) If the profit is $8\frac{1}{3}$% of the cost of production, find the profit made on 1500
tapes.

6. If you are using a calculator, do both parts of this question.
   If you are not using a calculator, do part (a) only.
   (a) Find the volume of a sphere of diameter 6 cm, giving your answer correct to
       2 significant figures.
       [Volume of a sphere $= \frac{4}{3} \pi r^3$. Take $\pi = 3 \cdot 14$]
   (b) Calculate $\dfrac{(4 \cdot 67 - 1 \cdot 93)^2}{1 \cdot 02 + 3 \cdot 98 \times 6 \cdot 4}$ writing the answer correct to 1 decimal place.

7. Mr. Endall earns £10 095 per year. His tax allowances are:
   higher personal allowance   £2145
   additional personal allowance   £ 650
   dependent relative allowance   £ 100
   (a) Find Mr. Endall's taxable income.
   (b) If income tax is calculated as 30% of the first £11 250 of taxable income,
       find the tax payable.
   (c) Calculate Mr. Endall's monthly salary, after deduction of income tax.

8. In 1970, the Index of Furniture Prices was 116, and the price of a three-piece
   suite was £197·20.
   (a) If the index was 124 in 1972, find the cost of the three-piece suite.
   (b) Calculate the index number corresponding to a price of £408 in 1976.

9. Two identical circular discs are cut from a rectangular sheet of metal, as shown.

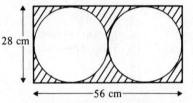

28 cm

—56 cm—

   (a) Calculate the area of waste metal around the discs (i.e. the shaded area).
   (b) Express the area of waste metal as a fraction of the area of one disc.

   [Area of circle $= \pi r^2$. Take $\pi = \frac{22}{7}$]

10. The flowchart shows how to calculate the cost of photocopies.

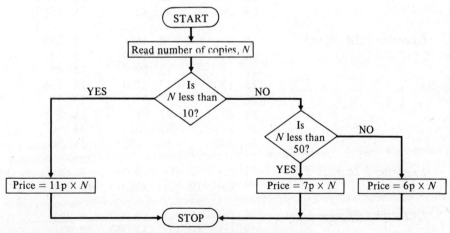

Calculate the cost of the following number of photocopies:
(a) 1   (b) 4   (c) 15   (d) 100   (e) 50

Exercise 1.1a page 1   1. (c)   2. (b)   3. (a)   4. (a)   5. (c)   6. (b)   7. (b)   8. (c)
9. (a)   10. (b)   11. (c)   12. (a)   13. (b)   14. (c)   15. (a)

Exercise 1.1b page 1   1. (a)   2. (a)   3. (b)   4. (c)   5. (a)   6. (a)   7. (b)   8. (b)
9. (b)   10. (c)   11. (a)   12. (c)   13. (c)   14. (b)   15. (a)

Exercise 1.1c page 2   1. (b)   2. (b)   3. (c)   4. (c)   5. (b)   6. (c)   7. (b)   8. (c)
9. (a)   10. (a)   11. (b)   12. (a)   13. (b)   14. (b)   15. (c)

Exercise 1.1d page 2   1. (a)   2. (b)   3. (b)   4. (a)   5. (c)   6. (a)   7. (a)   8. (b)
9. (a)   10. (c)   11. (a)   12. (a)   13. (a)   14. (b)   15. (a)

Exercise 1.1e page 3
1. 105, 45   2. 177 cm, 109 cm   3. 178 m, 315 m   4. 1900, 5450
5. 533 m, 453 m   6. 21, 8   7. 139, 333   8. 253 m, 481 m
9. 176 m, 451 m   10. 4720, 4884   11. 3290 m, 635 m 12. 25 m, 78 m, 21 ▸
13. 73, 43, 30   14. 169, 244, 75   15. 382, 391, 9

Exercise 1.2a page 4
1. 70, 700   2. 140, 1400   3. 250, 2500   4. 370, 3700
5. 100, 1000   6. 700, 7000   7. 2200, 22 000   8. 4000, 40 000
9. 1000, 10 000   10. 7860, 78 600   11. 80   12. 16
13. 3200   14. 100   15. 49   16. 100
17. 120   18. 5000   19. 900   20. 300

Exercise 1.2b page 4
1. 660, 66   2. 980, 98   3. 1050, 105   4. 1200, 120
5. 2000, 200   6. 12 500, 1250   7. 30 100, 3010   8. 100, 10
9. 20 000, 2000   10. 1000, 100   11. 72   12. 360
13. 28   14. 100   15. 6500   16. 100
17. 18 000   18. 400   19. 3000   20. 90 000

Exercise 1.2c page 4
1. 120   2. 260   3. 420   4. 540   5. 2240
6. 3260   7. 4500   8. 7680   9. 1940   10. 10 000
11. 280   12. 1050   13. 1330   14. 1820   15. 3500
16. 7840   17. 14 560 18. 9730   19. 24 500 20. 42 000
21. 2400   22. 3300   23. 5100   24. 8700   25. 14 100
26. 18 000 27. 36 300 28. 45 000 29. 210 000 30. 900 000
31. 540   32. 9   33. 7500   34. 20   35. 9
36. 200   37. 42   38. 5670   39. 21   40. 70

Exercise 1.2d page 5
1. 6   2. 15   3. 12   4. 200   5. 250
6. 340   7. 4000   8. 600   9. 360   10. 5400
11. 15   12. 19   13. 80   14. 140   15. 220
16. 1200   17. 1700   18. 114   19. 480   20. 2500
21. 2   22. 5   23. 11   24. 30   25. 70
26. 62   27. 89   28. 8   29. 126   30. 250
31. 14   32. 300   33. 90   34. 300   35. 40
36. 200   37. 1250   38. 50   39. 70 000 40. 2100

Exercise 1.2e page 5   1. (c)   2. (b)   3. (a)   4. (b)   5. (b)   6. (a)   7. (b)   8. (b)
9. (b)   10. (c)   11. (c)   12. (b)

Exercise 1.2f page 6   1. (b)   2. (c)   3. (a)   4. (b)   5. (c)   6. (a)   7. (b)   8. (a)
9. (a)   10. (c)   11. (b)   12. (a)

Exercise 1.2g page 6
1. (b) **2.** (c) **3.** (b) **4.** (c) **5.** (b) **6.** (a) **7.** (c) **8.** (c)
9. (c) **10.** (a) **11.** (a) **12.** (b)

Exercise 1.2h page 7
1. 84 kg, 28 kg **2.** 15 kg, 75 kg **3.** 144 km, 18 km **4.** 18, 90, 38
5. 168 kg, 24 kg **6.** 3240, 540 **7.** 1440, 120 **8.** 1170, 39
9. 2175, 145 **10.** 720, 30 **11.** 702, 13 **12.** 756, 36
13. 48 000 km, 1500 km **14.** 720 000, 30 **15.** 17, 4760

Exercise 1.3a page 8
1. 5	**2.** 3	**3.** 10	**4.** 10	**5.** 3
6. 8	**7.** 0	**8.** 36	**9.** 28	**10.** 3
11. 6	**12.** 14	**13.** 18	**14.** 4	**15.** 6
16. 11	**17.** 18	**18.** 14	**19.** 36	**20.** 32
21. 3	**22.** 1	**23.** 4	**24.** 20	**25.** 8
26. 30	**27.** 0	**28.** 2	**29.** 2	**30.** 10
31. 4	**32.** 12	**33.** 13	**34.** 8	**35.** 1
36. 15	**37.** 7	**38.** 264	**39.** 28	**40.** 161

Exercise 1.3b page 8
1. 42	**2.** 62	**3.** 86	**4.** 92	**5.** 61
6. 1	**7.** 12	**8.** 6	**9.** 61	**10.** 0
11. 14	**12.** 12	**13.** 65	**14.** 17	**15.** 28
16. 16	**17.** 22	**18.** 1	**19.** 0	**20.** 267
21. 18	**22.** 25	**23.** 34	**24.** 78	**25.** 492
26. 14	**27.** 72	**28.** 65	**29.** 153	**30.** 135
31. 8	**32.** 9	**33.** 3	**34.** 4	**35.** 7
36. 11	**37.** 51	**38.** 31	**39.** 93	**40.** 4

Exercise 1.4a page 8
1. 2, 4, 6, 8, 10 **2.** 5, 10, 15, 20, 25
3. 8, 16, 24, 32, 40 **4.** 7, 14, 21, 28, 35
5. 11, 22, 33, 44, 55 **6.** 20, 40, 60, 80, 100
7. 30, 60, 90, 120, 150 **8.** 60, 120, 180, 240, 300
9. 15, 30, 45, 60, 75 **10.** 25, 50, 75, 100, 125
11. 16, 32, 48, 64, 80 **12.** 18, 36, 54, 72, 90
13. 14, 28, 42, 56, 70 **14.** 13, 26, 39, 52, 65
15. 24, 48, 72, 96, 120 **16.** 21, 42, 63, 84, 105
17. 45, 90, 135, 180, 225 **18.** 36, 72, 108, 144, 180
19. 51, 102, 153, 204, 255 **20.** 72, 144, 216, 288, 360

Exercise 1.4b page 9
1. 1, 3 **2.** 1, 2, 4, 8 **3.** 1, 2, 5, 10
4. 1, 2, 3, 4, 6, 12 **5.** 1, 3, 5, 15 **6.** 1, 2, 3, 6, 9, 18
7. 1, 2, 3, 5, 6, 10, 15, 30 **8.** 1, 3, 9, 27 **9.** 1, 2, 3, 4, 6, 8, 12, 24
10. 1, 2, 4, 8, 16, 32 **11.** 1, 3, 5, 9, 15, 45 **12.** 1, 2, 4, 5, 8, 10, 20, 40
13. 1, 2, 3, 6, 9, 18, 27, 54 **14.** 1, 2, 3, 6, 7, 14, 21, 42
15. 1, 2, 3, 4, 5, 6, 10, 12, 15, 20, 30, 60 **16.** 1, 2, 3, 4, 6, 8, 12, 16, 24, 48
17. 1, 3, 7, 9, 21, 63
18. 1, 2, 3, 4, 6, 7, 12, 14, 21, 28, 42, 84 **19.** 1, 2, 3, 6, 11, 22, 33, 66
20. 1, 2, 3, 4, 6, 8, 9, 12, 18, 24, 36, 72

186   Answers

Exercise 1.4c page 9

1. 1, 2
2. 1, 5
3. 1, 2, 3, 6
4. 1, 7
5. 1, 11
6. 1, 2, 4, 5, 10, 20
7. 1, 5, 7, 35
8. 1, 2, 4; square number
9. 1, 2, 5, 10, 25, 50
10. 1, 5, 25; square number
11. 1, 2, 4, 8, 16; square number
12. 1, 2, 11, 22
13. 1, 2, 7, 14
14. 1, 13
15. 1, 2, 4, 7, 14, 28
16. 1, 3, 7, 21
17. 1, 7, 49; square number
18. 1, 2, 3, 4, 6, 9, 12, 18, 36; square number
19. 1, 3, 17, 51
20. 1, 2, 5, 7, 10, 14, 35, 70

Exercise 1.4d page 9

1. 1, 17; prime number
2. 1, 3, 11, 33
3. 1, 2, 13, 26
4. 1, 29; prime number
5. 1, 3, 13, 39
6. 1, 3, 19, 57
7. 1, 37; prime number
8. 1, 7, 13, 91
9. 1, 3, 29, 87
10. 1, 53; prime number
11. 1, 3, 37, 111
12. 1, 97; prime number
13. 1, 61; prime number
14. 1, 67; prime number
15. 1, 7, 17, 119
16. 1, 103, prime number
17. 1, 73; prime number
18. 1, 3, 41, 123
19. 1, 101; prime number
20. 1, 3, 9, 13, 39, 117

Exercise 1.4e page 9

1. 2×5
2. 3×5
3. $2 \times 2 \times 7$
4. $2 \times 2 \times 2 \times 3$
5. $2 \times 2 \times 3 \times 3$
6. $2 \times 2 \times 2 \times 5$
7. $3 \times 3 \times 5$
8. $2 \times 3 \times 3 \times 3$
9. $2 \times 5 \times 7$
10. $3 \times 3 \times 7$
11. $2 \times 3 \times 13$
12. $2 \times 2 \times 3 \times 5$
13. $2 \times 3 \times 11$
14. $2 \times 2 \times 2 \times 2 \times 7$
15. $2 \times 2 \times 2 \times 2 \times 2 \times 3$
16. $2 \times 3 \times 3 \times 3 \times 3$
17. $2 \times 7 \times 11$
18. $2 \times 2 \times 3 \times 3 \times 5$
19. $2 \times 2 \times 2 \times 3 \times 7$
20. $2 \times 2 \times 2 \times 3 \times 3 \times 3$

Exercise 1.4f page 10

1. 2   2. 6   3. 6   4. 9   5. 14   6. 24   7. 18   8. 21
9. 15   10. 16   11. 6   12. 4   13. 14   14. 16   15. 8   16. 6
17. 12   18. 13   19. 16   20. 4

Exercise 1.4g page 10

1. 24   2. 30   3. 24   4. 18   5. 36   6. 60   7. 30   8. 40
9. 45   10. 48   11. 24   12. 36   13. 40   14. 60   15. 36   16. 60
17. 144   18. 120   19. 90   20. 144

Exercise 1.4h page 10

1. (b)   2. (a)   3. (c)   4. (c)   5. (a)   6. (c)   7. (a)   8. (a)
9. (b)   10. (a)

Exercise 1.5a page 11

1. $\frac{7}{10}$
2. $10, \frac{4}{100}$
3. $500, \frac{9}{10}$
4. $\frac{3}{10}, \frac{5}{1000}$
5. $\frac{3}{100}, \frac{6}{1000}$
6. $30, \frac{5}{10}, \frac{7}{1000}$
7. $5, \frac{3}{10}, \frac{2}{100}$
8. $80, \frac{4}{10}$
9. $5, \frac{2}{10}, \frac{2}{1000}$
10. $200, \frac{3}{1000}$

Exercise 1.5b page 11

1. 37·6   2. 31·2   3. 35·85   4. 72·11   5. 87·834
6. 43·185   7. 38·16   8. 61·385   9. 16·935   10. 24·995
11. 1·4   12. 1·8   13. 3·85   14. 3·72   15. 3·75
16. 7·15   17. 4·38   18. 18·85   19. 106·65   20. 41·89
21. (b)   22. (c)   23. (a)   24. (a)   25. (b)
26. (b)   27. (c)   28. (a)   29. (b)   30. (a)

Exercise 1.5c page 12

1. 24, 240	**2.** 122, 1220	**3.** 37·5, 375	**4.** 153·6, 1536
5. 21·35, 213·5	**6.** 185·76, 1857·6	**7.** 8·5, 85	**8.** 7, 70
9. 2·368, 23·68	**10.** 0·139, 1·39	**11.** 36	**12.** 4·5
13. 290	**14.** 3·2	**15.** 100	**16.** 90
17. 0·8	**18.** 100	**19.** 0·04	**20.** 0·0015

Exercise 1.5d page 13

1. 2·53, 0·253	**2.** 3·816, 0·3816	**3.** 0·625, 0·0625	**4.** 0·735, 0·0735
5. 3·6, 0·36	**6.** 6, 0·6	**7.** 2·04, 0·204	**8.** 10·03, 1·003
9. 0·085, 0·0085	**10.** 0·0032, 0·00032	**11.** 0·42	**12.** 5·1
13. 0·36	**14.** 480	**15.** 100	**16.** 0·6
17. 30	**18.** 100	**19.** 0·009	**20.** 0·012

Exercise 1.5e page 13

1. 3·6	**2.** 1·6	**3.** 15·2	**4.** 34·4	**5.** 37·5
6. 15·12	**7.** 28·08	**8.** 61·8	**9.** 1·92	**10.** 21·76
11. 35·1	**12.** 14·72	**13.** 0·135	**14.** 0·126	**15.** 0·14
16. 0·0288	**17.** 0·027	**18.** 7·2	**19.** 0·9	**20.** 0·39

Exercise 1.5f page 13

1. 2·22	**2.** 2·76	**3.** 6·02	**4.** 2·496	**5.** 3·3
6. 0·7572	**7.** 46·35	**8.** 19·44	**9.** 10·38	**10.** 0·5525
11. 0·8192	**12.** 0·648	**13.** 3·248	**14.** 7·815	**15.** 0·0984
16. 0·0904	**17.** 0·062	**18.** 0·063	**19.** 0·0951	**20.** 0·07712
21. 2·875	**22.** 14·212	**23.** 12·81	**24.** 32·568	**25.** 9·87
26. 9·112	**27.** 83·46	**28.** 96·66	**29.** 1199·08	**30.** 906·88

Exercise 1.5g page 13

1. 0·3	**2.** 0·2	**3.** 0·24	**4.** 0·14	**5.** 0·143
6. 0·182	**7.** 0·272	**8.** 0·245	**9.** 0·252	**10.** 0·221
11. 0·086	**12.** 0·083	**13.** 0·019	**14.** 0·0124	**15.** 0·042
16. 0·093	**17.** 0·045	**18.** 0·042	**19.** 0·013	**20.** 0·011

Exercise 1.5h page 13

1. 21·2	**2.** 9·3	**3.** 11·7	**4.** 1·5	**5.** 10·9
6. 18	**7.** 13	**8.** 14	**9.** 14	**10.** 65
11. 630	**12.** 730	**13.** 920	**14.** 6500	**15.** 1·9
16. 1·4	**17.** 12	**18.** 27	**19.** 140	**20.** 130
21. 3	**22.** 11	**23.** 11	**24.** 5	**25.** 110
26. 2·6	**27.** 6·2	**28.** 2·15	**29.** 0·023	**30.** 28·5

Exercise 1.5i page 14

1. 32	**2.** 48	**3.** 136	**4.** 344	**5.** 504
6. 610	**7.** 120·4	**8.** 10·4	**9.** 1·6	**10.** 0·015
11. 306	**12.** 513	**13.** 567	**14.** 1215	**15.** 1944
16. 3681	**17.** 361·8	**18.** 34·2	**19.** 5·4	**20.** 0·225
21. 1160	**22.** 3120	**23.** 3800	**24.** 7080	**25.** 14 240
26. 25 680	**27.** 1236	**28.** 284	**29.** 1·2	**30.** 0·136
31. 44	**32.** 3·2	**33.** 216	**34.** 13·1	**35.** 40
36. 385	**37.** 200	**38.** 2·4	**39.** 300	**40.** 2·1

Exercise 1.5j page 14

1. 0·4	**2.** 0·12	**3.** 1·76	**4.** 0·224	**5.** 80·4
6. 0·84	**7.** 0·024	**8.** 0·00284	**9.** 0·0008	**10.** 0·0036
11. 0·6	**12.** 0·21	**13.** 1·71	**14.** 0·156	**15.** 80·6
16. 0·69	**17.** 0·014	**18.** 0·00234	**19.** 0·0008	**20.** 0·0077
21. 0·2	**22.** 0·131	**23.** 0·159	**24.** 0·43	**25.** 0·081
26. 0·396	**27.** 0·0013	**28.** 0·000142	**29.** 0·00162	**30.** 0·000098
31. 0·08	**32.** 20	**33.** 0·42	**34.** 96	**35.** 0·066
36. 30	**37.** 0·0064	**38.** 500	**39.** 0·0092	**40.** 1·98

Exercise 1.6a page 15

1. 14·99	**2.** 2·92	**3.** 2·42	**4.** 4·3	**5.** 2·8
6. 3	**7.** 6·4	**8.** 14·95	**9.** 4·12	**10.** 2·3
11. 2·48	**12.** 2·2	**13.** 5·22	**14.** 0·49	**15.** 6·82
16. 9·8	**17.** 10·2	**18.** 1	**19.** 0·5	**20.** 0·1
21. 2·3	**22.** 0·5	**23.** 14·55	**24.** 2	**25.** 3·54
26. 4	**27.** 9·9	**28.** 1·3	**29.** 5·9	**30.** 4·8
31. 0·54	**32.** 1·33	**33.** 4·1	**34.** 5·3	**35.** 5·34
36. 0·8	**37.** 7·6	**38.** 1·2	**39.** 7·9	**40.** 0

Exercise 1.6b page 15

1. 5·6	**2.** 10·2	**3.** 7·4	**4.** 23·4	**5.** 15·33
6. 2·8	**7.** 8·4	**8.** 6·76	**9.** 3·8	**10.** 3·32
11. 8·04	**12.** 8·5	**13.** 10·3	**14.** 14·4	**15.** 9·2
16. 10·9	**17.** 1·9	**18.** 4·5	**19.** 10·1	**20.** 0·1
21. 12·5	**22.** 1·2	**23.** 15·5	**24.** 8·4	**25.** 22·77
26. 9	**27.** 3·5	**28.** 1·33	**29.** 2·4	**30.** 3·45
31. 6·5	**32.** 3·6	**33.** 32	**34.** 3·6	**35.** 22·8
36. 5	**37.** 0·84	**38.** 7·2	**39.** 108·4	**40.** 16·1

Exercise 1.7a page 16

1. 2·64	**2.** 1·34	**3.** 17·6	**4.** 42·8	**5.** 1·734
6. 1·563	**7.** 13·6	**8.** 25·38	**9.** 8·05	**10.** 24·0
11. 5·10	**12.** 4·51	**13.** 27·1	**14.** 9·51	**15.** 6·30
16. 3·90	**17.** 5·10	**18.** 14·80	**19.** 14·0	**20.** 2·00
21. 0·68	**22.** 91·8	**23.** 0·74	**24.** 4·0	**25.** 5·6
26. 13·62	**27.** 1·472	**28.** 0·05	**29.** 10·0	**30.** 1·9

Exercise 1.7b page 16

1. 26	**2.** 6	**3.** 6	**4.** 1	**5.** 138
6. 12	**7.** 1	**8.** 36	**9.** 4	**10.** 105
11. 4·8	**12.** 12·8	**13.** 0·8	**14.** 7·9	**15.** 16·8
16. 0·1	**17.** 6·4	**18.** 1·3	**19.** 4·0	**20.** 4·3
21. 4760	**22.** 950	**23.** 6320	**24.** 1000	**25.** 130
26. 370	**27.** 70	**28.** 4100	**29.** 140	**30.** 1790

Exercise 1.7c page 16

1. 4	**2.** 3	**3.** 3	**4.** 2	**5.** 3
6. 4	**7.** 2	**8.** 1	**9.** 2	**10.** 3
11. 3	**12.** 4	**13.** 3·2	**14.** 7·6	**15.** 13·4
16. 17·2	**17.** 31	**18.** 37	**19.** 15	**20.** 20
21. 3570	**22.** 4280	**23.** 5500	**24.** 20·0	**25.** 4·0
26. 10·0	**27.** 0·064	**28.** 0·0072	**29.** 0·009	**30.** 0·008

Exercise 1.7d page 17

1. 18	**2.** 24	**3.** 28	**4.** 12	**5.** 36
6. 50	**7.** 30	**8.** 60	**9.** 10	**10.** 8
11. 3	**12.** 3	**13.** 2	**14.** 3	**15.** 2
16. 2	**17.** 2	**18.** 5	**19.** 50	**20.** 40
21. (a)	**22.** (b)	**23.** (a)	**24.** (a)	**25.** (a)
26. (b)	**27.** (a)	**28.** (a)	**29.** (b)	**30.** (a)

Exercise 1.7e page 17

1. 2	**2.** 2	**3.** 5	**4.** 1	**5.** 2
6. 0·1	**7.** 10	**8.** 0·1	**9.** 50	**10.** 0·01
11. 2	**12.** 20	**13.** 0·5	**14.** 18	**15.** 20
16. 0·5	**17.** 300	**18.** 400	**19.** 8	**20.** 10

Exercise 1.8a page 18

1. 3·57 m	**2.** 53·29 m	**3.** 37·6 m	**4.** 0·49 m	**5.** 0·6 m
6. 0·09 m	**7.** 5·276 m	**8.** 0·752 m	**9.** 0·08 m	**10.** 0·007 m
11. 9·137 km	**12.** 0·83 km	**13.** 3372 m	**14.** 2490 m	**15.** 19600 m
16. 345 cm	**17.** 920 cm	**18.** 5936 mm	**19.** 8210 mm	**20.** 7900 mm

Exercise 1.8b page 18

1. 1·328 g	**2.** 0·536 g	**3.** 0·78 g	**4.** 0·09 g	**5.** 0·008 g
6. 1·5 kg	**7.** 0·59 kg	**8.** 0·03 kg	**9.** 0·002 kg	**10.** 1·32 t
11. 0·8 t	**12.** 4536 mg	**13.** 8980 mg	**14.** 3400 mg	**15.** 5260 g
16. 8500 g	**17.** 700 g	**18.** 3710 kg	**19.** 5600 kg	**20.** 300 kg

Exercise 1.8c page 18

1. 3·278 l	**2.** 8·25 l	**3.** 9·3 l	**4.** 6·035 l	**5.** 5·02 l
6. 1332 ml	**7.** 7600 ml	**8.** 755 ml	**9.** 320 ml	**10.** 100 ml
11. 0·47 l	**12.** 0·3 l	**13.** 0·025 l	**14.** 1·47 l	**15.** 9000 ml
16. 84 000 ml	**17.** 500 ml	**18.** 8 ml	**19.** 186 l	**20.** 370 l

Exercise 1.8d page 18

1. 20p	**2.** 15p	**3.** 18p	**4.** 21p	**5.** £1·20
6. 75p	**7.** 36p	**8.** 3p	**9.** 15p	**10.** 26p
11. 16p	**12.** 15p	**13.** £2·94	**14.** £0·75	**15.** £11·76

PART 2

Exercise 2.1a page 19

1. $\frac{2}{12}$ **2.** $\frac{10}{16}$ **3.** $\frac{2}{3}$ **4.** $\frac{9}{13}$ **5.** $\frac{1}{3}$ **6.** $\frac{2}{3}$ **7.** $\frac{1}{2}$ **8.** $\frac{1}{3}$

9. $\frac{1}{4}$ **10.** $\frac{2}{3}$ **11.** $\frac{3}{4}$ **12.** $\frac{3}{4}$ **13.** $\frac{3}{5}$ **14.** $\frac{6}{7}$ **15.** $\frac{2}{3}$

Exercise 2.1b page 19

1. $1\frac{1}{2}$ **2.** $1\frac{1}{3}$ **3.** $1\frac{3}{4}$ **4.** $2\frac{1}{2}$ **5.** $2\frac{1}{3}$ **6.** $2\frac{4}{7}$ **7.** $1\frac{1}{2}$ **8.** $1\frac{1}{3}$

9. $\frac{5}{4}$ **10.** $\frac{6}{5}$ **11.** $\frac{10}{7}$ **12.** $\frac{7}{5}$ **13.** $\frac{5}{2}$ **14.** $\frac{13}{4}$ **15.** $\frac{25}{4}$ **16.** $\frac{7}{2}$

17. $\frac{11}{3}$ **18.** $\frac{38}{5}$ **19.** $\frac{26}{7}$ **20.** $\frac{23}{3}$

Exercise 2.1c page 19

1. $\frac{2}{5}$ **2.** $\frac{3}{5}$ **3.** $\frac{7}{10}$ **4.** $\frac{1}{2}$ **5.** $1\frac{1}{3}$ **6.** 1 **7.** $1\frac{4}{7}$ **8.** $1\frac{1}{4}$

9. $\frac{1}{3}$ **10.** $\frac{2}{5}$ **11.** $\frac{3}{7}$ **12.** $\frac{3}{4}$ **13.** $\frac{3}{5}$ **14.** $\frac{1}{3}$ **15.** $\frac{3}{8}$

Exercise 2.1d page 20

1. $\frac{8}{15}$ **2.** $\frac{7}{12}$ **3.** $\frac{3}{10}$ **4.** $\frac{11}{12}$ **5.** $\frac{7}{8}$ **6.** $\frac{1}{2}$ **7.** $\frac{1}{3}$ **8.** $\frac{3}{4}$

9. $\frac{1}{2}$ **10.** $\frac{9}{16}$ **11.** $\frac{1}{4}$ **12.** $\frac{1}{10}$ **13.** $\frac{1}{6}$ **14.** $\frac{1}{20}$ **15.** $\frac{5}{12}$ **16.** $\frac{1}{8}$

17. $\frac{1}{3}$ **18.** $\frac{1}{6}$ **19.** $\frac{7}{12}$ **20.** $\frac{1}{6}$

Exercise 2.1e page 20

1. $3\frac{1}{2}$ **2.** $4\frac{8}{15}$ **3.** $5\frac{7}{12}$ **4.** $5\frac{3}{10}$ **5.** $7\frac{7}{8}$ **6.** $3\frac{3}{10}$ **7.** $5\frac{5}{12}$ **8.** $6\frac{1}{10}$

9. $3\frac{1}{2}$ **10.** $2\frac{3}{4}$ **11.** $3\frac{1}{3}$ **12.** $4\frac{2}{5}$ **13.** $1\frac{1}{4}$ **14.** $1\frac{1}{6}$ **15.** $1\frac{3}{8}$ **16.** $5\frac{1}{10}$

17. $5\frac{1}{4}$ **18.** $\frac{5}{6}$ **19.** $1\frac{3}{5}$ **20.** $2\frac{1}{2}$ **21.** $\frac{2}{3}$ **22.** $2\frac{5}{6}$ **23.** $\frac{29}{30}$ **24.** $1\frac{5}{12}$

25. $2\frac{8}{9}$

Exercise 2.1f page 20

1. $\frac{1}{6}$ **2.** $\frac{1}{10}$ **3.** $\frac{3}{8}$ **4.** $\frac{3}{14}$ **5.** $\frac{8}{15}$ **6.** $\frac{1}{5}$ **7.** $\frac{2}{7}$ **8.** $\frac{2}{7}$

9. $\frac{3}{8}$ **10.** $\frac{15}{22}$ **11.** $\frac{4}{39}$ **12.** $\frac{1}{14}$ **13.** $\frac{3}{40}$ **14.** $\frac{2}{3}$ **15.** $\frac{3}{4}$

Exercise 2.1g page 20

1. $2\frac{11}{12}$ **2.** $2\frac{1}{12}$ **3.** $6\frac{1}{4}$ **4.** $2\frac{11}{12}$ **5.** $3\frac{9}{10}$ **6.** $3\frac{1}{3}$ **7.** $2\frac{2}{3}$ **8.** $4\frac{1}{2}$

9. $12\frac{1}{2}$ **10.** 30 **11.** $2\frac{6}{7}$ **12.** $8\frac{5}{8}$ **13.** 12 **14.** 12 **15.** $2\frac{1}{2}$

Exercise 2.1h page 21

1. $\frac{3}{4}$ 2. $1\frac{2}{5}$ 3. $1\frac{1}{15}$ 4. $1\frac{1}{14}$ 5. $\frac{25}{36}$ 6. $\frac{3}{5}$ 7. $\frac{4}{9}$ 8. $\frac{6}{11}$

9. $\frac{2}{3}$ 10. $\frac{1}{2}$ 11. 4   12. 4   13. 3   14. 2   15. $\frac{11}{12}$ 16. $\frac{11}{18}$

17. $2\frac{2}{3}$ 18. $1\frac{1}{2}$ 19. $\frac{4}{9}$ 20. $1\frac{3}{5}$

Exercise 2.1i page 21

1. (b)   2. (c)   3. (b)   4. (a)   5. (c)   6. (c)   7. (b)   8. (a)

9. (b)   10. (c)   11. $1\frac{1}{5}$ 12. $1\frac{3}{5}$ 13. $\frac{1}{3}$ 14. $\frac{1}{6}$ 15. $\frac{1}{10}$ 16. $\frac{7}{15}$

17. $\frac{9}{20}$ 18. $\frac{2}{3}$ 19. (a)   20. (a)   21. (b)   22. (b)   23. (a)   24. (a)

25. (c)   26. (b)   27. (c)   28. (b)   29. (b)   30. (a)

Exercise 2.2a page 22

1. $\frac{7}{10}$ 2. $\frac{1}{2}$ 3. $\frac{2}{5}$ 4. $\frac{9}{10}$ 5. $\frac{9}{20}$ 6. $\frac{13}{50}$

7. $\frac{17}{20}$ 8. $\frac{8}{25}$ 9. $\frac{29}{50}$ 10. $\frac{18}{25}$ 11. $\frac{7}{8}$ 12. $\frac{13}{40}$

13. $\frac{5}{8}$ 14. $\frac{19}{40}$ 15. $\frac{3}{100}$ 16. $\frac{2}{25}$ 17. $\frac{1}{20}$ 18. $\frac{7}{100}$

19. $\frac{11}{200}$ 20. $\frac{3}{125}$ 21. $\frac{31}{500}$ 22. $\frac{11}{400}$ 23. $\frac{1}{16}$ 24. $\frac{3}{1000}$

25. $\frac{3}{500}$ 26. $\frac{1}{500}$ 27. $\frac{7}{2000}$ 28. $\frac{3}{625}$ 29. $3\frac{1}{5}$ 30. $8\frac{1}{4}$

31. $12\frac{4}{25}$ 32. $41\frac{1}{50}$ 33. $5\frac{9}{500}$ 34. $13\frac{3}{250}$ 35. $2\frac{1}{80}$ 36. $8\frac{7}{400}$

37. $14\frac{1}{250}$ 38. $11\frac{1}{200}$ 39. $16\frac{1}{400}$ 40. $21\frac{3}{2000}$

Exercise 2.2b page 23

1. 0·375   2. 0·125   3. 0·25   4. 0·15   5. 0·55   6. 0·35

7. 0·65   8. 0·95   9. 0·6   10. 0·2   11. 0·8   12. 0·275

13. 0·225   14. 0·425   15. 0·525   16. 0·1875   17. 0·6875   18. 0·14

19. 0·18   20. 0·62

Exercise 2.2c page 23

1. $0·\dot{6}$ 2. $0·8\dot{3}$ 3. $0·1\dot{6}$ 4. $0·41\dot{6}$ 5. $0·58\dot{3}$ 6. $0·91\dot{6}$

7. $0·4\dot{5}$ 8. $0·8\dot{1}$ 9. $0·2\dot{7}$ 10. $0·6\dot{3}$ 11. $0·2\dot{6}$ 12. $0·7\dot{3}$

13. $0·4\dot{6}$ 14. $0·8\dot{6}$ 15. $0·5\dot{3}$ 16. $0·2\dot{3}$ 17. $0·3\dot{6}$ 18. $0·4\dot{3}$

19. $0·2\dot{7}$ 20. $0·3\dot{8}$

Exercise 2.2d page 23

1. 0·714   2. 0·857   3. 0·571   4. 0·286   5. 0·143   6. 0·429

7. 0·538   8. 0·769   9. 0·692   10. 0·846   11. 0·476   12. 0·095

13. 0·762   14. 0·529   15. 0·235   16. 0·176   17. 0·353   18. 0·158

19. 0·263   20. 0·368

Exercise 2.2e page 23

1. $\frac{5}{8}$, 0·625   2. $\frac{7}{20}$, 0·35   3. $\frac{9}{10}$, 0·9   4. $\frac{39}{100}$, 0·39

5. $\frac{1}{10}$, 0·1   6. $\frac{9}{10}$, 0·9   7. $\frac{49}{100}$, 0·49   8. $\frac{1}{10}$, 0·1

9. $\frac{7}{10}$, 0·7   10. $\frac{3}{10}$, 0·3   11. (a) by $\frac{1}{100}$ or 0·01

12. (b) by $\frac{1}{100}$ or 0·01   13. (b) by $\frac{1}{100}$ or 0·01

14. (b) by $\frac{3}{50}$ or 0·06   15. (a) by $\frac{1}{25}$ or 0·04

Exercise 2.2f page 24

1. $0.48, 0.7, \frac{3}{4}, 1\frac{1}{5}, 2$ **2.** $\frac{3}{8}, 0.6, \frac{7}{10}, \frac{3}{4}, 0.85$

3. $\frac{2}{5}, \frac{5}{8}, 0.65, 0.76, 0.8$ **4.** $0.26, \frac{3}{5}, 0.71, \frac{3}{4}, \frac{7}{8}$

5. $1.2, 1\frac{7}{20}, 1.39, 1\frac{3}{5}, 1\frac{5}{8}$ **6.** $\frac{1}{2}, \frac{11}{20}, \frac{3}{4}, 1\frac{1}{8}, \frac{6}{5}$

7. $\frac{1}{4}, 0.28, \frac{3}{10}, \frac{3}{8}, \frac{3}{5}$ **8.** $\frac{2}{5}, \frac{9}{20}, \frac{1}{2}, 0.6, \frac{7}{10}$

9. $2, 2\frac{1}{4}, 2.3, 2\frac{2}{5}, 2.41$ **10.** $0.1, 0.11, 0.12, \frac{1}{8}, \frac{3}{20}$

11. $1, 0.8, \frac{13}{20}, \frac{3}{5}, \frac{1}{2}$ **12.** $0.81, \frac{4}{5}, \frac{3}{4}, 0.7, \frac{3}{8}$

13. $0.71, \frac{7}{10}, 0.69, \frac{17}{25}, \frac{5}{8}$ **14.** $\frac{9}{20}, \frac{11}{25}, 0.41, \frac{4}{10}, 0.39$

15. $\frac{9}{16}, \frac{11}{20}, 0.52, 0.51, \frac{1}{2}$ **16.** $1\frac{2}{5}, 1.3, 1.2, 1\frac{1}{10}, 1.02$

17. $0.29, \frac{7}{25}, \frac{27}{100}, 0.26, \frac{1}{4}$ **18.** $2.31, 2\frac{3}{10}, 2.21, 2\frac{1}{5}, 2\frac{9}{50}$

19. $\frac{4}{25}, \frac{7}{50}, \frac{11}{100}, 0.10, 0.01$ **20.** $\frac{9}{25}, 0.313, \frac{5}{16}, 0.31, \frac{3}{10}$

Exercise 2.3a page 25

1. 2 **2.** $3\frac{5}{6}$ **3.** $4\frac{5}{16}$ **4.** $\frac{9}{10}$ **5.** $\frac{1}{24}$

6. $\frac{16}{21}$ **7.** $\frac{17}{24}$ **8.** $\frac{1}{15}$ **9.** $3\frac{19}{20}$ **10.** $2\frac{1}{8}$

11. $\frac{7}{9}$ **12.** $4\frac{3}{8}$ **13.** $6\frac{13}{28}$ **14.** $\frac{7}{8}$ **15.** $1\frac{1}{3}$

16. $\frac{5}{16}$ **17.** $\frac{17}{60}$ **18.** $\frac{7}{15}$ **19.** $\frac{11}{12}$ **20.** $\frac{7}{15}$

Exercise 2.3b page 25

1. $5\frac{3}{4}$ **2.** $2\frac{1}{4}$ **3.** 14 **4.** $2\frac{3}{4}$ **5.** 6

6. $1\frac{1}{2}$ **7.** 6 **8.** $3\frac{19}{21}$ **9.** $3\frac{1}{12}$ **10.** $4\frac{5}{16}$

11. $8\frac{2}{5}$ **12.** $\frac{2}{3}$ **13.** $2\frac{1}{2}$ **14.** $\frac{3}{16}$ **15.** $11\frac{5}{12}$

16. $18\frac{1}{10}$ **17.** $7\frac{1}{18}$ **18.** $5\frac{11}{12}$ **19.** $\frac{10}{77}$ **20.** $\frac{2}{3}$

Exercise 2.4 page 26

1. $10\,l$ **2.** $18, 6$ **3.** $56, 8$

4. £0.68, £1.87   **5.** $24, 32$ **6.** $36, 9$

7. $36, 60$ **8.** £1.35, £3.15   **9.** 574 kg, 164 kg

10. $176, 16$ **11.** $3\frac{1}{2}$ **12.** $4\frac{1}{2}$

13. $4\frac{1}{2}$ **14.** 6 **15.** 40

16. 18 **17.** $3\frac{1}{2}\,l$ **18.** $4\frac{1}{8}\,l$

19. 6 km, $3\frac{1}{3}$ km   **20.** $9\frac{1}{3}$ kg, $2\frac{1}{3}$ kg

Exercise 2.5a page 27

1. 1:4   **2.** 1:6   **3.** 1:3   **4.** 2:5   **5.** 2:3   **6.** 3:4

7. 5:6   **8.** 4:7   **9.** 5:2   **10.** 8:3   **11.** 7:4   **12.** 9:5

13. 12:7   **14.** 3:20   **15.** 2:15   **16.** 3:50   **17.** 9:20   **18.** 1:8

19. 3:40   **20.** 2:25   **21.** 3:8   **22.** 20:3   **23.** 15:4   **24.** 50:9

25. 40:3   **26.** 25:2   **27.** 3:4   **28.** 7:8   **29.** 3:7   **30.** 1:6

31. 4:11   **32.** 12:5   **33.** 9:7   **34.** 15:2   **35.** 9:4   **36.** 5:8

37. 3:20   **38.** 5:7   **39.** 9:2   **40.** 12:5

Exercise 2.5b page 27

1. 2:3   **2.** 5:6   **3.** 3:10   **4.** 4:7   **5.** 9:16   **6.** 5:3

7. 8:7   **8.** 9:2   **9.** 7:5   **10.** 4:3

Exercise 2.5c page 28

1. £27 and £63   **2.** £24 and £60   **3.** £20 and £25

4. 25 and 35   **5.** 18 and 30 kg   **6.** £30 and £36

7. £30, £40 and £50   **8.** £16, £24 and £32   **9.** £24, £36 and £72

10. £7, £14 and £35   **11.** 10, 15 and 35   **12.** 5, 15 and 25

13. 8, 40 and 48 kg   **14.** £50, £40 and £20   **15.** 12, 36 and 72

Exercise 2.5d page 28

1. £45, £50 and £65
2. £32, £40 and £48
3. £45, £39 and £24
4. £64, £44 and £32
5. £55, £35 and £30
6. 21, 24 and 27 kg
7. 135, 165 and 240 kg
8. £2·50, £3 and £4·50
9. 56p, 64p and 80p
10. $3\frac{1}{2}$, $5\frac{1}{2}$ and 6 t

Exercise 2.5e page 29

1. £1, 20p
2. £1·10, 66p
3. £63, £27
4. 91p, £3·90, £1·17, £2·60
5. 1 h 50 min, 33 min
6. 30, 9
7. 28, 10
8. 44 m², 12 m²
9. 32, 22, 10, 9
10. Ochiltree, Maybole
11. 180 g, 300 g
12. £4·41, £12·60
13. 25·2 t, 100·8 t
14. £1·20, 1100
15. £12·36, 8
16. £1·95, 17
17. 24 min, 8 min, 32 min
18. £16, £20·80, £8·80
19. £1540, £968, £495
20. £7·50, £13·75, £6·25

Exercise 2.5f page 30

1. 90 min, 36 min
2. 4 h 10 min, 1 h 15 min
3. 50 h, 18 h
4. 24°, 16°
5. 24°, 15°
6. 20 min, 8 min
7. 15 min, 6 min
8. 12 cm, 8 cm
9. 36 min, 48 min
10. 30 min, 42 min
11. 60, 80
12. 20 days, 8 days, 3 days
13. 10 h, 3 h 20 min
14. 10 days, 3 days, 1 day
15. 10 km, 20 km, 8 km
16. £0·90
17. 9 s, 25 s
18. 11 days, 2 days

Exercise 2.6a page 31

1. 1:60
2. 1:200 000
3. 1:40 000
4. 1:150 000
5. 1:20 000
6. 1:500
7. 1:3000
8. 1:600 000
9. 1:20 000
10. 1:100 000

Exercise 2.6b page 31

1. 8 m
2. 56 m
3. 20 km
4. 90 m
5. 200 m
6. 3 km
7. 20 m
8. 10 m
9. 1·17 m
10. 14 km

Exercise 2.6c page 32

1. 20 cm
2. 2 cm
3. 30 cm
4. 17 cm
5. 2 cm
6. 0·5 cm
7. 6 cm
8. 5·6 cm
9. 0·15 cm
10. 2 cm

Exercise 2.7a page 32

1. $\frac{9}{100}$
2. $\frac{3}{20}$
3. $\frac{9}{20}$
4. $\frac{1}{5}$
5. $\frac{7}{10}$
6. $\frac{8}{25}$
7. $\frac{17}{25}$
8. $\frac{14}{25}$
9. $\frac{21}{50}$
10. $\frac{3}{50}$
11. $1\frac{2}{5}$
12. $1\frac{4}{5}$
13. $1\frac{3}{10}$
14. $2\frac{1}{10}$
15. $2\frac{3}{4}$
16. $3\frac{7}{20}$
17. $2\frac{9}{25}$
18. $3\frac{2}{25}$
19. $4\frac{1}{20}$
20. $4\frac{13}{20}$
21. 0·15
22. 0·35
23. 0·9
24. 0·8
25. 0·6
26. 0·05
27. 0·16
28. 0·72
29. 0·44
30. 0·58
31. 1·25
32. 1·56
33. 1·7
34. 2·3
35. 2·25
36. 3·4
37. 2·12
38. 3·04
39. 4·08
40. 4·01

Exercise 2.7b page 33

1. $\frac{7}{40}$
2. $\frac{3}{16}$
3. $\frac{19}{30}$
4. $\frac{8}{15}$
5. $\frac{5}{24}$
6. $\frac{5}{12}$
7. $\frac{17}{40}$
8. $\frac{1}{18}$
9. $\frac{11}{18}$
10. $\frac{7}{32}$
11. $1\frac{1}{40}$
12. $1\frac{19}{40}$
13. $1\frac{7}{30}$
14. $2\frac{1}{24}$
15. $4\frac{11}{40}$
16. $3\frac{1}{12}$
17. $3\frac{2}{15}$
18. $2\frac{1}{15}$
19. $1\frac{7}{24}$
20. $1\frac{5}{32}$
21. 0·625
22. 0·875
23. 0·325
24. 0·475
25. 0·5625
26. 0·8125
27. 0·9375
28. 0·6̇
29. 0·1̇
30. 0·8̇3̇
31. 1·125
32. 1·375
33. 1·075
34. 2·225
35. 1·3125
36. 3·0625
37. 2·03125
38. 3·3̇
39. 4·03̇
40. 2·16̇

Exercise 2.7c page 33

1. (a)
2. (b)
3. (b)
4. (c)
5. (a)
6. (b)
7. (a)
8. (b)
9. (a)
10. (b)

Exercise 2.7d page 34	**1.** (b)	**2.** (a)	**3.** (c)	**4.** (a)	**5.** (b)
	6. (a)	**7.** (a)	**8.** (b)	**9.** (b)	**10.** (c)
	11. (a)				

Exercise 2.7e page 34

1. 120	**2.** 240	**3.** 74

4. 600, 360, 960, 480   **5.** 1080, 1020, 1104, 1176, 1140   **6.** £81 000

7. 2375	**8.** 45	**9.** 80%
10. 25%	**11.** 75%	**12.** 15%

13. 40%, 25%, 30%, 5%   **14.** $66\frac{2}{3}\%, 26\frac{2}{3}\%, 6\frac{2}{3}\%$ **15.** 20%, 25%, 50%, 5%

PART 3

Exercise 3.1a page 35	**1.** 162	**2.** 191	**3.** 134	**4.** 161	**5.** 214
	6. 116	**7.** 197	**8.** 148	**9.** 201	**10.** 99
	11. 72	**12.** 29	**13.** 43	**14.** 29	**15.** 29
	16. 63	**17.** 38	**18.** 79	**19.** 52	**20.** 48
	21. 2378	**22.** 2627	**23.** 1692	**24.** 1378	**25.** 738
	26. 1218	**27.** 2958	**28.** 2838	**29.** 3900	**30.** 2134
	31. 3·25	**32.** 2·25	**33.** 2·875	**34.** 2·625	**35.** 7·75
	36. 1·75	**37.** 3·25	**38.** 3·625	**39.** 6·125	**40.** 2·0625

Exercise 3.1b page 36	**1.** 830	**2.** 720	**3.** 940	**4.** 880	**5.** 680
	6. 780	**7.** 900	**8.** 792	**9.** 848	**10.** 539
	11. 453	**12.** 718	**13.** 455	**14.** 280	**15.** 190
	16. 610	**17.** 310	**18.** 280	**19.** 790	**20.** 422
	21. 178	**22.** 683	**23.** 289	**24.** 171	**25.** 419
	26. 33 600	**27.** 114 700	**28.** 105 800	**29.** 77 900	**30.** 172 800
	31. 97 200	**32.** 63 237	**33.** 124 124	**34.** 94 804	**35.** 59 396
	36. 89 397	**37.** 185 556	**38.** 61 845	**39.** 3·75	**40.** 1·875
	41. 6·625	**42.** 3·25	**43.** 5·785	**44.** 2·75	**45.** 2·625
	46. 2·125	**47.** 3·625	**48.** 3·25	**49.** 4·25	**50.** 3·75

Exercise 3.1c page 37	**1.** 13·2	**2.** 16·2	**3.** 30·3	**4.** 36·4	**5.** 38·1
	6. 47·8	**7.** 50·7	**8.** 68·9	**9.** 72·8	**10.** 70·7
	11. 4·8	**12.** 6·9	**13.** 11·8	**14.** 11·9	**15.** 9·1
	16. 15·1	**17.** 18·8	**18.** 19·8	**19.** 17·3	**20.** 18·8
	21. 31·98	**22.** 53·94	**23.** 32·67	**24.** 70·84	**25.** 157·56
	26. 92·38	**27.** 94·34	**28.** 293·04	**29.** 200·07	**30.** 257·4
	31. 1·75	**32.** 7·875	**33.** 2·25	**34.** 2·5	**35.** 4·75
	36. 6·5	**37.** 5·5	**38.** 2·25	**39.** 5·25	**40.** 2·125

Exercise 3.1d page 37	**1.** 10·96	**2.** 21·03	**3.** 27·68	**4.** 31·85	**5.** 51·54
	6. 61·75	**7.** 62·59	**8.** 69·42	**9.** 83·69	**10.** 55·95
	11. 5·17	**12.** 3·21	**13.** 9·94	**14.** 6·93	**15.** 13·24
	16. 14·07	**17.** 22·03	**18.** 20·94	**19.** 37·21	**20.** 42·32
	21. 42·4948	**22.** 35·7588	**23.** 39·0208	**24.** 34·9036	**25.** 48·4376
	26. 60·6464	**27.** 83·7183	**28.** 67·4584	**29.** 81·8892	**30.** 39·9228
	31. 6·25	**32.** 8·375	**33.** 9·875	**34.** 4·75	**35.** 9·25
	36. 9·125	**37.** 7·875	**38.** 4·788	**39.** 2·625	**40.** 6·375

194   Answers

Exercise 3.1e page 39

1. 26·9	2. 7	3. 77	4. 50·7	5. 3·75
6. 3·79	7. 8·2	8. 3·02	9. 3	10. 5·16
11. 15·96	12. 4·2	13. 9·75	14. 10·5	15. 13·2
16. 20·12	17. 5·34	18. 0·74	19. 5·27	20. 1·11
21. 8·68	22. 0·04	23. 4·01	24. 20·24	25. 29·4
26. 11·75	27. 0·345	28. 5016	29. 11·78	30. 414
31. 4·96	32. 13·72	33. 9·8	34. 9·5	35. 0·48
36. 0·8	37. 19·5	38. 3·8	39. 8·92	40. 4·2
41. 7·21	42. 13·152	43. 0·025	44. 18·5	45. 4·16
46. 12·5	47. 0·682	48. 2·56	49. 7·36	50. 9·19

Exercise 3.2a page 39

1. 0·25	2. 0·75	3. 0·125	4. 0·3125	5. 0·375
6. 0·4	7. 0·175	8. 0·15	9. 0·16	10. 0·1875
11. 0·875	12. 0·8	13. 0·45	14. 0·425	15. 0·1125
16. 0·15	17. 0·54	18. 0·28125	19. 0·1375	20. 0·65
21. 0·625	22. 0·075	23. 0·3375	24. 0·275	25. 0·35
26. 0·53	27. 0·34	28. 0·125	29. 0·0252	30. 0·018
31. 0·264	32. 0·0304	33. 0·725	34. 0·4375	35. 0·7375
36. 0·1776	37. 0·8125	38. 0·56	39. 0·42	40. 0·1375
41. 0·096	42. 0·248	43. 0·9375	44. 0·25	45. 0·05625
46. 0·028	47. 0·0912	48. 0·02112	49. 0·0424	50. 0·03536

Exercise 3.2b page 40

1. 1·25	2. 2·25	3. 3·5	4. 4·375	5. 2·4375
6. 1·625	7. 5·4	8. 3·16	9. 10·025	10. 5·1875
11. 8·25	12. 2·875	13. 4·225	14. 11·06	15. 6·125
16. 3·76	17. 1·6875	18. 5·6	19. 3·425	20. 4·42
21. 6·81	22. 12·15625	23. 9·1404	24. 2·784	25. 1·12
26. 7·104	27. 88·5625	28. 21·0224	29. 11·1325	30. 13·1368
31. 6·475	32. 9·9375	33. 2·8125	34. 1·02016	35. 32·0648
36. 14·346	37. 7·1075	38. 11·168	39. 12·575	40. 48·85
41. 182·296	42. 17·775	43. 9·275	44. 4·00736	45. 76·2112
46. 16·532	47. 3·0952	48. 19·054	49. 4·764	50. 9·248

Exercise 3.2c page 42

1. 0·3	2. 0·875	3. 0·325	4. 0·225	5. 29
6. 9·108	7. £6·18	8. £63·50	9. £5·44	10. £425
11. £18·60	12. £31·14	13. 0·25	14. 2·5	15. 2
16. 4	17. 0·2	18. 0·125	19. 480 t	20. 45 g
21. 2500 l	22. 15·75 m	23. 29·25 g	24. 81·6 t	25. 8
26. 1·2	27. 125	28. 0·025	29. 0·504	30. 40
31. 0·625	32. 0·2	33. 95·5	34. 24·75	35. 106·15
36. 250·25	37. £0·80	38. £18·20	39. £25·74	40. £387·80
41. £2367·75	42. £267·05	43. 2	44. 0·75	45. 0·11
46. 0·08	47. 4	48. 0·25	49. 0·0625	50. 5·1

Exercise 3.3a page 43

1. 2·19	2. 1·94	3. 4·31	4. 1·56	5. 3·06
6. 1·43	7. 1·52	8. 1·28	9. 2·47	10. 3·14
11. 2·35	12. 1·42	13. 3·23	14. 2·37	15. 1·56
16. 4·43	17. 3·72	18. 1·26	19. 2·40	20. 3·18
21. 1·13	22. 2·27	23. 3·47	24. 6·14	25. 4·58
26. 2·74	27. 3·56	28. 1·77	29. 1·13	30. 2·30
31. 1·53	32. 2·41	33. 3·35	34. 2·17	35. 4·29
36. 3·45	37. 2·51	38. 3·25	39. 1·53	40. 4·49

Exercise 3.3b page 44

1. 2·13	**2.** 2·74	**3.** 0·91	**4.** 1·22	**5.** 2·44
6. 3·66	**7.** 0·61	**8.** 1·52	**9.** 4·57	**10.** 3·05
11. 10·67	**12.** 13·72			

Exercise 3.3c page 44

1. £1·56	**2.** £3·06	**3.** £2·19	**4.** £5·44	**5.** £4·12
6. £1·97	**7.** £4·66	**8.** £3·44	**9.** £2·84	**10.** £3·53
11. £3·09	**12.** £4·56			

Exercise 3.3d page 45

1. 3·281	**2.** 4·156	**3.** 2·469	**4.** 3·594	**5.** 1·406
6. 3·541	**7.** 4·364	**8.** 6·278	**9.** 3·618	**10.** 5·255
11. 4·373	**12.** 2·149	**13.** 1·934	**14.** 2·134	**15.** 2·208
16. 3·583	**17.** 1·347	**18.** 2·720	**19.** 5·352	**20.** 6·148
21. 1·513	**22.** 3·331	**23.** 1·292	**24.** 4·957	**25.** 3·178
26. 6·347	**27.** 2·284	**28.** 1·555	**29.** 4·398	**30.** 3·140
31. 2·953	**32.** 1·724	**33.** 6·638	**34.** 1·287	**35.** 4·661
36. 4·359	**37.** 6·543	**38.** 3·284	**39.** 1·271	**40.** 5·560

Exercise 3.3e page 45

1. 6·437 km	**2.** 11·265 km	**3.** 3·219 km	**4.** 9·656 km
5. 14·484 km	**6.** 12·874 km	**7.** 38·623 km	**8.** 33·795 km
9. 19·312 km	**10.** 57·935 km	**11.** 8·046 km	**12.** 24·140 km
13. 43·451 km	**14.** 28·967 km	**15.** 45·060 km	

Exercise 3.3f page 46

1. 8·165 kg	**2.** 9·526 kg	**3.** 3·175 kg	**4.** 5·443 kg
5. 0·907 kg	**6.** 4·082 kg	**7.** 6·350 kg	**8.** 2·722 kg
9. 3·629 kg	**10.** 7·258 kg	**11.** 1·361 kg	**12.** 12·701 kg

Exercise 3.3g page 46

1. 13·634 *l*	**2.** 18·179 *l*	**3.** 31·814 *l*	**4.** 72·717 *l*
5. 54·538 *l*	**6.** 36·358 *l*	**7.** 81·806 *l*	**8.** 27·269 *l*
9. 9·090 *l*	**10.** 95·441 *l*	**11.** 1·136 *l*	**12.** 1·704 *l*
13. 4·545 *l*	**14.** 3·409 *l*	**15.** 2·840 *l*	

Exercise 3.3h page 47

1. 2·31	**2.** 3·56	**3.** 4·06	**4.** 5·19	**5.** 3·44
6. 2·17	**7.** 1·38	**8.** 3·54	**9.** 3·41	**10.** 4·23
11. 6·18	**12.** 2·38	**13.** 1·55	**14.** 3·69	**15.** 2·84
16. 1·57	**17.** 3·49	**18.** 2·32	**19.** 4·15	**20.** 1·80
21. 31·9	**22.** 54·4	**23.** 13·1	**24.** 25·6	**25.** 60·6
26. 15·4	**27.** 24·8	**28.** 18·8	**29.** 12·9	**30.** 14·4
31. 139	**32.** 253	**33.** 351	**34.** 123	**35.** 560
36. 4310	**37.** 2520	**38.** 2170	**39.** 1350	**40.** 3030

Exercise 3.3i page 48

1. 0·170	**2.** 0·385	**3.** 0·158	**4.** 0·263	**5.** 0·615
6. 0·571	**7.** 0·417	**8.** 0·667	**9.** 0·889	**10.** 0·583
11. 0·333	**12.** 0·278	**13.** 0·556	**14.** 0·222	**15.** 0·083
16. 0·353	**17.** 0·190	**18.** 0·778		

Exercise 3.3j page 48

1. 0·0538	**2.** 0·519	**3.** 0·0492	**4.** 0·316	**5.** 0·0331
6. 0·348	**7.** 0·212	**8.** 0·0840	**9.** 0·0149	**10.** 0·0722
11. 0·160	**12.** 0·319	**13.** 0·0650	**14.** 0·131	**15.** 0·0368
16. 0·0811	**17.** 0·0519	**18.** 0·0723		

Exercise 4.1a page 49
1. £480, £1920   2. £90, £510   3. £36, £364   4. £20, £230
5. £42, £308   6. £24, £126   7. £15, £110   8. £12, £68
9. £45, £135   10. £17, £68   11. £4·80, £55·20   12. £2·70, £51·30
13. £3·78, £59·22   14. £1·10, £20·90   15. 60p, £14·40

Exercise 4.1b page 49
1. £0·75, £14·25   2. £1·25, £11·25   3. £4·32, £67·68
4. £0·87, £6·38   5. £29·68, £712·32   6. £0·13, £6·37
7. £0·34, £3·91   8. £4·06, £53·94   9. £243, £2997
10. £7·56, £76·44   11. £12·60, £39·90   12. £12, £84
13. £0·20, £1·05   14. £0·58, £6·67   15. £8·25$\frac{1}{2}$, £55·24$\frac{1}{2}$

Exercise 4.1c page 50
1. £20, £100   2. £0·32, £6·72   3. £9, £159
4. £0·33, £8·58   5. £97, £4947   6. £114, £3914
7. £41·60, £561·60   8. £40, £290   9. £1·17, £7·67
10. £8·40, £848·40   11. £13, £273   12. £72, £552
13. £14, £70   14. £2·10, £86·10   15. £7·50, £127·50

Exercise 4.2 page 51
1. £6   2. £9   3. £50, £44, £37, £31, £41   4.£24, £16·80, £19·20
5. £9, £14·40, £12·70   6. £180, £235, £305, £363
7. £450, £625 , £822·50, £513·75   8. £117, £257, £291, £295·40
9. £109, £565, £21·25, £712   10. £570, £1021, £1336, £2003·08

Exercise 4.3a page 52
1. £15, £60, £210   2. £18, £72, £192   3. £30, £90, £340
4. £56, £112, £462   5. £15, £60, £180   6. £45, £135, £375
7. £34, £102, £262   8. £31·50, £126, £336
9. £37·80, £75·60, £390·60   10. £11·25, £22·50, £147·50
11. £21, £84, £434   12. £14, £56, £336   13. £34, £102, £527
14. £27, £81, £441   15. £7, £35, £147   16. £8, £32, £288
17. £15, £75, £235   18. £40·50, £162, £612
19. £10·80, £21·60, £156·60   20. £14·40, £28·80, £388·80

Exercise 4.3b page 52
1. £30, £45, £295   2. £48, £60, £380   3. £72, £162, £612
4. £18, £24, £174   5. £12, £30, £180   6. £32, £56, £376
7. £6, £21, £141   8. £60, £200, £950   9. £15, £40, £290
10. £36, £66, £786   11. £36, £114, £564

Exercise 4.3c page 53
1. 5%   2. 5%   3. 3%   4. 3%   5. 3%

Exercise 4.4a page 53
1. £518·40, £218·40   2. £529, £129   3. £752·64, £152·64
4. £665·50, £165·50   5. £672·80, £172·80   6. £278·48, £78·48
7. £463·05, £63·05   8. £1123·60, £123·60   9. £237·62, £37·62
10. £540·80, £40·80

Exercise 4.4b page 54
1. £699·84, £99·84   2. £332·75, £82·75   3. £1825·05, £625·05
4. £842·70, £92·70   5. £1254·40, £254·40   6. £1211·04, £311·04
7. £3906·25, £2306·25   8. £829·44, £349·44   9. £742·63, £42·63
10. £15 994·16, £10 394·16

Exercise 4.4c page 54
1. £405   2. £512   3. £180·50   4. £1687·50
5. £1440·60   6. £25   7. £441·80   8. £4423·68
9. £5648·16   10. £117 482·12

Exercise 4.5a page 55
1. £120, £12   2. £117, £23   3. £80, £8
4. £256, £36   5. £404, £49   6. £309·40, £9·40
7. £327·40, £42·40   8. £102, £14   9. £380, £50
10. £180·10, £22·10

Exercise 4.5b page 56
1. £102, £12   2. £63, £3   3. £81, £6   4. £147. £12
5. £252, £12   6. £504, £54   7. £402, £52   8. £2655, £155
9. £945, £45   10. £1968, £168

Exercise 4.5c page 56
1. £7   2. £3   3. £8   4. £6   5. £2
6. £72   7. £19   8. £152·50   9. £15·50   10. £23·80

Exercise 4.6a page 57
1. +3p 2. +2p 3. +1p 4. −1p 5. +2p 6. +2p 7. +4p 8. +6p
9. +10p 10. −6p 11. −2p 12. −20p 13. +12p 14. −4p 15. +60p 16. −30p
17. +4p 18. +6p 19. +3p 20. +6p

Exercise 4.6b page 58
1. $+33\frac{1}{3}\%$ 2. $+25\%$ 3. $+12\frac{1}{2}\%$ 4. $-16\frac{2}{3}\%$ 5. $+33\frac{1}{3}\%$ 6. $+4\%$

7. $+11\frac{1}{9}\%$ 8. $+11\frac{1}{9}\%$ 9. $+25\%$ 10. $-16\frac{2}{3}\%$ 11. $-6\frac{2}{3}\%$ 12. $-13\frac{1}{3}\%$

13. $+11\frac{1}{9}\%$ 14. $-16\frac{2}{3}\%$ 15. $+12\frac{1}{2}\%$ 16. -20% 17. $+50\%$ 18. $+11\frac{1}{9}\%$

19. $+20\%$ 20. $+150\%$

Exercise 4.6c page 59
1. +25%   2. +20%   3. −10%   4. +5%   5. −15%
6. +12%   7. +24%   8. −100%   9. $+12\frac{1}{2}\%$ 10. +6%

Exercise 4.6d page 60
1. £8   2. £6·25   3. £34·14   4. £0·77   5. £36·60
6. £0·22   7. £0·54   8. £4·50   9. £2·31   10. $22\frac{1}{2}$ p
11. £1·12   12. £2·53   13. £0·21   14. £4·25   15. $25\frac{1}{2}$ p

PART 5

Exercise 5.1 page 61
1. £5·20   2. £12   3. £20   4. £46·92   5. £50·60

Exercise 5.2a page 62
1. 2263, 674   2. 1865, 905   3. 455, 1204   4. 621, 1542
5. 534, 1636   6. 627, 516   7. 713, 429
8. 2321, 762, 548   9. 533, 1747, 2455   10. 1893, 815, 749, 3086

Exercise 5.2b page 62
1. £40·86   2. £42·84   3. £13·41   4. £18·54   5. £25·56   6. £23·40
7. £32·76   8. £29·88   9. £43·56   10. £55·80   11. £29·16   12. £21·60

Exercise 5.2c page 63
1. £45·30   2. £12·96   3. £11·42
4. £39·80   5. £37·38, £10·54   6. £12·52, £9·00
7. £33·20, £34·74   8. £30·56, £11·64   9. £36·94, £12·96
10. £10·98, £9·44   11. £16·92, £31·22   12. £33·64, £49·92
13. £48·38, £15·16, £11·86, £43·32   14. £40·68, £14·06, £13·18, £33·20
15. £46·18, £14·28, £12·08, £38·26

Exercise 5.3a page 63
1. £100   2. £180   3. £200   4. £126   5. £270
6. £336   7. £252   8. £297   9. £360   10. £176
11. £181·50   12. £178·20   13. £518·16   14. £251·25   15. £529·65

Exercise 5.3b page 64

1. £360	**2.** £450	**3.** £240	**4.** £500	**5.** £250	**6.** £550
7. £420	**8.** £340	**9.** £375	**10.** £286	**11.** £512	**12.** £336
13. £254	**14.** £192	**15.** £394	**16.** £542	**17.** £285	**18.** £325

Exercise 5.3c page 64

1. 75p	**2.** 80p	**3.** 60p	**4.** 70p	**5.** 90p	**6.** 100p
7. 110p	**8.** 104p	**9.** 105p	**10.** 102p		

Exercise 5.3d page 65

1. 86p, £1000   **2.** 97p, £12 000   **3.** 63p, £2000
4. 59p, £24 000   **5.** 88p, £3000   **6.** 103p, £15 000
7. 105p, £28 000   **8.** 109p, £27 000   **9.** 107p, £750
10. 111p, £13 500

Exercise 5.3e page 65

1. 103p, £2400, £1287·50   **2.** £136·08   **3.** 98p, £270·48

Exercise 5.4a page 65

1. £48	**2.** £54	**3.** £42	**4.** £49·60	**5.** £59·20	**6.** £63
7. £46·20	**8.** £58·80	**9.** £48·30	**10.** £55·44	**11.** £45	**12.** £46·80
13. £57·60	**14.** £52·20	**15.** £37·44	**16.** £52·80		

Exercise 5.4b page 65

1. £1·65, £2·20   **2.** £1·86, £2·48   **3.** £2·55, £3·40   **4.** £1·98, £2·64
5. £2·10, £2·80   **6.** £2·04, £2·72   **7.** £1·45, £2·61   **8.** £2·00, £3·60
9. £1·80, £3·24   **10.** £1·60, £2·88

Exercise 5.4c page 66

1. £62·40	**2.** £61·75	**3.** £60·50	**4.** £65·28	**5.** £73·44	**6.** £55·80
7. £56·10	**8.** £62·40	**9.** £65·10	**10.** £65·54		

Exercise 5.4d page 66

1. £3000	**2.** £3840	**3.** £4260	**4.** £3300	**5.** £3792	**6.** £4128
7. £5016	**8.** £3516	**9.** £3846	**10.** £3414	**11.** £3723	**12.** £3303
13. £4635	**14.** £3489	**15.** £4473			

Exercise 5.4e page 66

1. £300	**2.** £375	**3.** £267	**4.** £326	**5.** £319
6. £332	**7.** £278	**8.** £284	**9.** £340·25	**10.** £364·25
11. £290·75	**12.** £376·75	**13.** £240·50	**14.** £345·50	**15.** £336·50

Exercise 5.4f page 67

1. 1st by £4	**2.** 2nd by £8	**3.** 1st by £2	**4.** 2nd by £1
5. 1st by £4	**6.** 1st by £5	**7.** 2nd by £12	**8.** 2nd by £3

Exercise 5.5a page 67

1. £1500	**2.** £2250	**3.** £2550	**4.** £2450	**5.** £2250	**6.** £3180
7. £2610	**8.** £3190	**9.** £1730	**10.** £2080		

Exercise 5.5b page 68

1. £1375, £3425   **2.** £3520, £1530   **3.** £2915, £4685
4. £3520, £5980   **5.** £1555, £2645   **6.** £1375, £1867
7. £1520, £3000   **8.** £2930, £5490   **9.** £1875, £23
10. £2005, £5651

Exercise 5.5c page 69

1. £504, £42, £200   **2.** £624, £52, £250
3. £408, £34, £184   **4.** £264, £22, £160
5. £888, £74, £355   **6.** £768, £64, £282
7. £804, £67, £312   **8.** £432, £36, £210
9. £324, £27, £176   **10.** £1272, £106, £482
11. £2856, £238, £719   **12.** £1404, £117, £987
13. £2916, £243, £798   **14.** £912, £76, £362·50
15. £1578, £131·50, £513·50   **16.** £3455
17. £4544   **18.** £4463

Answers

Exercise 5.6a page 69
1. £104   2. £615   3. £145   4. £2520   5. £1962·50
6. £840, £16 800   7. £1522·50, £38 062·50
8. £328·10, £9843   9. £3712·50, £92 812·50
10. £232·05, £4641   11. £2059·60, £1441·72
12. £178·50, £107·10   13. £2312·40, £1040·58
14. £237·60, £130·68, £3267   15. £278·20, £1391, £5109

Exercise 5.6b page 70
1. £9, £0·75   2. £18, £1·50   3. £15·60, £1·30
4. £30, £2·50   5. £93·60, £7·80   6. £30, £2·50
7. £231, £19·25   8. £102, £8·50   9. £30, £2·50
10. £35·40, £2·95   11. £142·32, £11·86   12. £133·68, £11·14
13. £24·72, £2·06   14. £61·20, £5·10   15. £276, £23
16. £63·84, £5·32   17. £14·88, £1·24   18. £20·88, £1·74
19. £64·92, £5·41   20. £95·64, £7·97

Exercise 5.6c page 71
1. £84   2. £90   3. £54   4. £36   5. £63
6. £90   7. £48   8. £48   9. £74   10. £84
11. £84   12. £43·70   13. £67·50   14. £42   15. £114·70

Exercise 5.7a page 72
1. $392   2. $294   3. $49   4. Fr. 364   5. Fr. 2730
6. Fr. 1274   7. DM 288   8. DM 768   9. DM 960   10. SF 162
11. SF 378   12. Ptas. 2436   13. Ptas. 3276   14. BF 7500   15. BF 4320

Exercise 5.7b page 72
1. £300   2. £250   3. £75   4. £60   5. £400
6. £160   7. £50   8. £350   9. £125   10. £15
11. £35   12. £18   13. £24   14. £115   15. £64

Exercise 5.7c page 72
1. Britain, £25   2. France, £40   3. Switzerland, 5p
4. Britain, 10p   5. Spain, 25p   6. Britain, 50p
7. Germany, 25p
8. Spain, £2200; Germany, £2250; Britain, £2400; Switzerland, £2450; U.S.A. £2500; Belgium, £2600; France, £2700
9. France, 50p; Belgium, 57p; Britain, 60p; Spain, 63p; Switzerland, 65p
10. Spain, £4.50; U.S.A. £4.75; Switzerland, £4.80; Belgium, £4.90; .Britain, £5; France, £5.20; Germany, £5.25
11. £30·50   12. $484·61, £28·50   13. DM984·96
14. £256   15. £17·20, £136·15, £24·40, £0·30

Exercise 5.8a page 74
1. 6 h   2. 5 h   3. 10 h   4. 6 h 15 min
5. 4 h 20 min   6. 1 h 5 min   7. 5 h 12 min   8. 1 h 38 min
9. 18 min   10. 2 h 6 min   11. 59 min   12. 2 h 45 min
13. 2 h 47 min   14. 5 h 55 min   15. 3 h 46 min   16. 1 h 22 min
17. 5 h 47 min   18. 17 min   19. 9 h 49 min   20. 4 h 57 min

Exercise 5.8b page 74
1. 10 h   2. 3 h 30 min   3. 14 h   4. 14 h 30 min
5. 9 h 25 min   6. 3 h 55 min   7. 15 h 50 min   8. 31 h
9. 13 h 20 min   10. 2 h 12 min   11. 37 h 37 min   12. 29 h 25 min
13. 36 h 32 min   14. 12 h 45 min   15. 84 h 55 min

Exercise 5.8c page 75

1. 14.00 h	**2.** 17.00 h	**3.** 18.00 h	**4.** 14.15 h	**5.** 14.40 h
6. 17.20 h	**7.** 17.32 h	**8.** 18.30 h	**9.** 18.48 h	**10.** 19.50 h
11. 19.05 h	**12.** 16.35 h	**13.** 20.00 h	**14.** 20.30 h	**15.** 21.15 h
16. 22.00 h	**17.** 22.24 h	**18.** 23.45 h	**19.** 23.05 h	**20.** 08.00 h
21. 06.00 h	**22.** 09.15 h	**23.** 08.42 h	**24.** 07.35 h	**25.** 06.08 h
26. 09.03 h	**27.** 10.00 h	**28.** 10.15 h	**29.** 11.55 h	**30.** 11.32 h

Exercise 5.8d page 75

1. 1.00 p.m.	**2.** 4.00 p.m.	**3.** 7.00 p.m.	**4.** 1.30 p.m.	**5.** 1.50 p.m.
6. 7.35 p.m.	**7.** 7.18 p.m.	**8.** 4.25 p.m.	**9.** 4.05 p.m.	**10.** 6.40 p.m.
11. 6.12 p.m.	**12.** 3.55 p.m.	**13.** 2.36 p.m.	**14.** 9.00 p.m.	**15.** 9.45 p.m.
16. 11.00 p.m.	**17.** 11.30 p.m.	**18.** 10.45 p.m.	**19.** 10.06 p.m.	**20.** 9.00 a.m.
21. 7.00 a.m.	**22.** 8.45 a.m.	**23.** 9.53 a.m.	**24.** 8.56 a.m.	**25.** 7.02 a.m.
26. 8.09 a.m.	**27.** 11.00 a.m.	**28.** 11.25 a.m.	**29.** 10.50 a.m.	**30.** 10.36 a.m.

Exercise 5.8e page 75

1. 1 h 45 min   **2.** 1 h 35 min   **3.** 1 h 15 min   **4.** 34 min
5. 2 h 25 min   **6.** 3 h 30 min   **7.** 1 h 50 min   **8.** 1 h 45 min
9. 45 min   **10.** 2 h 45 min
11. 1 h 12 min, 1 h 45 min, 2 h 0 min, 2 h 40 min
12. 43 min, 2 h 14 min, 2 h 29 min, 3 h 35 min
13. 2 h 12 min, 2 h 40 min, 3 h 45 min, 6 h 0 min
14. (a) 1.35 p.m., 2.15 p.m., 40 min   (b) 26 min
    (c) 8.28 a.m., 8.50 a.m., 22 min, 6.25 p.m., 55 min
    (d) 5.50 p.m.   (e) 14 min, 3 h 41 min
15. (a) 5 h 10 min, 5 h 55 min, 5 h 10 min, 7.05 p.m.
    (b) 08.28 h, 08.45 h, 17 min, 2 h 30 min, 12.25 h, 40 min, 12.42 p.m.
    (c) 33 min, 6.03 p.m.
16. (a) 9.45 p.m., 35 min   (b) 15 min
    (c) 8.15 a.m., 8.45 a.m., 30 min, 4.20 p.m., 4.50 p.m.
    (d) 2.15 p.m., 25 min   (e) 1.05 p.m., 3 h 35 min

Exercise 5.8f page 78

1. 33, 35	**2.** 32, 34	**3.** 46, 48	**4.** 64, 66	**5.** 79, 81
6. 70, 72	**7.** 64, 66	**8.** 38, 40	**9.** 45, 47	**10.** 104, 106
11. 102, 104	**12.** 322, 324	**13.** 146, 148	**14.** 390, 392	**15.** 454, 456
16. 436, 438	**17.** 442, 444	**18.** 75, 77	**19.** 423, 425	**20.** 880, 882

Exercise 5.8g page 78

1. yes	**2.** no	**3.** yes	**4.** yes	**5.** yes
6. no	**7.** yes	**8.** yes	**9.** yes	**10.** no
11. no	**12.** no	**13.** yes	**14.** no	**15.** no
16. yes	**17.** no	**18.** yes	**19.** yes	**20.** yes

Exercise 5.9a page 78

1. 632 km/h	**2.** 488 km/h	**3.** 720 km/h	**4.** 864 km/h	**5.** 632 km/h
6. 40 km/h	**7.** 24 km/h	**8.** 40 km/h	**9.** 42 km/h	**10.** 44 km/h
11. 64 km/h	**12.** 48 km/h	**13.** 52 km/h	**14.** 60 km/h	**15.** 52 km/h
16. 76 km/h	**17.** 68 km/h	**18.** 63 km/h	**19.** 68 km/h	**20.** 75 km/h

Exercise 5.9b page 79

1. 90 km/h	**2.** 36 km/h	**3.** 50 km/h	**4.** 48 km/h	**5.** 99 km/h
6. 117 km/h	**7.** 24 km/h	**8.** 30 km/h	**9.** 85 km/h	**10.** 36 km/h

Exercise 5.9c page 79

1. 5 h	**2.** 19 h	**3.** 16 h	**4.** 146 h
5. 16 h	**6.** 15 h 30 min	**7.** 6 h 30 min	**8.** 12 h 30 min
9. 3 h 15 min	**10.** 2 h 45 min	**11.** 5 h 20 min	**12.** 3 h 12 min
13. 6 h 40 min	**14.** 16 h 12 min	**15.** 2 h 10 min	**16.** 7 h 40 min
17. 3.2 s	**18.** 2.25 s	**19.** 82 min	**20.** 1.25 s

Exercise 5.9d page 80	**1.** 240 km	**2.** 852 km	**3.** 780 km	**4.** 770 km	**5.** 672 km
	6. 245 km	**7.** 441 km	**8.** 2580 km	**9.** 475 km	**10.** 2071 km
	11. 6160 km	**12.** 145 km	**13.** 306 km	**14.** 3564 km	**15.** 726 km
	16. 3763 m	**17.** 106 cm	**18.** 9639 m	**19.** 12·6 km	**20.** 2028 m

Exercise 5.10 page 80

1. £82·80, £51·75, £124·20
2. £3·60, £8·40, £7
3. £50, £90, £350, £50
4. £6, £13, £15·50, £11
5. yes, no, yes, yes
6. £30, £45, £55, £47·50, £30
7. £50, £50, £135, £150, £290
8. 111, 1110, 11011, 100000, 1101101
9. 11 025, 231 525, 4 862 025
10. 16p, 18p, 36p, 18p

PART 6

Exercise 6.1a page 84

1. 4	**2.** 25	**3.** 81	**4.** 400	**5.** 3600
6. 4900	**7.** 160 000	**8.** 640 000	**9.** 10 000	**10.** 4 000 000
11. 25 000 000	**12.** 36 000 000	**13.** 0·25	**14.** 0·64	**15.** 0·01
16. 0·0009	**17.** 0·0004	**18.** 0·0036	**19.** $\frac{1}{16}$	**20.** $\frac{1}{49}$
21. $\frac{1}{144}$	**22.** $\frac{9}{16}$	**23.** $\frac{4}{25}$	**24.** $\frac{16}{121}$	**25.** $1\frac{11}{25}$
26. $1\frac{7}{9}$	**27.** $1\frac{9}{16}$	**28.** $2\frac{7}{9}$	**39.** $4\frac{21}{25}$	**30.** $11\frac{1}{9}$

Exercise 6.1b page 84

1. 2	**2.** 4	**3.** 7	**4.** 8	**5.** 6
6. 10	**7.** 9	**8.** 12	**9.** 11	**10.** 1
11. $\frac{1}{3}$	**12.** $\frac{1}{8}$	**13.** $\frac{1}{11}$	**14.** $\frac{3}{5}$	**15.** $\frac{7}{8}$
16. $\frac{5}{12}$	**17.** $2\frac{1}{2}$	**18.** $2\frac{2}{3}$	**19.** $1\frac{3}{4}$	**20.** $1\frac{2}{5}$

Exercise 6.1c page 84

1. 30	**2.** 50	**3.** 60	**4.** 900	**5.** 700
6. 500	**7.** 110	**8.** 1000	**9.** 200	**10.** 800
11. 0·3	**12.** 0·2	**13.** 0·6	**14.** 0·4	**15.** 0·8
16. 0·7	**17.** 0·5	**18.** 0·9	**19.** 0·05	**20.** 0·08
21. 0·04	**22.** 0·01	**23.** 0·12	**24.** 14	**25.** 16
26. 18	**27.** 21	**28.** 27		

Exercise 6.1d page 85

1. 1, 4	**2.** 4, 9	**3.** 4, 9	**4.** 1, 4	**5.** 9, 16
6. 9, 16	**7.** 16, 25	**8.** 16, 25	**9.** 9, 16	**10.** 25, 36
11. 16, 25	**12.** 36, 49	**13.** 36, 49	**14.** 49, 64	**15.** 36, 49
16. 49, 64	**17.** 64, 81	**18.** 16, 25	**19.** 1, 4	**20.** 1, 4
21. 9, 16	**22.** 9, 16	**23.** 16, 25	**24.** 9, 16	**25.** 16, 25
26. 36, 49	**27.** 36, 49	**28.** 49, 64	**29.** 25, 36	**30.** 64, 81

Exercise 6.1e page 85

1. 2	**2.** 3	**3.** 4	**4.** 3	**5.** 4
6. 4	**7.** 5	**8.** 4	**9.** 5	**10.** 6
11. 5	**12.** 6	**13.** 7	**14.** 7	**15.** 7
16. 7	**17.** 8	**18.** 8	**19.** 2	**20.** 1
21. 3	**22.** 3	**23.** 3	**24.** 3	**25.** 4
26. 4	**27.** 5	**28.** 5	**29.** 6	**30.** 8

Exercise 6.1f page 86

1. 20	**2.** 30	**3.** 60	**4.** 50	**5.** 40
6. 80	**7.** 70	**8.** 60	**9.** 90	**10.** 200
11. 1000	**12.** 300	**13.** 500	**14.** 600	**15.** 700
16. 800	**17.** 400	**18.** 400	**19.** 0·3	**20.** 0·4
21. 0·8	**22.** 0·6	**23.** 0·7	**24.** 0·9	**25.** 0·04
26. 0·03	**27.** 0·06	**28.** 0·05	**29.** 0·08	**30.** 0·04

Exercise 6.2a page 86

1. 27	**2.** 64	**3.** 125
4. 343	**5.** 729	**6.** 1000
7. 1728	**8.** 8000	**9.** 1 000 000
10. 27 000 000	**11.** 0·001	**12.** 0·008
13. 0·125	**14.** 0·216	**15.** 0·729
16. 0·000 000 027	**17.** $\frac{8}{27}$	**18.** $\frac{27}{125}$
19. $\frac{216}{343}$	**20.** $\frac{729}{1000}$	**21.** $\frac{64}{1331}$
22. $\frac{27}{64}$	**23.** $3\frac{3}{8}$	**24.** $11\frac{25}{64}$

Exercise 6.2b page 86

1. 2	**2.** 1	**3.** 4	**4.** 10	**5.** 6
6. 7	**7.** $\frac{1}{2}$	**8.** $\frac{1}{4}$	**9.** $\frac{1}{3}$	**10.** $\frac{1}{6}$
11. $\frac{1}{5}$	**12.** $\frac{2}{3}$	**13.** $\frac{3}{4}$	**14.** $\frac{4}{5}$	**15.** $\frac{5}{6}$
16. $1\frac{1}{3}$	**17.** $1\frac{2}{3}$	**18.** $1\frac{1}{4}$	**19.** $2\frac{1}{2}$	**20.** $1\frac{1}{5}$

Exercise 6.2c page 87

1. 20	**2.** 40	**3.** 50	**4.** 300	**5.** 1000
6. 70	**7.** 200	**8.** 600	**9.** 0·2	**10.** 0·4
11. 0·5	**12.** 0·6	**13.** 0·1	**14.** 0·7	**15.** 0·02
16. 0·01	**17.** 0·05	**18.** 0·03	**19.** 0·06	**20.** 0·07
21. 8	**22.** 14	**23.** 18	**24.** 16	**25.** 24

Exercise 6.2d page 87

1. 1, 8	**2.** 1, 8	**3.** 8, 27	**4.** 8, 27	**5.** 1, 8
6. 8, 27	**7.** 27, 64	**8.** 27, 64	**9.** 64, 125	**10.** 64, 125
11. 27, 64	**12.** 64, 125	**13.** 125, 216	**14.** 64, 125	**15.** 27, 64
16. 8, 27	**17.** 27, 64	**18.** 64, 125	**19.** 1, 8	**20.** 1, 8
21. 8, 27	**22.** 1, 8	**23.** 8, 27	**24.** 8, 27	**25.** 8, 27
26. 27, 64	**27.** 27, 64	**28.** 27, 64	**29.** 64, 125	**30.** 125, 216

Exercise 6.2e page 88

1. 1	**2.** 2	**3.** 2	**4.** 3	**5.** 2
6. 3	**7.** 4	**8.** 3	**9.** 4	**10.** 4
11. 5	**12.** 4	**13.** 5	**14.** 5	**15.** 4
16. 3	**17.** 4	**18.** 5	**19.** 1	**20.** 2
21. 1	**22.** 2	**23.** 3	**24.** 2	**25.** 3
26. 4	**27.** 3	**28.** 4	**29.** 2	**30.** 1

Exercise 6.2f page 88

1. 30	**2.** 20	**3.** 40	**4.** 60	**5.** 20
6. 30	**7.** 20	**8.** 40	**9.** 10	**10.** 200
11. 300	**12.** 200	**13.** 300	**14.** 200	**15.** 100
16. 400	**17.** 500	**18.** 400	**19.** 0·2	**20.** 0·2
21. 0·3	**22.** 0·4	**23.** 0·3	**24.** 0·5	**25.** 0·02
26. 0·02	**27.** 0·03	**28.** 0·04	**29.** 0·05	**30.** 0·04

Exercise 6.3a page 89

1. 8×10^5	**2.** $3 \cdot 6 \times 10^5$	**3.** $5 \cdot 48 \times 10^5$
4. 5×10^4	**5.** $3 \cdot 5 \times 10^4$	**6.** $2 \cdot 24 \times 10^4$
7. 9×10^3	**8.** $7 \cdot 5 \times 10^3$	**9.** $2 \cdot 84 \times 10^3$
10. $1 \cdot 563 \times 10^3$	**11.** 7×10^2	**12.** $2 \cdot 9 \times 10^2$
13. $3 \cdot 42 \times 10^2$	**14.** $1 \cdot 863 \times 10^2$	**15.** 8×10^1
16. $3 \cdot 6 \times 10^1$	**17.** $2 \cdot 98 \times 10^1$	**18.** 4×10^0
19. $8 \cdot 2 \times 10^0$	**20.** $7 \cdot 36 \times 10^0$	**21.** $1 \cdot 23 \times 10^0$
22. $4 \cdot 274 \times 10^2$	**23.** $1 \cdot 1692 \times 10^3$	**24.** $4 \cdot 0006 \times 10^3$
25. $1 \cdot 111\ 111 \times 10^6$	**26.** $2 \cdot 020\ 206 \times 10^6$	**27.** $1 \cdot 000\ 0001 \times 10^7$
28. $4 \cdot 006\ 0007 \times 10^7$	**29.** $1 \cdot 86 \times 10^5$	**30.** $1 \cdot 000\ 01 \times 10^1$

Exercise 6.3b page 89

1. 700 000	**2.** 825 000	**3.** 966 300	**4.** 500 100	**5.** 80 000
6. 63 000	**7.** 75 060	**8.** 6000	**9.** 3910	**10.** 5376
11. 450	**12.** 938	**13.** 883·7	**14.** 620·6	**15.** 50
16. 72·8	**17.** 15·32	**18.** 50·86	**19.** 9·51	**20.** 4·537
21. 1·6	**22.** 10·6	**23.** 482	**24.** 400 700	**25.** 2700 000
26. 991 000		**27.** 10 200 000		**28.** 94 360 000
29. 842 000 000		**30.** 96 300 000 000		

Exercise 6.3c page 89

1. $5 \cdot 6 \times 10^{-1}$	**2.** $8 \cdot 19 \times 10^{-1}$	**3.** $7 \cdot 04 \times 10^{-1}$
4. 2×10^{-2}	**5.** $3 \cdot 5 \times 10^{-2}$	**6.** $6 \cdot 01 \times 10^{-2}$
7. 4×10^{-4}	**8.** $5 \cdot 7 \times 10^{-4}$	**9.** 8×10^{-5}
10. $9 \cdot 65 \times 10^{-5}$	**11.** $7 \cdot 94 \times 10^{-4}$	**12.** 7×10^{-7}
13. $8 \cdot 21 \times 10^{-7}$	**14.** $3 \cdot 6 \times 10^{-1}$	**15.** $9 \cdot 1 \times 10^{-2}$
16. $4 \cdot 23 \times 10^{-4}$	**17.** $6 \cdot 241 \times 10^{-1}$	**18.** $7 \cdot 9 \times 10^{-3}$
19. $8 \cdot 05 \times 10^{-4}$	**20.** 7×10^{-2}	**21.** $6 \cdot 2 \times 10^{-1}$
22. $9 \cdot 09 \times 10^{-2}$	**23.** $6 \cdot 04 \times 10^{-1}$	**24.** 3×10^{-4}
25. $6 \cdot 9 \times 10^{-2}$	**26.** $7 \cdot 92 \times 10^{-5}$	**27.** $4 \cdot 3 \times 10^{-2}$
28. $3 \cdot 7 \times 10^{-5}$	**29.** $8 \cdot 4 \times 10^{-3}$	**30.** $6 \cdot 8 \times 10^{-1}$

Exercise 6.3d page 90

1. 0·004	**2.** 0·062	**3.** 0·000 071
4. 0·000 000 403	**5.** 0·000 635	**6.** 0·74
7. 0·000 0089	**8.** 0·000 0198	**9.** 0·0124
10. 0·000 000 022	**11.** 0·000 003	**12.** 0·0005
13. 0·009 091	**14.** 0·000 000 101	**15.** 0·000 019
16. 0·007 97	**17.** 0·141	**18.** 0·000 624
19. 0·081 82	**20.** 0·000 000 000 473	**21.** 5·6
22. 0·000 734	**23.** 0·0649	**24.** 0·008 91
25. 0·000 432	**26.** 0·705	**27.** 0·000 066
28. 0·004 083	**29.** 0·0173	**30.** 0·000 0389

Exercise 6.3e page 90

1. $6 \cdot 42 \times 10^2$	**2.** $3 \cdot 5 \times 10^{-3}$	**3.** $1 \cdot 76 \times 10^1$
4. $9 \cdot 1 \times 10^{-1}$	**5.** $3 \cdot 74 \times 10^5$	**6.** $4 \cdot 5 \times 10^{-3}$
7. $7 \cdot 9 \times 10^{-3}$	**8.** $1 \cdot 394 \times 10^2$	**9.** $2 \cdot 76 \times 10^0$
10. $8 \cdot 4 \times 10^{-1}$	**11.** $7 \cdot 6 \times 10^{-4}$	**12.** $9 \cdot 7 \times 10^3$
13. $1 \cdot 0004 \times 10^5$	**14.** $1 \cdot 019 \times 10^{-1}$	**15.** $1 \cdot 4 \times 10^0$
16. 5000	**17.** 0·04	**18.** 630 000
19. 7·91	**20.** 0·000 481	**21.** 6000
22. 0·000 007 15	**23.** 89 300	**24.** 5600
25. 0·000 074	**26.** 0·008	**27.** 151
28. 0·428	**29.** 32	**30.** 0·000 523

Exercise 6.4b page 91 **1.** 196 **2.** 315 **3.** 82 **4.** 37 **5.** 26 **6.** 14 **7.** 109 **8.** 242 **9.** 186 **10.** 61 **11.** (a) **12.** (c) **13.** (a) **14.** (a) **15.** (b) **16.** (a) **17.** (c) **18.** (b) **19.** (b) **20.** (a)

Exercise 6.4c page 91
1. 2303 **2.** 4211 **3.** 3220 **4.** 2040 **5.** 101110
6. 100111 **7.** 11001 **8.** 10011 **9.** 10332 **10.** 12233
11. 3000 **12.** 3205 **13.** 2010 **14.** 421 **15.** 3212
16. 1220 **17.** 3120 **18.** 1160 **19.** 520 **20.** 323
21. 102 **22.** 300 **23.** 159 **24.** E0 **25.** 1000
26. 50T

Exercise 6.4d page 92
1. 110001 **2.** 101010 **3.** 110100 **4.** 221 **5.** 124
6. 131 **7.** 355 **8.** 425 **9.** 330 **10.** 234
11. 302 **12.** 76 **13.** 10001 **14.** 2130 **15.** 10000
16. 107 **17.** 65 **18.** 131 **19.** 66 **20.** 101

Exercise 6.4e page 93
1. 111 **2.** 1001 **3.** 1010 **4.** 222 **5.** 221
6. 211 **7.** 241 **8.** 410 **9.** 1012 **10.** 136
11. 123 **12.** 101 **13.** 1004 **14.** 1012 **15.** 143
16. 10010 **17.** 1101 **18.** 1021 **19.** 343 **20.** 955

Exercise 6.4f page 93
1. 101 **2.** 10 **3.** 110 **4.** 210 **5.** 120
6. 11 **7.** 301 **8.** 123 **9.** 203 **10.** 310
11. 330 **12.** 213 **13.** 131 **14.** 14 **15.** 411
16. 616 **17.** 11 **18.** 1 **19.** 186 **20.** 301

Exercise 6.4g page 93
1. 1000 **2.** 1111 **3.** 100011 **4.** 1001 **5.** 1111
6. 12222 **7.** 4244 **8.** 11411 **9.** 23414 **10.** 20544
11. 20343 **12.** 1011011

Exercise 6.5a page 94
1. 5·5, 4·5 m **2.** 3·5, 2·5 cm **3.** 14·5, 13·5 g
4. 25·5, 24·5 kg **5.** 8·1, 7·9 cm **6.** 17·1, 16·9 g
7. 16·2, 15·8 cm **8.** 12·2, 11·8 m **9.** 30·5, 29·5 mg
10. 26·5, 25·5 cm **11.** 6·55, 6·45 *l* **12.** 9·85, 9·75 g
13. 25·55, 25·45 cm **14.** 42·65, 42·55 mm **15.** 37·61, 37·59 kg
16. 14·71, 14·69 *l* **17.** 48·21, 48·19 g **18.** 173·51, 173·49 mg
19. 63·42, 63·38 N **20.** 82·02, 81·98 *l*

Exercise 6.5b page 94
1. 2·5, 1·5 mm **2.** 14·5, 13·5 cm **3.** 8·5, 7·5 m*l*
4. 16·5, 15·5 cm **5.** 7·5, 6·5 mm **6.** 9·5, 8·5 mg
7. 5·45, 5·35 m **8.** 6·25, 6·15 cm **9.** 1·85, 1·75 *l*
10. 5·05, 4·95 m **11.** 1·25, 1·15 g **12.** 2·85, 2·75 cm
13. 6·925, 6·915 m **14.** 0·765, 0·755 kg **15.** 4·645, 4·635 g
16. 13·585, 13·575 *l* **17.** 21·505, 21·495 mg **18.** 7·805, 7·795 m*l*
19. 30·5, 29·5 m **20.** 20·5, 19·5 cm **21.** 100·5, 99·5 g
23. 4·05, 3·95 m **24.** 7·05, 6·95 g **25.** 5·305, 5·295 mg
25. 4·8625, 4·8615 *l* **26.** 8·7045, 8·7035 mg **27.** 12·7425, 12·7415 kg
28. 0·0845, 0·0835 cm **29.** 0·9735, 0·9725 g **30.** 12·1505, 12·1495 N

Exercise 6.5c page 95

1. 11, 9 m	**2.** 36, 34 m	**3.** 38, 36 m
4. 19, 17 cm	**5.** 36, 34 cm	**6.** 68, 66 m
7. 21, 19 l	**8.** 49, 47 mg	**9.** 6·9, 6·7 cm
10. 24·4, 24·2 m	**11.** 6·6, 6·4 cm	**12.** 2·8, 2·6 l
13. 43·3, 43·1 kg	**14.** 51·0, 50·8 g	**15.** 58·0, 57·8 mg
16. 132·1, 131·9 m	**17.** 16·55, 16·53 cm	**18.** 17·42, 17·40 g
19. 18·41, 18·39 m	**20.** 17·69, 17·67 cm	**21.** 42·46, 42·44 g
22. 15·98, 15·96 km	**23.** 2·66, 2·64 mg	**24.** 7·76, 7·74 l
25. 24, 22 g	**26.** 19·2, 19·0 l	**27.** 18·5, 15·5 g
28. 38·5, 35·5 m	**29.** 16·05, 15·75 m	**30.** 8·45, 8·15 km

31. 34 cm   **32.** 26·2, 25·8 m   **33.** 4·485 cm   **34.** 65 l **35.** 3·66, 3·64 m

Exercise 6.5d page 95

1. 29·25, 19·25 m²	**2.** 68·75, 51·75 cm²
3. 40·25, 26·25 mm²	**4.** 63·75, 48·75 cm²
5. 194·75, 165·75 m²	**6.** 26·25, 14·25 m²
7. 116·25, 94·25 cm²	**8.** 106·75, 73·75 m²
9. 432·25, 378·25 cm²	**10.** 57·75, 42·75 km²
11. 9·1375, 8·5075 cm²	**12.** 26·3925, 25·2925 mm²
13. 27·4625, 26·3025 m²	**14.** 53·5275, 52·0375 m²
15. 30·2325, 29·1125 cm²	**16.** 30·1125, 28·9325 mm²
17. 10·3275, 9·6775 cm²	**18.** 45·4725, 44·1325 m²
19. 81·7075, 79·8975 mm²	**20.** 35·0975, 33·9075 km²
21. 0·331375, 0·270675 cm²	**22.** 0·090525, 0·079125 cm²
23. 0·189225, 0·146825 m²	**24.** 0·143325, 0·047725 km²
25. 8·75, 3·75 m²	**26.** 63·44, 56·64 mm²
27. 57·51, 54·51 cm²	**28.** 21·75, 6·75 m²
29. 0·7544, 0·6864 m²	**30.** 1·1644, 1·0764 mm²
31. 36·75 cm²	**32.** 18 cm, 19·25 cm²

Exercise 6.5e page 96

1. 3, 1 m	**2.** 6, 4 cm	**3.** 3, 1 g
4. 36, 34 km	**5.** 46, 44 m	**6.** 18, 16 cm
7. 84, 82 cm	**8.** 19, 17 g	**9.** 27, 25 l
10. 261, 259 l	**11.** 3·7, 3·5 g	**12.** 5·5, 5·3 mm
13. 3·3, 3·1 l	**14.** 23·3, 23·1 g	**15.** 9·5, 9·3 km
16. 60·1, 59·9 N	**17.** 3·5, 3·3 l	**18.** 7·7, 7·5 kg
19. 81·5, 81·3 mg	**20.** 0·7, 0·5 m	**21.** 17 t

22. 154·4 l **23.** 98·01, 97·99 m   **24.** 11 kg   **25.** 1 km

Exercise 6.5f page 96

1. 3, 1·4	**2.** 3·57, 2·56	**3.** 2·6, 1·57
4. 5·67, 3	**5.** 4·45, 3·62	**6.** 2·23, 1·8
7. 2·18, 1·84	**8.** 11, 7·57	**9.** 4·56, 3·55
10. 4·71, 3·44	**11.** 8·23, 7·78	**12.** 1·67, 0·6
13. 5·10, 4·90	**14.** 2·07, 1·93	**15.** 8·07, 7·93
16. 6·05, 5·95	**17.** 5·08, 4·93	**18.** 3·02, 2·98
19. 11·24, 10·76	**20.** 41·33, 39·44	**21.** 80

22. 17   **23.** 4·06, 3·94 cm   **24.** 8·09 m/s   **25.** 49

Exercise 7.1a page 97
1. 37 500 cm² 2. 275 mm² 3. 4·65 cm² 4. 1·25 m² 5. 1·5 cm²
6. 6750 cm² 7. 1·27 cm² 8. 41 000 cm² 9. 0·0062 m² 10. 0·001 764 m²
11. 4·8 h 12. 210 000 m² 13. 8250 h 14. 41 km² 15. 6·5 km²
16. 4·7 h 17. 700 m² 18. 38·45 km² 19. 49 h 20. 63 290 m²

Exercise 7.1b page 97
1. (a) 144 cm², 50 cm (b) 198 cm², 58 cm (c) 700 mm², 110 mm
   (d) 960 mm², 128 mm (e) 300 m², 74 m (f) 360 m², 78 m
2. (a) 400 cm², 80 cm (b) 256 cm², 64 cm (c) 1·96 m², 5·6 m
   (d) 3·61 m², 7·6 m (e) 324 mm², 72 mm (f) 625 mm², 100 mm²
3. (a) 8 ha (b) 12 ha (c) 6 ha (d) 9 ha (e) 4·5 ha (f) 0·39 ha

Exercise 7.1c page 98
1. (a) 5 cm (b) 10 cm (c) 15 mm (d) 3 m (e) 11 mm
   (f) 12 m (g) 19 cm (h) 24 mm (i) 2·5 km (j) 3·8 m
2. (a) 8 cm (b) 9 cm (c) 6 m (d) 17 cm (e) 13 mm
   (f) 12 mm (g) 11 km (h) 21 m (i) 4·5 cm (j) 4·4 mm

Exercise 7.1d page 98
1. 20 cm 2. 30 mm 3. 16 m 4. 19 m 5. 38 cm
6. 55 cm 7. 108 cm 8. 256 mm 9. 1·6 km 10. 0·81 km
11. 0·68 m 12. 0·24 mm

Exercise 7.1e page 99
1. 13 cm 2. 17 cm 3. 19 mm 4. 23 mm 5. 15 m
6. 36 cm 7. 44 cm 8. 60 cm 9. 4·8 m 10. 6·4 m

Exercise 7.1f page 99
1. 24 2. 80 3. 30, £8·40 4. 150, £18 5. 40
6. 48 7. 24 8. 500 9. 1500 10. 2500

Exercise 7.1g page 100
1. 120 cm² 2. 272 cm² 3. 528 cm² 4. 720 cm² 5. 25 m²

Exercise 7.2a page 101
1. 350 cm² 2. 304 cm² 3. 300 mm² 4. 300 mm² 5. 90 cm²
6. 150 cm² 7. 45 m² 8. 25 cm² 9. 150 cm² 10. 57 cm²

Exercise 7.2b page 101
1. (a) 5 cm (b) 20 cm (c) 4 mm (d) 3 mm (d) 4 cm
   (f) 11 m (g) 8 m (h) 9 cm (i) 7 cm (j) 5 mm
2. (a) 3 cm (b) 5 cm (c) 3 cm (d) 5 mm (e) 7 mm
   (f) 40 mm (g) 7 cm (h) 2·5 cm (i) 3·5 m (j) 7·2 cm

Exercise 7.2c page 102
1. 945 cm² 2. 2000 cm² 3. 274 cm² 4. 12 000 mm² 5. 3168 cm²
6. 609 cm² 7. 512 cm² 8. 24 cm² 9. 644 cm²

Exercise 7.3a page 103
1. 66 cm 2. 154 cm 3. 220 cm 4. 44 cm 5. 13·2 m
6. 17·6 m 7. 26·4 m 8. 462 mm 9. 330 mm 10. 286 mm
11. 15·7 cm 12. 9·42 cm 13. 34·54 cm 14. 25·12 cm 15. 94·2 cm
16. 1256 mm 17. 1884 mm 18. 2826 mm 19. 4·71 m 20. 7·85 m

Exercise 7.3b page 103
1. 56 cm, 28 cm 2. 84 cm, 42 cm 3. 112 mm, 56 mm
4. 98 mm, 49 mm 5. 126 mm, 63 mm 6. 154 mm, 77 mm
7. 4·2 m, 2·1 m 8. 1·4 m, 0·7 m 9. 18·2 cm, 9·1 cm
10. 19·6 cm, 9·8 cm

Exercise 7.3c page 104

1. 1386 cm²	**2.** 154 cm²	**3.** 3850 mm²	**4.** 2464 mm²
5. 1·54 m²	**6.** 3·465 m²	**7.** $38\frac{1}{2}$ cm²	**8.** $9\frac{5}{8}$ cm²
9. $\frac{77}{200}$ m²	**10.** $6\frac{4}{25}$ m²	**11.** 12·56 cm²	**12.** 28·26 cm²
13. 78·5 cm²	**14.** 2826 mm²	**15.** 1256 mm²	**16.** 706·5 mm²
17. 1962·5 mm²	**18.** 7·065 m²	**19.** 0·785 m²	**20.** 19·625 m²

Exercise 7.3d page 104

1. 176 cm, 25   **2.** 44 cm, 75   **3.** 11 m, 45   **4.** 4·4 m, 7
5. 30·8 mm, 50 **6.** 9   **7.** (a) 15 400 mm²   (b) £5·28   **8.** £115·50

Exercise 7.3e page 105

1. 714 cm²   **2.** 2016 mm²   **3.** 1743 cm²   **4.** 3028 cm²
5. 203 cm²   **6.** 68 000 mm²   **7.** 21·98 cm²   **8.** 743·75 cm²

Exercise 7.4a page 106

1. 330 cm³   **2.** 720 cm³   **3.** 1200 cm³
4. 9 m³   **5.** 28 m³   **6.** 36 m³
7. 729 cm³   **8.** 1728 cm³   **9.** 216 000 mm³
10. 512 000 mm³   **11.** 72 m³, 72 000 l **12.** 84 m³, 84 000 l
13. 28 m³, 28 000 l **14.** 27 m³, 27 000 l **15.** 9 m³, 9000 l
16. 0·024 m³, 24 l **17.** 0·008 m³, 8 l **18.** 12 000 cm³, 12 l

Exercise 7.4b page 107

1. 8 cm	**2.** 15 cm	**3.** 80 mm	**4.** 81 cm	**5.** 31 m
6. 12 cm	**7.** 13 mm	**8.** 3·5 cm	**9.** 2·5 m	**10.** 8·2 cm

Exercise 7.4c page 107

1. 180 cm³	**2.** 154 cm³	**3.** 540 cm³	**4.** 1500 cm³
5. 75 000 mm³	**6.** 18 000 mm³	**7.** 12 000 mm³	**8.** 5400 mm³
9. 3 m³	**10.** 9 m³	**11.** 1848 cm³	**12.** 594 cm³
13. 1100 cm³	**14.** 7040 cm³	**15.** 8800 mm³	**16.** 9240 mm³
17. 70 400 mm³	**18.** 49 500 mm³	**19.** 3·08 m³	**20.** 1·1 m³

Exercise 7.4d page 107

1. 5 cm	**2.** 30 cm	**3.** 20 cm	**4.** 20 mm	**5.** 300 mm
6. 150 cm	**7.** 71 cm	**8.** 80 m	**9.** 7·5 cm	**10.** 42·5 mm

Exercise 7.4e page 108

1. 36	**2.** 60	**3.** 150	**4.** 250	**5.** 2160
6. 20 cm	**7.** 136 m³	**8.** 180 l, 12	**9.** 88 l, 160	**10.** 7040 m³, 80

Exercise 7.5a page 109

1. 66 cm³   **2.** 2970 cm³   **3.** 1570 cm³   **4.** 9420 cm³
5. 396 000 mm³   **6.** 148 500 mm³   **7.** 25 120 mm³   **8.** 0·77 m³
9. 0·154 m³   **10.** 0·1256 m³

Exercise 7.5b page 109

1. 113·04 cm³   **2.** 38 808 mm³   **3.** 3052 080 mm³
4. 904·32 cm³   **5.** 14·13 cm³   **6.** 4·851 cm³
7. $606\frac{3}{8}$ cm³   **8.** 381·51 cm³   **9.** 47·688 75 cm³
10. 1·766 25 cm³

Exercise 7.5c page 110

1. 1540 cm³   **2.** 452·16 cm³   **3.** 32 342 cm³   **4.** 17·584 cm³, 4·396 g
5. 22·608 l **6.** 0·14 l **7.** 7065 cm³   **8.** 4710 cm³, 12 cm
9. 80   **10.** 7007 cm³, 8408·4 g

Exercise 7.6a page 111

1. 248 cm²	**2.** 232 cm²	**3.** 168 m²	**4.** 552 cm²	**5.** 146 mm²
6. 600 cm²	**7.** 150 mm²	**8.** 216 m²	**9.** 864 cm²	**10.** 13·5 cm²

Exercise 7.6b page 112	**1.** 108 m²	**2.** 247 cm²	**3.** 272 cm²	**4.** 415 mm²	**5.** 161 cm²
	6. 20 mm²	**7.** 320 mm²	**8.** 4500 cm²	**9.** 9·8 m²	**10.** 72·2 cm²

Exercise 7.6c page 112

1. 942 cm²　　**2.** 3768 cm²　　**3.** 1884 m²　　**4.** 942 cm²
5. 28·26 m²　　**6.** 594 cm²　　**7.** 2813 cm²　　**8.** 2442 mm²
9. 23·54 cm²　　**10.** 3646·5 cm²　　**11.** 471 cm²　　**12.** 1281·12 cm²
13. 7762·08 m²　　**14.** 80 384 mm²　　**15.** 15 857 cm²

Exercise 7.6d page 113

1. 220 cm²　　**2.** 1760 cm²　　**3.** 44 m²　　**4.** 88 m²
5. 396 cm²　　**6.** 942 cm²　　**7.** 5024 cm²　　**8.** 69 080 mm²
9. 8949 cm²　　**10.** 39 250 mm²

Exercise 7.6e page 113

1. 50·24 mm²　**2.** 314 cm²　**3.** 154 m²　**4.** 1256 cm²　**5.** 346½ mm²
6. 616 cm²　**7.** 314 cm²　**8.** 1386 m²　**9.** 962·5 mm²　**10.** 5024 cm²

Exercise 7.7a page 114

1. 2 cm	**2.** 4 mm	**3.** 12 cm	**4.** 5 m	**5.** 14 cm
6. 4·5 cm	**7.** 60 cm	**8.** 10 m	**9.** 2 mm	**10.** 5 cm
11. 2 mm	**12.** 4 m	**13.** 7 cm	**14.** 10 cm	**15.** 20 cm

PART 8

Exercise 8.1a page 115

1. (a) 3°C, 13°C, 7°C, 3°C, 11°C, 8°C; (b) 11 a.m. to 12 noon, 5°C;
(c) 4 p.m. to 5 p.m., 4°C
2. (a) 238 V, 247 V, 250 V, 246 V, 241 V, 237 V; (b) 5 p.m. to 6 p.m., 13 V
3. 55 m, 53 m, 30 m, 47 m, 55 m, 31 m, 57·5 m, 48·5 m
4. (a) 751 mm, 759 mm, 757 mm, 761 mm, 756 mm, 755 mm; (b) Tues. to
Wed., 8 mm; (c) Thurs. to Fri., 6 mm
5. (a) 282·5 m, 255 m, 205 m; (b) first; (c) 100 m
6. (a) 14 cm², 30 cm², 53 cm², 77 cm²; (b) 18 cm, 27 cm, 31 cm, 38 cm
7. (a) 30 cm, 17 cm, 16 cm, 13·5 cm; (b) 15 cm, 12 cm, 30 cm
8. 77°C, 87°C, 94°C; (b) 36 s, 174 s

Exercise 8.2a page 119

1. (a) 54 km/h, 126 km/h, 198 km/h; (b) 10 m/s, 25 m/s, 45 m/s
2. (a) 45 km/h, 72 km/h, 99 km/h; (b) 2 s, 7 s, 10 s
3. (a) 0·4 A, 1·6 A, 2·8 A, 4·4 A; (b) 2 V, 5 V, 8 V, 10·5 V
4. (a) £1·60, £3·60, £4·80; (b) 30 km, 70 km, 110 km
5. (a) 0·75 cm, 6 cm, 7·5 cm; (b) 2 cm, 5 cm, 11 cm
6. (a) £12, £28, £64; (b) £100, £450, £750

Exercise 8.2b page 121

1. 28°, 81 m, 10°, $\frac{2}{3}$　　**2.** 32 cm, 350 g, 30 cm, $\frac{2}{25}$
3. 830 mm, 68°, 750 mm, 2·5　　**4.** 170 m, 13·5 km, 110 m, $\frac{1}{75}$
5. 1002·7 mm, 30°, 1002 mm, $\frac{1}{100}$ mm

Exercise 8.3a page 122

1. 5.30 p.m., 6.15 p.m., 6.30 p.m., 7.00 p.m., 7.15 p.m.
2. 8.15 p.m., 8.30 p.m., 9.15 p.m., 10.15 p.m., 10.30 p.m., 11.00 p.m.
3. 4.30 p.m., 4.45 p.m., 5.15 p.m., 5.30 p.m., 6.15 p.m., 7.00 p.m.
4. 2.30 p.m., 2.45 p.m., 3.30 p.m., 4.00 p.m., 4.15 p.m., 4.30 p.m.
5. 8.00 a.m., 8.30 a.m., 8.45 a.m., 9.15 a.m., 10.00 a.m., 10.45 a.m., 11.15 a.m.
6. 3.30 p.m., 4.15 p.m., 4.30 p.m., 4.45 p.m., 5.15 p.m.
7. 12.45 p.m., 1.15 p.m., 3.00 p.m., 3.30 p.m., 4.00 p.m.
8. 2.15 p.m., 2.45 p.m., 3.30 p.m., 3.45 p.m.

Exercise 8.3b page 124
1. 3.15 p.m.; 6 km; 15 min; 45 min; 18 km/h, 12 km/h, 20 km/h
2. 10.10 a.m.; 10.40 a.m.; 14 km; 21 km; 10 min; 1 h 30 min; 84 km/h; 63 km/h, 70 km/h
3. 8.40 a.m.; 9.50 a.m.; 12 km; 30 min; 30 km/h; 36 km/h, 24 km/h
4. 2.30 p.m.; 60 km; 2 h; 50 km/h; 40 km/h, 40 km/h
5. 7.30 p.m.; 2 h 30 min; 140 km; 75 m; 95 km/h; 80 km/h, 100 km/h

Exercise 8.3c page 126
1. 2.50 p.m., 5 km   2. 10.27 a.m., 9 km   3. 9.30 a.m., 25 km
4. 8.40 a.m, 50 km   5. 9.30 a.m, 9 km   6. 3.36 p.m., 30 km

PART 9

Exercise 9.2e page 140
1. $\frac{1}{2}, \frac{1}{3}, \frac{1}{6}$ 2. $\frac{1}{2}, \frac{2}{5}, \frac{1}{10}$ 3. $\frac{7}{15}, \frac{1}{3}, \frac{1}{5}$ 4. $\frac{3}{4}, \frac{1}{5}, \frac{1}{20}$
5. $\frac{2}{5}, \frac{1}{4}, \frac{1}{5}, \frac{3}{20}$

Exercise 9.2f page 141
1. (a) $\frac{1}{2}, \frac{1}{3}, \frac{1}{6}$ (b) 150, 100, 50
2. (a) $\frac{2}{5}, \frac{3}{10}, \frac{1}{5}, \frac{1}{10}$ (b) 200, 150, 100, 50
3. (a) $\frac{2}{3}, \frac{1}{4}, \frac{1}{12}$ (b) 160, 60, 20
4. (a) $\frac{5}{6}, \frac{1}{9}, \frac{1}{18}$ (b) 45, 6, 3
5. (a) $\frac{1}{3}, \frac{3}{10}, \frac{1}{5}, \frac{1}{6}$ (b) 20, 18, 12, 10
6. (a) $\frac{3}{8}, \frac{1}{3}, \frac{1}{6}, \frac{1}{8}$ (b) 45, 40, 20, 15
7. (a) $\frac{5}{12}, \frac{1}{3}, \frac{1}{6}, \frac{1}{12}$ (b) 775, 620, 310, 155
8. (a) $\frac{1}{2}, \frac{7}{20}, \frac{1}{10}, \frac{1}{20}$ (b) £20 000, £14 000, £4000, £2000
9. (a) $\frac{1}{3}, \frac{7}{24}, \frac{1}{4}, \frac{1}{8}$ (b) £20, £17·50, £15, £7·50

Exercise 9.3 page 143
1. 13   2. 78   3. 18   4. 16   5. 231
6. 12   7. 6   8. 5   9. 6   10. 6

Exercise 9.4a page 144
1. 3°C   2. 2   3. 54   4. 6   5. 43
6. 28   7. 8   8. 1H, 2T   9. 14   10. 27

Exercise 9.4b page 146
1. 6-8 mm   2. 5-9   3. 9-11   4. 2000-2999 *l*/h
5. 80-84 kg   6. 1·0-1·9 mm   7. 900-999   8. 2000-2999
9. 0-149   10. 10-19

Exercise 9.5a page 147
1. 32   2. 73   3. 22°C   4. 80 p   5. 23·4 s
6. 38 min   7. 29 p   8. 39 kg   9. 66 cm

Exercise 9.5b page 149
1. 1   2. 49   3. 38   4. 2   5. 3
6. 3   7. 5·6   8. 14   9. 0·25

Exercise 9.5c page 150
1. 165-169 cm   2. 5-9   3. 16-17   4. 20-29
5. 10-19   6. 15-19   7. 36-45 years

Exercise 9.6a page 151   **1.** 14   **2.** £57·47   **3.** 46 min   **4.** 2h 49 min **5.** 13 724 **6.** 84 p
7. 47 s   **8.** 7·5   **9.** 164 cm **10.** 181 cm

Exercise 9.6b page 152   **1.** 25 p   **2.** 14   **3.** 30   **4.** 5   **5.** 4   **6.** 20·4°C
7. 1·76   **8.** 0·32   **9.** 0·56   **10.** 761 mm Hg

Exercise 9.6c page 153   **1.** 9   **2.** 15   **3.** 12 cm   **4.** 13   **5.** 5·8 kg
6. 32   **7.** 7·2 cm   **8.** 14·4   **9.** 7·5   **10.** 36 min

Exercise 9.7a page 155   **1.** $\frac{2}{3}, \frac{1}{3}$ **2.** $\frac{1}{2}, \frac{1}{3}$ **3.** $\frac{3}{4}, \frac{1}{4}$ **4.** $\frac{1}{13}, \frac{3}{13}, \frac{9}{13}$
5. $\frac{5}{8}, \frac{3}{8}$ **6.** $\frac{5}{9}, \frac{4}{9}$ **7.** $\frac{3}{5}, \frac{4}{13}, \frac{2}{15}$ **8.** $\frac{3}{4}, \frac{1}{2}$
9. $\frac{1}{3}, \frac{5}{18}, \frac{2}{9}, \frac{1}{6}$ **10.** $\frac{1}{4}, \frac{1}{4}, \frac{1}{2}$

Exercise 9.7b page 156   **1.** $\frac{12}{13}, \frac{9}{13}, \frac{5}{13}, 1$ **2.** $\frac{5}{6}, \frac{1}{2}, \frac{2}{3}, \frac{1}{2}$ **3.** $\frac{11}{15}, \frac{3}{5}, \frac{2}{3}$ **4.** $\frac{2}{11}, \frac{7}{11}, \frac{6}{11}, \frac{9}{11}$
5. $\frac{1}{2}, \frac{3}{5}, \frac{9}{10}$ **6.** $\frac{4}{5}, \frac{17}{25}, \frac{13}{25}$ **7.** $\frac{4}{5}, \frac{8}{15}, \frac{2}{3}$ **8.** $\frac{11}{12}, \frac{3}{4}, \frac{11}{12}$
9. $\frac{13}{20}, \frac{17}{20}, \frac{13}{20}$ **10.** $\frac{19}{35}, \frac{24}{35}, 1, 0, \frac{24}{35}$

Exercise 9.7c page 157   **1.** $\frac{9}{25}, \frac{4}{25}, \frac{12}{25}, \frac{1}{3}, \frac{2}{15}, \frac{8}{15}$ **2.** $\frac{1}{64}, \frac{1}{4}, \frac{1}{4}, \frac{1}{16}, \frac{1}{28}, \frac{1}{32}, \frac{1}{28}$ **3.** $\frac{1}{18}, \frac{1}{36}, \frac{5}{18}, \frac{1}{4}, \frac{1}{6}, \frac{25}{36}$
4. $\frac{1}{4}, \frac{4}{9}, \frac{1}{9}, \frac{4}{9}, \frac{2}{5}, \frac{8}{15}$ **5.** $\frac{1}{16}, \frac{1}{16}, \frac{1}{4}, \frac{9}{16}, \frac{1}{4}, \frac{1}{16}$ **6.** $\frac{4}{25}, \frac{9}{100}, \frac{9}{100}, \frac{3}{25}, \frac{2}{15}$
7. $\frac{6}{25}, \frac{9}{25}, \frac{6}{25}, \frac{4}{25}, \frac{2}{15}$ **8.** $\frac{1}{12}, \frac{1}{12}, \frac{1}{4}, \frac{1}{3}, \frac{1}{6}$ **9.** $\frac{1}{30}, \frac{1}{10}, \frac{1}{15}, \frac{1}{5}, 0$
10. $\frac{3}{20}, \frac{7}{20}, \frac{1}{4}, \frac{3}{20}, \frac{1}{4}$

Exercise 9.8a page 159
1. 100, 110, 120, 140   **2.** 100, 150, 165, 200
3. 80, 100, 120, 160   **4.** 100, 160, 240, 480
5. 100, 140, 120, 144   **6.** 75, 100, 95, 115
7. 100, 120, 180, 240   **8.** 20, 100, 80, 150; 50%
9. 70, 75, 100, 110; 10%   **10.** 120, 90, 100, 70; 20%
11. 84, 100, 220, 228; 120%   **12.** 100, 110, 130, 155; 55%
13. 88, 100, 104, 92; 8%   **14.** 100, 110, 120, 140; 10%, 20%, 40%
15. 60, 64, 80, 100; 40%

Exercise 9.8b page 161
1. £30, £40, £18, £50   **2.** 75p, 45p, 39p, 90p
3. 20 t, 29 t, 11 t, 17 t   **4.** 5 t, 3 t, 1·5 t, 6·25 t
5. £0·80, £0·96, £0·60, £1   **6.** £1·74, £2·03, £3·19
7. £24, £27, £32   **8.** 72 h.p., 84 h.p., 96 h.p.
9. 2 mg, 2·75 mg, 3 mg   **10.** 23, 25, 27
11. 16 200 l, 18 600 l, 17 400 l **12.** 620, 880, 940
13. £2106, £2430, £2916   **14.** £297·50, £360·50, £554·75
15. 8505 t, 9585 t, 11 250 t

Exercise 9.8c page 162
1. 100, 200, 160, 240   **2.** 125, 100, 115, 130
3. 80, 100, 120, 128   **4.** 90, 75, 50, 100
5. 250, 190, 100, 145   **6.** 70, 80, 90, 100
7. 80, 90, 100, 120   **8.** 100, 116, 148, 168

Test Paper 1 *page 163*
1. $10 \cdot 37$, $39 \cdot 42$, $5 \cdot 74$, $5 \cdot 3$, $0 \cdot 02$, 15   2. $\frac{7}{30}$, $\frac{23}{40}$, $\frac{1}{16}$, $\frac{1}{40}$
3. $4 \cdot 27 \times 10^4$, $1 \cdot 63 \times 10^1$, $4 \cdot 4 \times 10^{-3}$
4. $5 \cdot 0$, $17 \cdot 9$ 5. $37 \cdot 5\%$, £$6 \cdot 57$, 600, £$0 \cdot 69$, 58_{ten}
6. D, D, D, E, D

Test Paper 2 *page 164*
1. $4 \cdot 87$, 611, $14 \cdot 6$, $0 \cdot 059$, $3 \cdot 14$, 0   2. $\frac{19}{60}$, $\frac{33}{40}$, $\frac{2}{3}$
3. $4 \cdot 7$, $0 \cdot 05$, 695   4. $4 \cdot 08 \times 10^3$, $1 \cdot 7 \times 10^5$, $4 \cdot 2 \times 10^{-4}$
5. $\frac{7}{25}$, $62\frac{1}{2}\%$, £$2 \cdot 45$, 300, £$7 \cdot 15$ 6. C, C, B, A, D

Test Paper 3 *page 165*
1. $1\frac{1}{12}$, $1\frac{1}{2}$, $1\frac{9}{11}$, $5\frac{1}{5}$, 4, $3\frac{3}{4}$ 2. £$0 \cdot 80$, $1\frac{9}{20}$, 65%, $12\frac{1}{2}\%$
3. $4 \cdot 78 \times 10^4$, $3 \cdot 9 \times 10^{-3}$, $5 \cdot 29 \times 10^0$
4. 9.30 p.m., 61 hours, yes, Thursday, 30 014
5. £$14 \cdot 55$, 20 minutes   6. D, B, B, C, B

Test Paper 4 *page 166*
1. 640, $1 \cdot 88$, $0 \cdot 8$, $2 \cdot 5$, $2 \cdot 5$, 45   2. $0 \cdot 4$, $\frac{19}{25}$, $237 \cdot 5\%$, £$36 \cdot 04$
3. 160 km, 165 km, $164 \cdot 5$ km   4. £$7 \cdot 13$, 35%, 72 km/h, 5 h 7 min
5. 80, $2\frac{1}{2}$, $0 \cdot 15$ 6. D, C, E, D, D

Test Paper 5 *page 167*
1. $8 \cdot 28$, 12 920, $5 \cdot 52$, $2 \cdot 55$, $132 \cdot 37$, 4
2. $\frac{29}{42}$, $11\frac{2}{3}$, $1\frac{1}{12}$ 3. $7 \cdot 43$, $0 \cdot 06$
4. $4 \cdot 27 \times 10^3$, $9 \cdot 8 \times 10^2$, $8 \cdot 4 \times 10^6$, 3×10^1
5. £$3 \cdot 30$, $64 \cdot 3\%$, 150, $0 \cdot 3$, £$8 \cdot 51$ 6. D, C, B, C, B

Test Paper 6 *page 168*
1. 5, $1 \cdot 69$, 3 510, $1 \cdot 4632$, $4 \cdot 72$, 52   2. $\frac{4}{9}$, $\frac{5}{12}$, $1\frac{1}{4}$
3. $\frac{23}{250}$, $4 \cdot 78 \times 10^4$, $9 \cdot 82$, $0 \cdot 041$, $0 \cdot 000\,723$
4. £$8 \cdot 10$, 160, $0 \cdot 375$ 5. 25 min, 29 h 58 min, 60%
6. C, C, D, C, D

Test Paper 7 *page 169*
1. $2\frac{1}{6}$, $6 \cdot 4$, $27 \cdot 3$, $4\frac{4}{5}$, 280, 6   2. Fr$234 \cdot 74$, £$179 \cdot 85$, £$41 \cdot 34$, £$12 \cdot 75$
3. $4 \cdot 7 \times 10^4$, $6 \cdot 04 \times 10^{-3}$, $1 \cdot 7 \times 10^{-2}$
4. 18 km/h, 12 cm, 20 seconds   5. $4 \cdot 7$, 18 000, $0 \cdot 060$, 47 210 km
6. A, B, C, C, E

Test Paper 8 *page 170*
1. $1\frac{11}{12}$, $\frac{1}{6}$, $\frac{7}{10}$, $1\frac{1}{10}$, $2\frac{9}{10}$, $3\frac{37}{40}$ 2. $0 \cdot 079$, $5 \cdot 4$, $0 \cdot 84$
3. 2 500, $0 \cdot 063$, $4 \cdot 736 \times 10^2$ 4. £$7 \cdot 03$, 1996, 79, 48%
5. $9\frac{1}{2}$ hours, 350 minutes, £$18 \cdot 65$, 8   6. D, D, C, D, D

Test Paper 9 *page 171*
1. 15, $5\frac{1}{2}$, 14, $\frac{11}{12}$, $1\frac{17}{24}$, $4\frac{1}{6}$ 2. $\frac{3}{8}$, $0 \cdot 19$, 85%, $0 \cdot 345$
3. $0 \cdot 046$, $0 \cdot 046$, $4 \cdot 65 \times 10^{-2}$ 4. 45 days, 1 h 12 min, $12\frac{1}{2}\%$, £$213 \cdot 90$
5. 1001000_{two}, 2200_{three}, 242_{five} 6. D, D, E, C, C

Test Paper 10 *page 172*
1. $6 \cdot 25$, 1 520, $0 \cdot 152$, 8, $1 \cdot 2$, $3 \cdot 44$ 2. $\frac{7}{8}$, $\frac{3}{40}$, $\frac{1}{2}$
3. $1 \cdot 07 \times 10^5$, $1 \cdot 91 \times 10^{-2}$, $2 \cdot 5 \times 10^3$
4. $146 \cdot 5$ km, 100 km, 150 km
5. 12020_{three}, £$103 \cdot 80$, £$105 \cdot 45$, 338 km, £$4 \cdot 75$
6. D, A, C, C, C

Test Paper 1 page 173
1. £11·60, 25%
2. £363·30, £34·60
3. £30, £864
4. 255 km, 01·05 hours
5. 2601 cm^2, 72 828 cm^3
6. 390 cm^3, 0·096
7. £32, £33·60, £42
8. £320 000, 146, 128
9. £18·34
10. 200, 28·6, Thursday, 4 days, $\frac{1}{5}$

Test Paper 2 page 176
1. £18·50
2. £45·60
3. 65 km/h, 18.09 hours
4. 3360, 48 000 cm^3
5. 45 km/h, 10.30 hours, 60 km/h, 60 km, 30 minutes, 12.00 hours, 90 km
6. 3·5 cm, 1·2
7. 43 hours, £79·12, £12·42, 5 hours
8. −, 161, 4·6, 3
9. $\frac{2}{5}$, $\frac{1}{10}$
10. £6·45, £23·80, £26·10, £47·50

Test Paper 3 page 178
1. £35·40
2. £342
3. 1173$_{\text{eight}}$, 59$_{\text{ten}}$
4. £46·08, £51·84
5. 1:35, 6 cm
6. 582·2 cm^3, 0·0328
7. £4·05, £7·32, £6·75, £8·10
8. £70·40, 15 800 km
9. $\frac{1}{4}$, $\frac{1}{6}$
10. 40 000, 1978, 30%, 45%, $\frac{3}{10}$

Test Paper 4 page 180
1. 1 400 km, 1 h 12 min
2. £11·88, £2·39
3. £1 518·40, £1 804·70
4. 1·875 km, 8 cm, 1 250 000 m^2
5. 314 cm^2, 2 512 cm^2, 125 600 cm^3
6. 10 mm, 13
7. £1200, £530
8. £5760, £6400, 120; 62, 75, 90
9. £8·79, £18·91, £15·04, £8·80
10. 40, 0, £1·45

Test Paper 5 page 182
1. 60 km/h, 5.25 p.m.
2. 1950, 15 days
3. £137·67, £51·03
4. £14·28
5. £6·30, £3·15, £2·52, £0·63: £1575
6. 110 cm^3, 0·3
7. £7 200, £2 160, £661·25
8. £210·80, 240
9. 336 cm^2, $\frac{6}{11}$
10. 11p, 44p, £1·05, £6, £3